The Chitosan Biomaterials: Advances and Challenges

The Chitosan Biomaterials: Advances and Challenges

Guest Editors

Lacramioara Popa
Mihaela Violeta Ghica
Cristina Elena Dinu-Pirvu

MDPI

Basel • Beijing • Wuhan • Barcelona • Belgrade • Novi Sad • Cluj • Manchester

Guest Editors
Lacramioara Popa
Physical and Colloidal Chemistry Department
"Carol Davila" University of Medicine and Pharmacy
Bucharest
Romania

Mihaela Violeta Ghica
Department of Physical and Colloidal Chemistry
"Carol Davila" University of Medicine and Pharmacy
Bucharest
Romania

Cristina Elena Dinu-Pirvu
Physical and Colloidal Chemistry Department
"Carol Davila" University of Medicine and Pharmacy
Bucharest
Romania

Editorial Office
MDPI AG
Grosspeteranlage 5
4052 Basel, Switzerland

This is a reprint of the Special Issue, published open access by the journal *International Journal of Molecular Sciences* (ISSN 1422-0067), freely accessible at: https://www.mdpi.com/journal/ijms/special_issues/Chitosan_Biomaterials.

For citation purposes, cite each article independently as indicated on the article page online and as indicated below:

Lastname, A.A.; Lastname, B.B. Article Title. *Journal Name* **Year**, *Volume Number*, Page Range.

ISBN 978-3-7258-5421-9 (Hbk)
ISBN 978-3-7258-5422-6 (PDF)
https://doi.org/10.3390/books978-3-7258-5422-6

Contents

About the Editors

Lacramioara Popa

Lacramioara Popa is a Professor, PhD, Dr. Habil. in Pharmacy at the Department of Physical and Colloidal Chemistry, Faculty of Pharmacy, Carol Davila University of Medicine and Pharmacy, Bucharest, Romania. She is also a Researcher at the Innovative Therapeutic Structures R&D Center and President of the Ethics and Quality Assurance Commission in the Faculty of Pharmacy at the same university. She has 32 years' experience in the integration of the Quality by Design strategy in the development of drug delivery systems, surface characterization of raw materials and pharmaceutical systems, and the application of biopolymer chitosan in medicine. She has published over 75 ISI articles and has an H index of 21.

Mihaela Violeta Ghica

Mihaela Violeta Ghica is a Professor, PhD, Dr. Habil. in Pharmacy at the Department of Physical and Colloidal Chemistry, Faculty of Pharmacy, Carol Davila University of Medicine and Pharmacy, Bucharest, Romania. She is also a Researcher at the Innovative Therapeutic Structures R&D Center at the same university. She has a master's degree in Physical Chemistry and Applied Radiochemistry from the Faculty of Chemistry, University of Bucharest. She has 23 years' experience in developing biomaterials for precision treatment, working on advanced topical formulations, statistical experimental design for drug delivery systems, and biotechnological process optimization. She has published over 100 ISI articles and has an H index of 25.

Cristina Elena Dinu-Pirvu

Cristina Elena Dinu-Pirvu is a Professor, PhD, Dr. Habil. in Pharmacy at the Department of Physical and Colloidal Chemistry, Faculty of Pharmacy, Carol Davila University of Medicine and Pharmacy, Bucharest, Romania. She is also the Director of the Innovative Therapeutic Structures R&D Center and Director of the Doctoral School at the same university. She has 32 years' experience in the design and development of bioactive compound delivery systems for personalized medicine and functionalized nanoparticulate systems for theranostic purposes, as well as valorization of the biopharmacological potential of medicinal and aromatic plants. She has published more than 100 ISI articles and has an H index of 27.

Preface

Chitosan, a biopolymer derived from chitin, continues to attract significant interest due to its unique combination of biological activity and chemical versatility. Due to its biocompatibility, biodegradability, antibacterial properties, and simplicity of modification, it is considered an exceptionally appropriate material for many applications in biomedicine, environmental science, and materials engineering.

This collection integrates the articles from the Special Issue titled *The Chitosan Biomaterials: Advances and Challenges* in the *International Journal of Molecular Sciences*. The papers highlight current research on Chitosan and explore its possibilities in fundamental and applied sciences.

This reprint examines the design and characterisation of chitosan-based hydrogels with adjustable mechanical and biological properties for wound healing and tissue regeneration; the development of Chitosan nanoparticles and nanocomposites for targeted drug delivery, antimicrobial treatments, and gene transfer; investigations into physicochemical modifications of chitosan to improve solubility, reactivity, and biological specificity; and experimental assessments of biocompatibility, antibacterial efficacy, and controlled release performance *in vitro* and *in vivo*.

These studies demonstrate how innovations in Chitosan processing and functionalization are addressing ongoing challenges and enabling creative solutions across several domains.

We express sincere thanks to all the authors and reviewers whose contributions influenced this Special Issue. This collection aims to be a valuable resource for researchers and practitioners looking to maximise the potential of Chitosan-based products.

Lacramioara Popa, Mihaela Violeta Ghica, and Cristina Elena Dinu-Pirvu
Guest Editors

International Journal of
Molecular Sciences

Editorial

Chitosan Biomaterials: Advances and Challenges

Lăcrămioara Popa, Mihaela Violeta Ghica * and Cristina-Elena Dinu-Pîrvu

Department of Physical and Colloidal Chemistry, Faculty of Pharmacy, University of Medicine and Pharmacy "Carol Davila", 020956 Bucharest, Romania; lacramioara.popa@umfcd.ro (L.P.); cristina.dinu@umfcd.ro (C.-E.D.-P.)

* Correspondence: mihaela.ghica@umfcd.ro

The purpose of this Special Issue was to review research focusing on the development of formulations based on chitosan or its derivatives together with other molecules, producing biomaterials with improved physicochemical properties and effects. The lack of drinking water, especially in disadvantaged countries, is an increasingly discussed topic and represents a concern at the level of the World Health Organization. One of the manuscripts is based on this topic and presents a water treatment method designed to ensure access to drinking water. The theme of the other manuscripts mainly concerns the development of systems based on a chitosan biopolymer, but they also discuss the preparation, evaluation, and effects of these systems in the bio-medical field.

The applicability of chitosan in the medical field (drug delivery systems and tissue engineering) and other fields (food, cosmetics, and agriculture) is due to its already-well-known properties. Chitosan is a natural polymer, obtained via the deacetylation of chitin. It is biodegradable, biocompatible [1,2], bioabsorbable, and non-toxic [3].

Concerns regarding the world population's access to drinking water have increased in recent years, and have led to new water treatment and filtration procedures. The method developed by Holmes and his team combined chitosan's ability to coagulate substances suspended in water, which subsequently flocculate, with the technique of water filtration through sand columns, thus removing the floaters that contain impurities in water, including bacteria and viruses. Chitosan concentrations of 3, 10, and 30 mg/L were used to pretreat the water. The highest retained amount of bacteria and viruses was recorded at the concentrations of 10 mg/L and 30 mg/L of chitosan. The best result in terms of turbidity was in the case of water pretreated with 10 mg/L chitosan followed by filtration using sand columns [4].

The properties of chitosan and its derivatives, which help to enhance drug permeation through the skin, have led to the development of pharmaceutical systems in association with other natural or synthetic molecules, such as films, scaffolds, membranes, and nanoparticles. The results of the evaluation of these systems have shown that they are hemocompatible, biocompatible, and do not cause skin irritation [5]. A chitosan biopolymer's main asset is its mucoadhesive property, because this increases the residence time of formulations at the oral, nasal, vaginal, or skin level, and thus the bioavailability of the active pharmaceutical ingredient is improved [6].

In the case of wound treatment, in addition to the mucoadhesive property, hemostatic and antimicrobial actions play an important role in the healing process. Chitosan is generally used together with other biomaterials, obtaining smart-materials that stimulate wound healing [5,7].

The antimicrobial action of chitosan and its derivatives can be increased through association with other molecules or plant extracts, and this can help to obtain a synergy between the properties of chitosan and plants with similar applicability in several therapeutic areas [8].

The study carried out by Azueta-Aguayo, P.H. and collaborators on scaffolds based on chitosan and ammonium hydroxide, for use in the field of tissue engineering, showed that their structures are sufficiently porous and can help to increase cell development. Regarding formulations based on chitosan and neutralized ammonium hydroxides, the thermal stability and elastic properties were superior. Furthermore, for the non-neutralized scaffolds, the direct cytotoxicity and proliferation tests showed that a greater number of cells adhered to the scaffolds and their viability increased after 48 h. In conclusion, scaffolds based on chitosan and neutralized ammonium hydroxide have the potential to be used in the medical field due to their non-toxic and biocompatible properties [3].

Peng and colleagues developed and evaluated chitosan-coated oxymatrine liposomes for inhalation administration based on the antiviral action of oxymatrine, especially regarding the respiratory syncytial virus (RSV). Due to the mucoadhesive properties of chitosan, the retention time at the lung level of liposomes covered with chitosan is higher, meaning an improved bioavailability of oxymatrine. Preliminary in vitro experiments showed that liposomes with oxymatrine coated with chitosan inhibited the proliferation of lethal RSV induced in mice. The research results demonstrated that chitosan-coated liposomes have the potential to be used as inhalation drug delivery systems, but studies are needed to optimize them [9].

A challenging topic studied lately has been the intranasal administration of drugs, due to its advantages. Many drug administration systems are based on chitosan or its derivatives, such as gels for the intranasal administration of antihistamines (loratadine and chlorpheniramine) [10]; insulin [11]; liposomes and nanoemulsions for the administration of a derivative of Dehydroepiandrosterone (DHEA) [12]; microemulsion with silymarin for the treatment of Parkinson's disease [13]; nanoparticles with galantamine [14] or ropinirole [15]; and nanocrystals with memantine for the treatment of Alzheimer's disease [16].

In this Special Issue, "Chitosan Biomaterials: Advances and Challenges", the published articles offer a significant contribution in studying the properties of chitosan as a biomaterial and are of interest to those researching the development and optimization of chitosan-based systems. Perspectives for the use of chitosan are still very broad, and its applications are still in the initial stages of development.

Acknowledgments: "Carol Davila" University of Medicine and Pharmacy, Bucharest, Romania, through contract no. 33PFE/30.12.2021.

Conflicts of Interest: The authors declare no conflict of interest.

References

1. Sukpaita, T.; Chirachanchai, S.; Pimkhaokham, A.; Ampornaramveth, R.S. Effect of Storage Time and Temperature on the Bioactivity of a Chitosan-Derived Epigenetic Modulation Scaffold. *Mar. Drugs* **2023**, *21*, 175. [CrossRef] [PubMed]
2. Román-Hidalgo, C.; Martín-Valero, M.J.; López-Pérez, G.; Villar-Navarro, M. Green Method for the Selective Electromembrane Extraction of Parabens and Fluoroquinolones in the Presence of NSAIDs by Using Biopolymeric Chitosan Films. *Membranes* **2023**, *13*, 326. [CrossRef] [PubMed]
3. Azueta-Aguayo, P.H.; Chuc-Gamboa, M.G.; Aguilar-Perez, F.J.; Aguilar-Ayala, F.J.; Rodas-Junco, B.A.; Vargas-Coronado, R.F.; Cauich-Rodriguez, J.V. Effects of Neutralization on the Physicochemical, Mechanical, and Biological Properties of Ammonium-Hydroxide-Crosslinked Chitosan Scaffolds. *Int. J. Mol. Sci.* **2022**, *23*, 14822. [CrossRef] [PubMed]
4. Holmes, E.B.; Oza, H.H.; Bailey, E.S.; Sobsey, M.D. Evaluation of Chitosans as Coagulants-Flocculants to Improve Sand Filtration for Drinking Water Treatment. *Int. J. Mol. Sci.* **2023**, *24*, 1295. [CrossRef] [PubMed]
5. Desai, N.; Rana, D.; Salave, S.; Gupta, R.; Patel, P.; Karunakaran, B.; Sharma, A.; Giri, J.; Benival, D.; Kommineni, N. Chitosan: A Potential Biopolymer in Drug Delivery and Biomedical Applications. *Pharmaceutics* **2023**, *15*, 1313. [CrossRef] [PubMed]
6. Kaur, M.; Sharma, A.; Puri, V.; Aggarwal, G.; Maman, P.; Huanbutta, K.; Nagpal, M.; Sangnim, T. Chitosan-Based Polymer Blends for Drug Delivery Systems. *Polymers* **2023**, *15*, 2028. [CrossRef] [PubMed]
7. Popa, L.; Ghica, M.V.; Popescu, R.; Irimia, T.; Dinu-Pîrvu, C.-E. Development and Optimization of Chitosan-Hydroxypropyl Methylcellulose In Situ Gelling Systems for Ophthalmic Delivery of Bupivacaine Hydrochloride. *Processes* **2021**, *9*, 1694. [CrossRef]
8. Khubiev, O.M.; Egorov, A.R.; Kirichuk, A.A.; Khrustalev, V.N.; Tskhovrebov, A.G.; Kritchenkov, A.S. Chitosan-Based Antibacterial Films for Biomedical and Food Applications. *Int. J. Mol. Sci.* **2023**, *24*, 10738. [CrossRef] [PubMed]
9. Peng, J.; Wang, Q.; Guo, M.; Liu, C.; Chen, X.; Tao, L.; Zhang, K.; Shen, X. Development of Inhalable Chitosan-Coated Oxymatrine Liposomes to Alleviate RSV-Infected Mice. *Int. J. Mol. Sci.* **2022**, *23*, 15909. [CrossRef] [PubMed]
10. Viswanadhan Vasantha, P.; Sherafudeen, S.P.; Rahamathulla, M.; Mathew, S.T.; Murali, S.; Alshehri, S.; Shakeel, F.; Alam, P.; Sirhan, A.Y.; Narayana Iyer, B.A. Combination of Cellulose Derivatives and Chitosan-Based Polymers to Investigate the Effect of

Permeation Enhancers Added to In Situ Nasal Gels for the Controlled Release of Loratadine and Chlorpheniramine. *Polymers* **2023**, *15*, 1206. [CrossRef] [PubMed]
11. Kaur, G.; Goyal, J.; Behera, P.K.; Devi, S.; Singh, S.K.; Garg, V.; Mittal, N. Unraveling the role of chitosan for nasal drug delivery systems: A review. *Carbohydr. Polym. Technol. Appl.* **2023**, *5*, 100316. [CrossRef]
12. Kannavou, M.; Karali, K.; Katsila, T.; Siapi, E.; Marazioti, A.; Klepetsanis, P.; Calogeropoulou, T.; Charalampopoulos, I.; Antimisiaris, S.G. Development and Comparative In Vitro and In Vivo Study of BNN27 Mucoadhesive Liposomes and Nanoemulsions for Nose-to-Brain Delivery. *Pharmaceutics* **2023**, *15*, 419. [CrossRef] [PubMed]
13. Imran, M.; Almehmadi, M.; Alsaiari, A.A.; Kamal, M.; Alshammari, M.K.; Alzahrani, M.O.; Almaysari, F.K.; Alzahrani, A.O.; Elkerdasy, A.F.; Singh, S.K. Intranasal Delivery of a Silymarin Loaded Microemulsion for the Effective Treatment of Parkinson's Disease in Rats: Formulation, Optimization, Characterization, and In Vivo Evaluation. *Pharmaceutics* **2023**, *15*, 618. [CrossRef] [PubMed]
14. Georgieva, D.; Nikolova, D.; Vassileva, E.; Kostova, B. Chitosan-Based Nanoparticles for Targeted Nasal Galantamine Delivery as a Promising Tool in Alzheimer's Disease Therapy. *Pharmaceutics* **2023**, *15*, 829. [CrossRef] [PubMed]
15. Mura, P.; Maestrelli, F.; Cirri, M.; Mennini, N. Multiple Roles of Chitosan in Mucosal Drug Delivery: An Updated Review. *Mar. Drugs* **2022**, *20*, 335. [CrossRef] [PubMed]
16. Saleh, M.A.; Mohamed, J.M.M.; Ruby, J.J.; Kanthiah, S.; Alanazi, Y.F.; Majrashi, K.A.; Alshahrani, S.M.; Eladl, M.A.; Alaryani, F.S.; El-Sherbiny, M.; et al. Preparation of Memantine-Loaded Chitosan Nanocrystals: In Vitro and Ex Vivo Toxicity Analysis. *Crystals* **2022**, *13*, 21. [CrossRef]

International Journal of
Molecular Sciences

MDPI

Article

Optimization of Multilayer Films Composed of Chitosan and Low-Methoxy Amidated Pectin as Multifunctional Biomaterials for Drug Delivery

Joanna Potaś [1], Agnieszka Zofia Wilczewska [2], Paweł Misiak [2], Anna Basa [3] and Katarzyna Winnicka [1,*]

[1] Department of Pharmaceutical Technology, Faculty of Pharmacy, Medical University of Białystok, Mickiewicza 2C, 15-222 Białystok, Poland; joanna.potas@umb.edu.pl
[2] Department of Polymers and Organic Synthesis, Faculty of Chemistry, University of Białystok, Ciołkowskiego 1K, 15-245 Białystok, Poland; agawilcz@uwb.edu.pl (A.Z.W.); p.misiak@uwb.edu.pl (P.M.)
[3] Department of Physical Chemistry, Faculty of Chemistry, University of Białystok, Ciołkowskiego 1K, 15-245 Białystok, Poland; abasa@uwb.edu.pl
* Correspondence: kwin@umb.edu.pl; Tel.: +48-85-748-56-15

Abstract: Polyelectrolyte multilayers (PEMs) based on polyelectrolyte complex (PEC) structures are recognized as interesting materials for manufacturing functionalized coatings or drug delivery platforms. Difficulties in homogeneous PEC system development generated the idea of chitosan (CS)/low-methoxy amidated pectin (LM PC) multilayer film optimization with regard to the selected variables: the polymer ratio, PC type, and order of polymer mixing. Films were formulated by solvent casting method and then tested to characterize CS/LM PC PECs, using thermal analysis, Fourier transform infrared spectroscopy (FTIR), turbidity, and zeta potential measurements. The internal structure of the films was visualized by using scanning electron microscopy. Analysis of the mechanical and swelling properties enabled us to select the most promising formulations with high uniformity and mechanical strength. Films with confirmed multilayer architecture were indicated as a promising material for the multifunctional systems development for buccal drug delivery. They were also characterized by improved thermal stability as compared to the single polymers and their physical mixtures, most probably as a result of the CS–LM PC interactions. This also might indicate the potential protective effect on the active substances being incorporated in the PEC-based films.

Keywords: polyelectrolyte complex; polyelectrolyte multilayer; chitosan; pectin; buccal drug delivery materials

1. Introduction

Materials based on polyelectrolyte complexes (polycomplexes, PECs) have been under detailed investigation over the last years as structures with great possibilities in multifunctional systems development. Multilayer PEC-based films are characterized by huge potential in tissue engineering and chemical or pharmaceutical technology. They are characterized by unique properties arising from combining polymers with different characteristics enriched with the special properties of PEC structures [1–3]. Preparation of PECs does not require any addition of toxic cross-linking agents but usually occurs after mixing aqueous solutions of polyelectrolytes. Since rapid interactions between oppositely charged polymers might lead to PECs' coacervation/precipitation, separation of insoluble PEC particles is one of the methods to preserve the homogeneity of the systems with regard to their further utilization [4]. Considering the number of factors that usually impact PECs' performance, such as polymer properties (molecular weight, charge distribution, pKa, solubility, chain flexibility), ionic strength, pH values affecting the polyelectrolytes' ionization degree, temperature, concentration of the interacting compounds, polymer ratio, and preparation technique, precise evaluation of the polyelectrolytes should always be performed prior to the highly controlled process of PEC formation [5–8]. There are numerous criteria of PEC classification; nevertheless, differentiation of polycomplexes on water-soluble non-stoichiometric and water-insoluble stoichiometric structures is most often invoked [5].

Polyelectrolyte multilayers (PEMs), composed of alternately deposited layers of anionic and cationic polyions separated with an interfacial layer of ionically interacting polymeric chains creating PECs, are recognized as particularly interesting materials for obtaining functionalized coatings and drug delivery platforms [1]. While polycomplex-based systems obtained by mixing the polymers with opposite charges might be treated as systems with low homogeneity as a result of easily precipitating PEC particles [4,5], the technique of gradual deposition of polyelectrolyte solutions might enable one to prepare uniform composites with usually well-preserved physicochemical performance of the individual components. It is particularly valid for those composites for which maintaining the original behavior of polymers is necessary for the multifunctional character of PEMs.

Among the plethora of polymers being utilized for PEC-based systems technology, chitosan (CS)—a linear polycation with positively charged amino groups [9]—has been widely investigated. The presence of amino groups is mainly responsible for the unique characteristics of CS, including mucoadhesive [10], antimicrobial [11], and anti-inflammatory or hemostatic properties [12]. The possibility of developing multifunctional CS-based materials with therapeutic effects supported by the polycation by itself was the main reason for designing buccal carriers for clotrimazole delivery. Pros and cons of high-molecular-weight CS and high-methoxy amidated pectin (HM PC) multilayer composites based on interpolymer complexation of the oppositely charged polymer chains showed the need for careful adjustment of the technological procedure of the films since many different factors turned out to shape the final PEC-based product. PC is, similarly to CS, a natural origin polysaccharide with an essentially linear structure composed of D-galacturonic acid units linked by α-(1, 4)-glycosidic bonds [13]. According to the degree of methyl esterification, high-methoxy (HM PC) ($\geq$50%) and low-methoxy amidated PC (LM PC) (<50%) can be distinguished. While HM PC jellifies in the presence of soluble solids at a pH around 3, LM PC requires calcium or other divalent ions for a gelation process. The mechanism of LM PC gelation strictly corresponds to the "egg–box" model, where calcium ions are ionically bound to carboxylic groups of galacturonan monomers [14]. Calcium reactivity is a parameter that describes the amount of calcium required for the highest PC gel strength at a certain concentration of soluble solids. Upon increasing the calcium content, PC gel becomes firmer up to a certain ion concentration value, beyond which the mechanical strength noticeably declines. PC is treated as an interesting candidate for pharmaceutical use [15–17], including for the technology of the films [18–21]. As an anionic biopolymer with carboxylic groups capable of ionic interactions with positively charged polyelectrolytes, PC has been under careful investigation for PECs formation [1]. The mucoadhesive properties of LM PC have shown its great potential for mucosal drug administration [16]. The development of CS/LM PC multilayers can be regarded as particularly beneficial for the improved biocompatibility of CS as well as the increased mechanical strength of the hybrid PEC materials as compared to the single-polymer-based systems [17,22]. Layer-by-layer deposition of the polyelectrolyte solutions using the solvent evaporation technique aimed to eliminate the risk of polycomplexes' precipitation since preliminary studies pointed out the high susceptibility of CS/PC blends to phase separation.

Taking into consideration the increasing potential of polycomplex materials with CS as novel and complex drug delivery platforms and our experience in PEC-based systems fabrication [23,24], the idea was to dive deeper into technological aspects of the interpolymer complexation process. Furthermore, the aim was to verify whether the selection of the internal and external stimuli that might potentially influence the films' characteristics can help develop optimized CS/PC materials. For this purpose, medium-molecular-weight CS and low-methoxy amidated PC with different calcium reactivity were utilized. In the framework of PEMs optimization, different PC types, polymer ratios, and orders of polyelectrolyte mixing were applied. The prepared films were analyzed with regard to CS/LM PC PECs' characterization. Thermal analysis, Fourier transform infrared spectroscopy (FTIR), turbidity, and zeta potential measurements were performed. The internal structure of the films was visualized by scanning electron microscopy (SEM), although visual observations were equally important since the films were morphologically diversified. Mechanical and swelling performance of the composites were also evaluated and became a valuable source of information about CS/LM PC behavior against the factors being modulated. Despite the attention given to the pharmaceutical assessment of PEMs, the paper focuses on searching for correlations between the intensity of ionic CS–PC interactions and the films' performance. The question is whether careful optimization of the interpolymer complexation process might be helpful in the development of highly functionalized PEMs, with potential application for buccal drug administration being the object of our special interests.

The increasing problem of oromucosal infections necessitates development of buccal products resistant to the eroding capacity of saliva. From the other side, the buccal mucosa might be treated as an alternative route of systemic drugs administration. Considering the specificity of the oral cavity, PEM development and optimization aims to obtain non-toxic, biodegradable films with satisfactory mechanical strength and retentivity. The structure of the systems composed of polymers with different behaviors upon contact with body fluids creates an opportunity for controllable release of a drug/drugs at the application site [25]. Selection of CS and LM PC—safe, mucoadhesive polymers particularly suited to mucosal administration, but with varied water-solubility and sensitivity to ions—can be recognized as highly beneficial for developing buccal films. It is expected that maintained physical integrity of the films in contact with saliva will enable prolonged delivery of the active substance as a consequence of the stable, slowly disintegrating structure of PEC, modulated by gradually swelling and eroding PC macromolecules.

2. Results and Discussion

2.1. Mechanical Properties

The mechanical strength of the films is important for their physical integrity during both technological and application procedures. Parameters such as tensile strength (σ_s), elongation at break (ε_s), Young's modulus (E), or tear resistance (TR) might be helpful in the evaluation process of the films. The σ_s and TR describe the stress needed for breaking of a film, and ε_s corresponds to the possible material deformation until it tears. E measures the susceptibility of a film to being deformed and might be indicated by the slope on the stress/strain curve drawn by Texture Analyzer [26,27]. In this study, great attention was paid to the mechanical analysis since visual differences in the physical performance of the formulations F1–F16 have been observed. Evaluation of the mechanical properties was regarded as crucial for the optimization of the buccal films. Satisfactory elasticity and strength of dosage forms are important with regard to both patient comfort and resistance to the saliva's eroding capacity when applied to the buccal mucosa [26].

The highest values of E were calculated for the films with an equal amount of CS and PC (F1–F4) (Figure 1, Table 1). To recognize potential differences resulting from either various polymer concentrations or the intensity of PECs formation at each polymer ratio, the mechanical behavior of F13–F16 was also evaluated. Despite different total polymer content, E values were comparable for both F1–F4 and F13–F16, especially for these formulations with the PC layer casted as the first one. Simultaneously, poorly repeatable measurements were noted for the films with the PC layer deposited first (F4), which might have arisen from the visually noticeable heterogeneity of these formulations. It might be then assumed that this order of polymer casting promoted the complexation of CS's amine groups with carboxylic groups of the polyanion. The films F5–F8—being mainly composed of PC—showed an inhomogeneous appearance after drying, probably as a result of less or more severe interactions between PC and easily penetrating CS chains. In fact, noticeable thickening and folds with air bubbles in the structure of the multilayers were observed for the F5–F8 formulations. The films with the ratio of CS to PC 10:1 were relatively uniform in color and transparency regardless of the polyelectrolyte mixing order. When it comes to the mechanical performance of the non-stoichiometric films, harder and more brittle formulations were obtained when CS was deposited as the first layer. On contrary, for the F1–F4 and F13–F16 films, the values of E were similar despite the preparation techniques used.

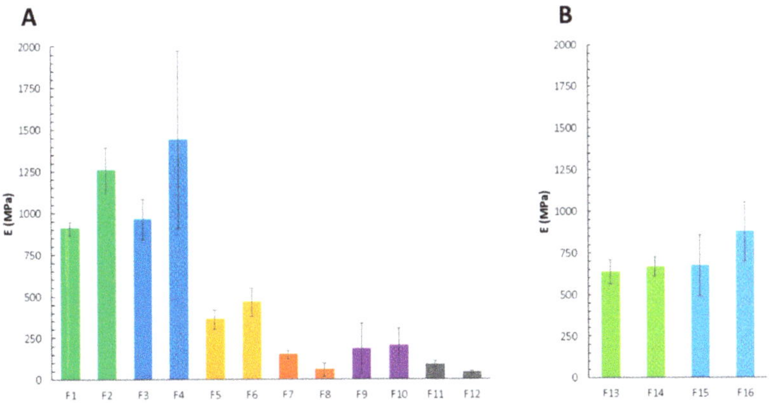

Figure 1. Mechanical properties of the films F1–F12 (**A**) and F13–F16 (**B**) expressed as Young's modulus (E) (mean ± standard deviation (SD), $n \geq 3$, $p < 0.05$). Formulations with the same polymer ratio and deposition order, but different LM PC type were indicated with the same color.

Table 1. Mechanical properties of the films F1–F16 expressed as tensile strength (σ_s) and tear resistance (TR) (mean ± SD, $n \geq 3$, $p < 0.05$).

Formulation	σ_s (N/mm^2)	TR (N)
F1	28.2 ± 0.0	27.2 ± 2.7
F2	16.6 ± 7.4	9.6 ± 2.5
F3	22.9 ± 6.1	24.6 ± 12.0
F4	44.4 ± 11.8	28.7 ± 2.9
F5	14.8 ± 3.6	10.0 ± 2.5
F6	13.7 ± 0.2	8.0 ± 0.3
F7	4.3 ± 1.9	4.0 ± 1.4
F8	1.5 ± 1.2	2.9 ± 0.7
F9	9.6 ± 4.3	7.8 ± 4.0
F10	8.4 ± 3.2	6.8 ± 3.5
F11	6.9 ± 1.2	5.7 ± 0.6
F12	6.7 ± 1.8	5.1 ± 1.8
F13	11.1 ± 3.2	6.1 ± 1.9
F14	9.8 ± 2.3	5.0 ± 0.6
F15	15.9 ± 9.5	6.0 ± 3.6
F16	10.0 ± 0.7	6.0 ± 0.4

Among all tested films, those with higher content of CS (F9–F12) showed higher values of ε_s, indicating outstanding elasticity compared to the others (Figure 2). However, due to the high standard deviation values, the main factors responsible for such mechanical behavior cannot be indicated. Only for films with PC 020 (F10, F12), the deposition order affected the values of ε_s—the films with PC deposited as the first one were more flexible compared to those with preliminary casted CS. The parameter values were comparable to the physicomechanical characteristics of the vaginal blend CS/PC films developed by Mishra et al. [28]; nevertheless, the great elasticity of these formulations was probably determined by the very high (30% (v/v)) concentration of glycerol as a plasticizer. Taking into consideration the previous observations on the correlation between the PECs formation process and the fragility of the films, the superiority of CS over PC content in F9–F12 might result in less intense polycation–polyanion interactions. This was confirmed by the significantly enhanced values of ε_s and more flexible films formation. It might be concluded that low participation of carboxylic groups in CS complexation resulting from both low PC content and relatively weak ionic strength of the polyanion affected the mechanical performance, being mostly "shaped" by free, not complexed, CS. Drying time can also influence the performance of the films. More pronounced water evaporation from the external CS layer—being exposed to a temperature for a longer time—resulted in lower elasticity. The impact of PEC structures on the mechanical behavior of polycation/polyanion materials was presented by Tejada's group on high-molecular-weight CS and high-methoxy amidated PC films as potential drug carriers for an antifungal miconazole nitrate [29]. Blend films consisting of a mixture of 3% (*w*/*v*) CS and 3% (*w*/*v*) PC solutions were more resistant to mechanical deformation as both the increased σ_s and reduced ε_s values suggested. The physicomechanical performance of PEC-based films highly corresponds to the polyelectrolytes' characteristics, especially their molecular weight [29–31].

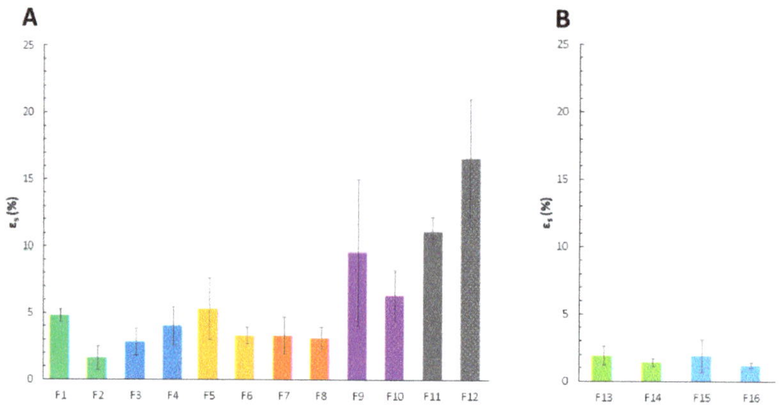

Figure 2. Mechanical properties of the films F1–F12 (**A**) and F13–F16 (**B**) expressed as elongation at break (ε_s) (mean ± SD, $n \geq 3$, $p < 0.05$). Formulations with the same polymer ratio and deposition order, but different LM PC type were indicated with the same color.

It has been clearly shown that the polymer content is not the only factor determining the mechanical strength of the films. The degree of interpolymer complexation is equally important since formulations with the same total amount of polyelectrolytes but various polymer ratios performed differently in the mechanical test. The composites F13–F16 with a high probability of stoichiometric PECs formation were noticeably more resistant to mechanical pressure compared to the delicate and more elastic F5–F12—films with low participation of electrostatically interacting amine–carboxylic groups, which is important for the enhanced strength of the films. As already mentioned, the impact of PEC particles on the development of more brittle and hard composites is well known in the literature [31,32].

2.2. Films Morphology

SEM analysis showed the layered structure of the films, composed of the relatively homogeneous and dense layer of CS and a noticeably less packed PC layer with visible delamination signs (Figure 3). Ionically interacting polymer chains were recognized as a thin, brighter area at the site of PC–CS contact; however, this was not observed for all analyzed formulations. Compared to F1 and F13, which consisted of two layers with similar thicknesses, F2–F4 showed an unevenly divided layered structure despite identical polymer content in both main layers of CS and PC. Cross-sections of F13 turned out to be full of cracks, with areas of inhomogeneously placed PECs particles. The PC-oriented layer of CS was illustrated for F15–F16. The horizontally directed structure of PC resulted in the formation of a significantly thinner polyanion layer compared to CS [13]. It was noticed that in the case of the composites with the PC layer distributed as the second one (F1–F2 and F13–F14), use of PC 010 increased the thickness and elasticity of the films. With comparison to F9–F12 characterized by the ratio of CS to PC 10:1 (*w*/*w*), films with dominant content of PC (CS to PC 1:10, *w*/*w*) were noticeably thinner, probably as a result of the linear structure of PC. The F3–F4 films were characterized visually as very delicate films with low elasticity, which might be a result of the presence of a relatively thick interlayer of PEC, as shown in Figure 4. Regardless of PC type, deposition of CS on a pre-jellified PC layer resulted in intense interpolymer complexation with a visible layer of CS/PC PEC. This is highly correlated with the results obtained in the mechanical analysis (Section 3.1), although it cannot be compared with the performance of the CS/sodium hyaluronate films developed by Kononova et al. [33]. The fact that deposition of an anionic sodium hyaluronate on the CS surface resulted in more ordered and denser PEC interlayer formation compared to the reverse addition order might indicate the individual and unique character of each polycomplex structure depending inter alia on the polyanion used.

According to the investigations made by Schlenoff's and Fares's groups [34,35], the importance of the site diffusion mechanism in the formation of the multilayers should be highlighted at this point. Rather than relying on the interdiffusion of polyions molecules throughout an entire film in PEMs fabrication, the aforementioned research groups pointed out the role of changes within its extrinsic sites composed of polyelectrolytes monomers balanced by counterions. The analysis of F5–F8 films showed very interesting results with regard to the potential mechanism of PEMs' formation. While the distribution of PC on firstly deposed CS led to thin film formation with, as anticipated, the dominant layer of PC being the result of a ten-fold larger amount of the polyanion versus polycation (F5–F6), the reverse order of polymer mixing led to almost equally divided bilayer composites (F7–F8). We suspect that strong electrostatic interactions between PC and CS added in the second step favored tight loading of linear chains of PC according to the polymer-dominated diffusion mechanism [34,35].

The order of polyelectrolyte deposition turned out to be crucial for the intensity of CS–PC ionic interactions. As the SEM images showed (Figure 3), reorganization of both CS and PC chains toward the polymer contact surface was observed; nevertheless, it was mainly characteristic for F8. The F7–F8 films were visually inhomogeneous, with the presence of wrinkles and areas of PEC precipitates placed throughout the whole films, which was also illustrated by the SEM analysis. The films F5–F6 had inhomogeneous cross-sections with a noticeably dominant layer of PC divided from a thin coat of CS with a visible PEC layer. The presence of air bubbles affected the non-uniform morphology of the films. Very interesting observations were made for F9–F12. These composites were homogeneous in the visual evaluation; nevertheless, deposition of PC of each type in the second step (F9–F10) resulted in the polyelectrolytes mixing across the whole width of the films and therefore the monolayer architecture. In the SEM analysis, a very dynamic process of 2% CS interpenetration throughout ten-fold less concentrated PC was illustrated.

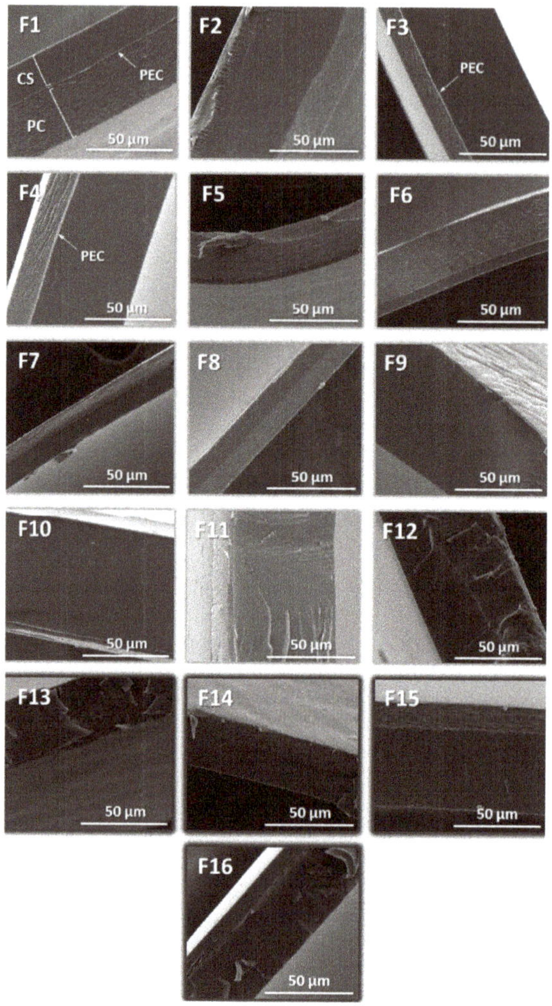

Figure 3. Representative SEM images of F1–F16 films under 3000× magnification.

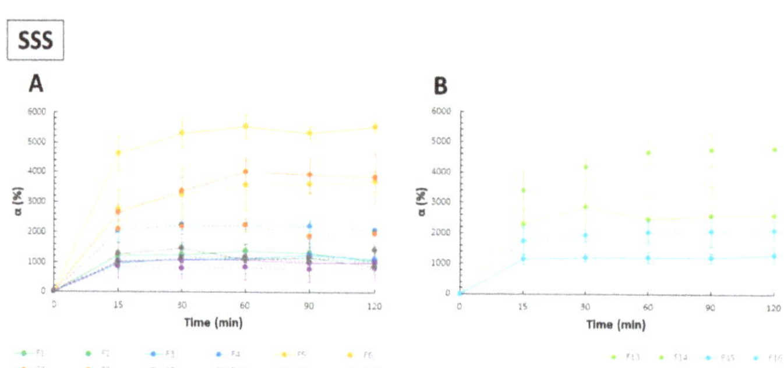

Figure 4. Swelling behavior of F1–F12 (**A**) and F13–F16 (**B**) films in the simulated saliva solution (SSS) (mean ± SD, $n = 3$).

2.3. Swelling Capacity

Considering the presence of calcium ions in different biological fluids [26], their interaction with PC chains leading to externally simulated gelation provides a tool for tailored and prolonged delivery of active substances at the application site. Although the susceptibility of the polyanion to other positive ions (mainly magnesium and potassium) cannot be omitted, calcium is particularly effective in PC gelation by forming bridges between carboxylic groups of two neighboring PC chains and divalent ions, which are additionally supported by hydrogen bonds.

The highest values of α in SSS were noted for the films with the ratio of CS to PC 1:10, which might have suggested the main role of PC in swelling and disintegration processes (Figure 4). Films F9–F12, being mainly composed of CS, showed significantly lower values of α, and no important differences were noted for both types of PC. Complexation of carboxylic groups of PC with amine groups of CS limited the capability of the polyanion for the SSS uptake, which can be regarded as characteristic for PECs-based materials. As stoichiometric PECs were known for their dense and rigid structure, a limited swelling process was repeatedly observed [32,36]. On the contrary, non-stoichiometric PECs, including those with PC as an anionic component, were mainly characterized by a high capacity of water uptake due to the formation of macroporous film structure [37,38]. While F1–F2 and F13–F14 showed varied swelling performance, mass changes upon contact with SSS noted for F3–F4 and F15–F16 were comparable (analogically prepared formulations with the same polymer ratio). For composites with predominant PC, more easily jellifying PC 020 hampered water entrance into the film matrix; nevertheless, the values of α were still markedly higher as compared to the other polymer ratios. Furthermore, films with the lowest values of thickness recorded by SEM (F5–F8 and F13–F14) showed great swelling capacity, which might be caused by the high porosity and "elasticity" of the PC layer despite the initially condensed linear structure of the polymer. In comparison to the swelling behavior of PEC-based films with high-molecular-weight CS and HM PC, which were the object of our previous studies [24], the use of LM PC resulted in obtaining relatively stable and slowly disintegrating films beneficial for prolonged drug delivery. In addition, buccal dosage forms are a group of products for which high retentivity is necessary due to the eroding properties of the saliva.

Except for F3–F4, no significant variations in water absorption were noted between the formulations with different PC type (Figure 5). The swelling performance of F1–F2 and F9–F12 were comparable regardless of the medium utilized, while the other composites were characterized by significantly weaker water entrance as compared to the SSS absorption. According to the above, it might be assumed that the high attendance of PC chains in the interpolymer complexation affected the limited ability of the polymer to jellify and easily swell the matrix formation.

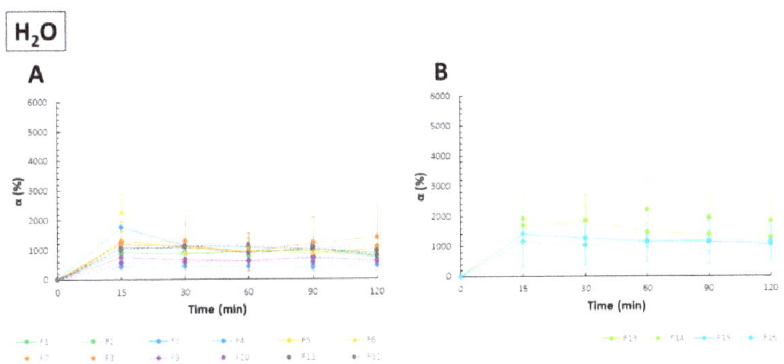

Figure 5. Swelling behavior of F1–F12 (**A**) and F13–F16 (**B**) films in purified water (mean ± SD, $n = 3$).

2.4. Thermal Characteristics

The weight loss was observed at the range of 50–150 °C for the "pure" polymers and their physical mixtures and was related to the elimination of moisture (Figure 6). Gradual degradation of the samples was recorded in the thermogravimetric analysis (TGA). The "pure" polyelectrolytes have undergone a two-step and three-step decomposition process for CS and both PCs, respectively. The first one was related to the water evaporation, with subsequent depolymerization of polymer chains [39]. For the polymer blends with the ratio of CS to PC 1:1, TG curves corresponded to the degradation processes characteristic of the single components. Simultaneously, thermograms of the physical mixtures with a predominance of either polycation or polyanion were very similar to those recorded for "pure" CS/PC.

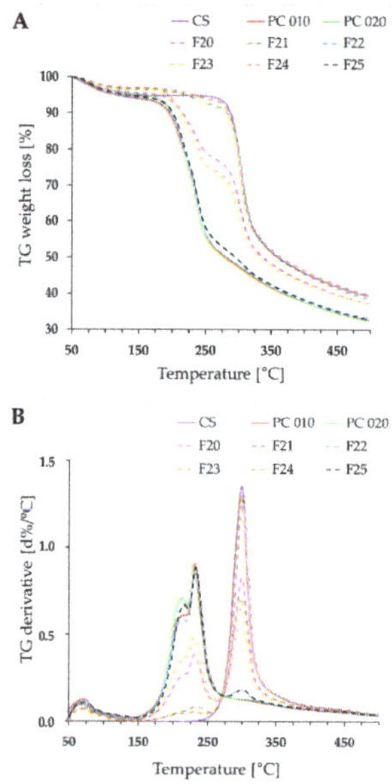

Figure 6. TGA profiles (**A**,**B**) of CS, PC 010, PC 020, and their physical mixtures in the polymer ratios corresponding to the films composition (F20—CS:PC 010 1:1, F21—CS:PC 010 10:1, F22—CS:PC 010 1:10, F23—CS:PC 020 1:1, F24—CS:PC 020 10:1, F25—CS:PC 020 1:10).

Calorimetric curves for the physical mixtures (Figure 7, F20–F25) and "pure" compounds were consistent. Any differences were correlated with the varied polymer content. Differential scanning calorimetry (DSC) thermograms for single polymers and their physical mixtures presented wide endothermic peaks in the range of 50–120 °C, probably related to the evaporation of the absorbed moisture [40]. They were not observed for the developed films, probably as a result of the drying process. The preparation of the films led to the formation of products characterized by new thermal properties. Analysis of the pervaporation membranes consisting of CS and hyaluronic acid in the ratio of 1:1 (w/w) performed by Kononova's research group also indicated the enhanced thermal stability of the PEC-based product [33]. No differences in the performance of both examined sides of the films were noted. The sharp endothermic peak at approximately 150 °C, probably corresponding to the melting point of PC 010/020, was absent for all analyzed films [15,16], indicating the possible transformation of PC into an amorphous form during the preparation of the films. The DSC curve recorded for CS was typical of an amorphous form of the compound [41]. The degradation process of both CS and PC was recorded above 200 °C; nevertheless, noticeably slower decomposition of the films was observed as compared to the "pure" polymers or their physical mixtures. This might be related to interpolymer bonding formation.

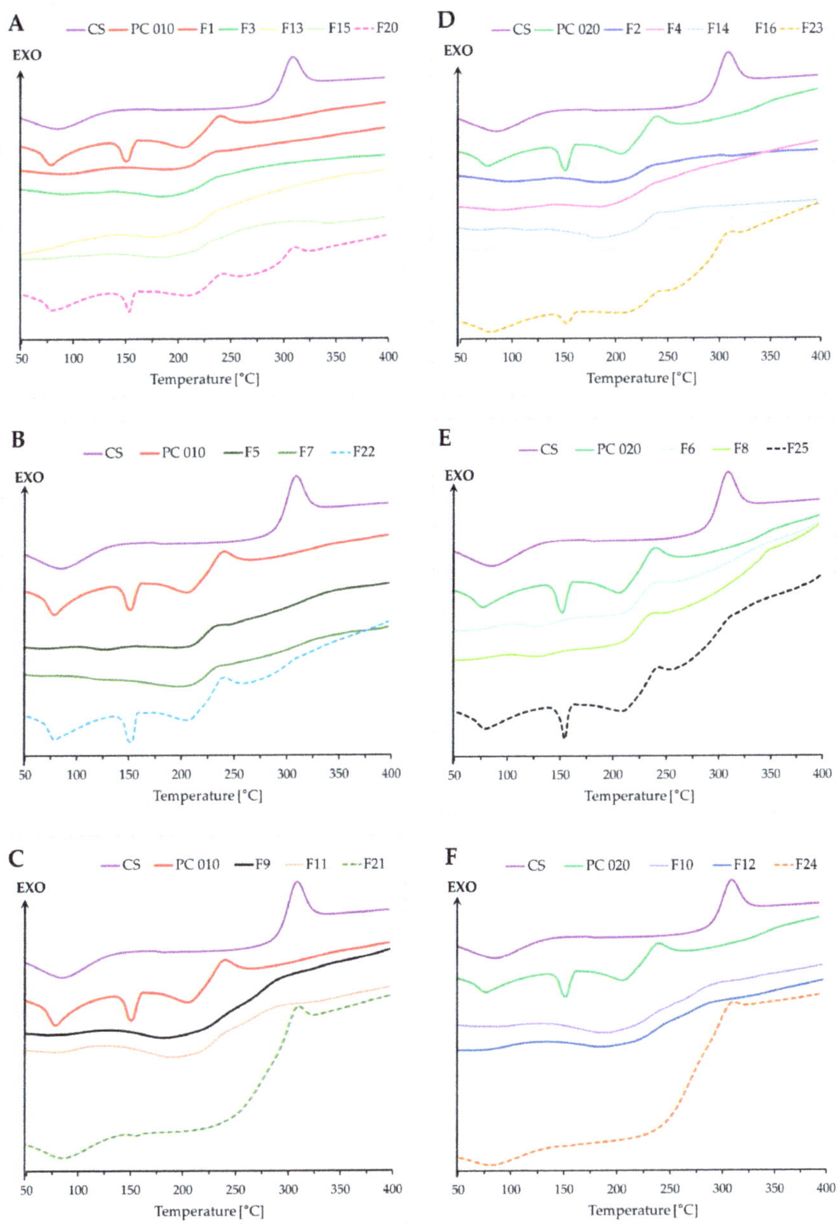

Figure 7. DSC curves for "pure" polyelectrolytes (CS, PC 010, PC 020), the films F1–F16, and the corresponding physical mixtures (F20–F25) (**A–F**).

2.5. Turbidity

PC type and the sequence of polyelectrolytes mixing affected the intensity of PECs precipitation and the morphology of the particles formed (Figure 8). For mixtures with PC 010, a more intense interpolymer complexation was recorded when the polyanion was added first. The highest values of turbidity were noted for 1:2 and 1:1 CS:PC 010 weight ratios. When a different order of polymer deposition was applied, the strongest precipitation was recorded for 1:5 and 1:2 ratios. The use of PC 020 gave the opposite outcomes. The addition of PC 020 to CS solution resulted in visually more intense polycomplexes precipitation compared to the reverse preparation technique; nevertheless, those observations were not confirmed by significantly enhanced turbidity values. The different appearance of CS/PC 020 particles/coacervates might be responsible for the misleading effect of the increased fraction of PECs formed. It cannot be excluded that the high sensitivity of PC 020 to di- but also monovalent ions might result in the hampered complexation by CS. By gradually adding PC to the CS solution, more efficient polymeric interactions might have occurred by replacing counterions with amine groups of CS that were in excess. The highest turbidity was recorded for the ratio of CS to PC 020 1:2 with the polycation added in the first place. It might be concluded that external factors, such as the order of polymer mixing or the polymer ratio, influenced the process of CS/PC PECs formation. The pH changes arising from the increasing amount of acidic CS solution resulted in a slight pH decline to the values corresponding to "pure" CS solution (2.3 ± 0.3). Nevertheless, it should be underlined that the turbidity test does not allow for precise quantitative detection of the phase separation since the ability of light scattering highly depends on the morphological features of the dispersed phase [8]. According to this, observed variations in the turbidity might be affected by both the intensity of PECs' precipitation/coacervation but also structural differences of the polycomplexes formed.

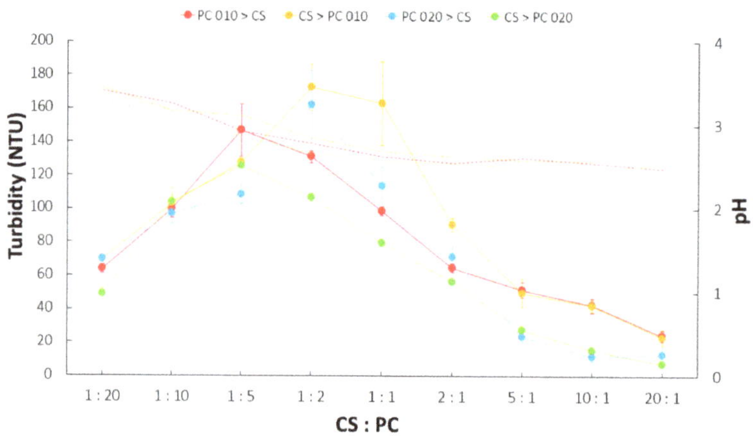

Figure 8. Turbidity (continuous line) and pH fluctuations (dashed line) recorded for CS/PC mixtures with different polymer ratio, polyelectrolyte deposition, and PC type (mean ± SD, $n = 3$).

2.6. Zeta Potential

The zeta potential is the charge density on the surface of suspended particles and colloids. The values over 60 mV indicate excellent stability of dispersions, those above 30 mV indicate physical stability, and those below 20 mV are characteristic for limited stability. In addition, a zeta potential lower than 5 mV induces an agglomeration process [42]. In the case of PECs, neutralization of the charge highly corresponds to the enhanced aggregation processes and the lowest solubility at this point.

In this study, changes in the zeta potential values were similar regardless of PC type and the applied order of polyelectrolytes mixing (Figure 9). What was interesting was that

CS addition resulted in very high zeta potential, indicating excellent stability of the PECs dispersions [42]. While the point of charge neutralization was observed for the ratio of CS to PC 010 close to 1:5, the addition of CS to PC 020 led to stoichiometric PECs preparation at the point between the 1:10 and 1:5 ratios. Zeta potential values recorded for samples based on PC 020 were inconsistent, since an addition of CS to PC solution resulted in more rapid stoichiometric PEC formation as the turbidity test confirmed. Simultaneously, the range of CS:PC weight ratios 1:10–1:5 was recognized as relevant to charge neutralization.

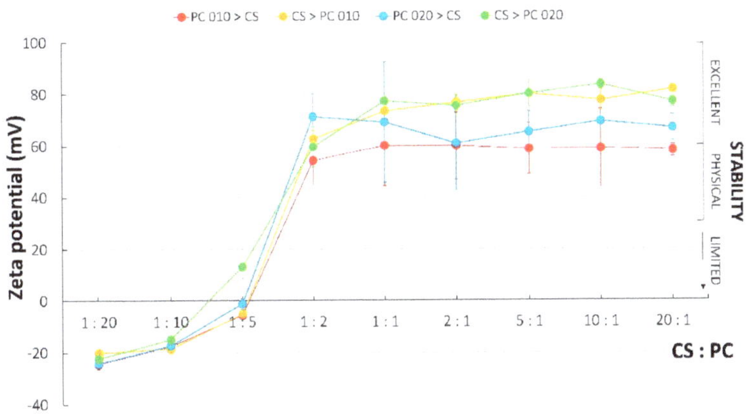

Figure 9. Changes of zeta potential noted for CS/PC mixtures with different polymer ratio and polyelectrolyte deposition. The range of zeta potential being characteristic of excellent, physical, and limited stability was also marked [42] (mean ± SD, n = 3).

2.7. FTIR Assay

"Pure" polyelectrolytes, films, and corresponding physical mixtures were assessed by FTIR (Figure S1). As previously observed [24], spectra of the physical mixtures were dominated by signals coming from PC, especially for the CS/PC 010 combinations. Among bands characteristic of the polyanion, C=O stretching of methylated carboxyl groups (1733 cm^{-1} and 1730 cm^{-1} for PC 010 and PC 020, respectively) and symmetric stretching vibrations of carboxylic groups (1421 cm^{-1} and 1424 cm^{-1} for PC 010 and PC 020, respectively) were recorded. A signal coming from N-H stretching of amine groups was noted as identifying CS at a wavenumber of 1650 cm^{-1}. The physical mixtures of CS and PC showed bands characteristic for both components without shifts, which could indicate potential interpolymer interactions. For comparison, shifts of the bands specific for both amine and carboxylic groups observed for all analyzed films might be a result of the polycomplexes formation [21,24,43]. While ten-fold higher PC content resulted in a "well-preserved" PC layer in F5–F6 films, with less affected bands characteristic for the polyanion, dominance of CS in the composites F9–F12 resulted in equally modified surfaces of the films. An exception to those observations was the FTIR spectrum recorded for F8, which revealed a very similar bands distribution for both tested sides.

3. Materials and Methods

3.1. Materials

Medium-molecular-weight CS 80/50 (degree of deacetylation: 77.6–82.5%, viscosity of 1% solution in 1% acetic acid: 31–70 mPa·s, molecular weight: 80–200 kDa) was obtained from Heppe Medical CS GmbH (Haale, Germany). Low-methoxy amidated PC with low- (a type CF 010, degree of esterification 34%, degree of amidation 17%, pH 4.2 for 2.5% solution in distilled water at 20 °C) and high-calcium reactivity (a type CF 020, degree of esterification 31%, degree of amidation 19%, pH 4.1 for 2.5% solution in distilled water at 20 °C) were kindly gifted by Herbstreith & Fox and GmbH & Co. KG (Neuenbürg,

Germany). Potassium dihydrogen phosphate, sodium chloride, calcium chloride dihydrate, and 85% lactic acid were purchased from Chempur (Piekary Śląskie, Poland). Glycerol was obtained from Fagron (Kraków, Poland).

SSS composed of potassium dihydrogen phosphate (1.63 mg/mL), sodium chloride (2.32 mg/mL), and calcium chloride dihydrate (0.22 mg/mL) with pH 6.2 adjusted by disodium hydrogen phosphate addition (according to Nair et al. with modification [25]) was used for the research analyses.

3.2. Methods

3.2.1. Preparation of the Films

Initially, twelve formulations (F1–F12) based on medium-molecular-weight CS and LM PC were prepared. As illustrated in Figure 10, the followed differentiating factors were applied: calcium reactivity of PC, polymer ratio, and order of polyelectrolyte deposition. PCs with low- (PC 010) and high-calcium reactivity (PC 020) but with an identical amount of free carboxyl groups were selected among the variety of PC types available on the market. The selection of a PC that is sensitive to calcium ions resulted from the huge potential of the polymer in the development of smart drug delivery carriers. Three different polymer ratios of CS to PC (10:1, 1:1, and 1:10, *w*/*w*) were assessed. For this purpose, 0.2% and 2% solutions of the polyelectrolytes, with an addition of 0.5% (*w*/*w*) glycerol as both plasticizer and chain interpenetration supporting agent, were utilized. CS solutions were prepared by gradual polymer addition to the mixture of 1% (*w*/*w*) lactic acid and glycerol heated to about 40 °C. PC solutions, as dispersions of the polyelectrolyte in water with the plasticizer, were manufactured at room temperature by using a magnetic stirrer. Since many scientific reports devoted to PECs pointed out the importance of the polymer mixing order concerning the polycomplexes' characteristics, varied deposition of CS/PC layers was applied (Table 2). To eliminate potential disparities in the performance of the films being a result of a different polymer content, additional F13–F16 films with the polymer ratio 1:1 (*w*/*w*) corresponding to the ratio in F1–F4 and with total polymer content identical to F5–F12 were developed (Table 2). This step was crucial for critical evaluation of the films concerning the impact of interpolymer complexation on the composites' physicomechanical behavior. The multilayers were prepared according to the previously optimized method [24], then wrapped with aluminum foil and stored at room temperature. Then, 15 g of an adequate polymer solution was placed on a Petri dish in each step of polyelectrolyte deposition by using an analytical balance. Then, they were stored in a fridge to remove air bubbles generated during the mixing processes of the polymer solutions. Films were subsequently dried in the oven at 35–40 °C for about 24 h for pre-gelation of the casted polymer layer.

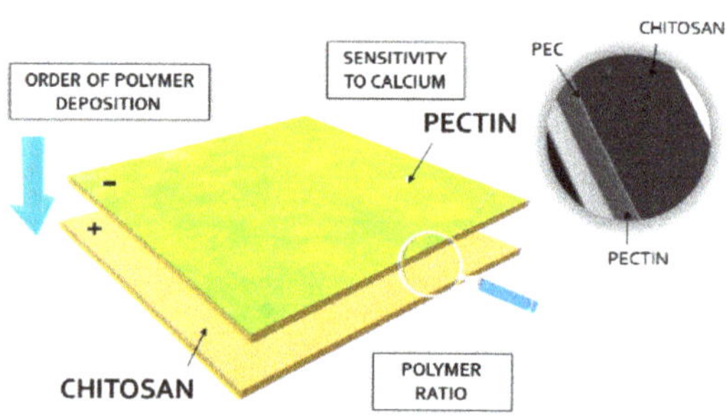

Figure 10. The scheme of the film preparation technique with an indication of the factors being optimized.

Table 2. Composition of F1–F16 films.

Formulation	CS:PC	I Layer	II Layer
F1 [1]	1:1	2%CS	2%PC010
F2 [1]		2%CS	2%PC020
F3 [1]		2%PC010	2%CS
F4 [1]		2%PC020	2%CS
F5	1:10	0.2%CS	2%PC010
F6		0.2%CS	2%PC020
F7		2%PC010	0.2%CS
F8		2%PC020	0.2%CS
F9	10:1	2%CS	0.2%PC010
F10		2%CS	0.2%PC020
F11		0.2%PC010	2%CS
F12		0.2%PC020	2%CS
F13 [1]	1:1	1.1%CS	1.1%PC010
F14 [1]		1.1%CS	1.1%PC020
F15 [1]		1.1%PC010	1.1%CS
F16 [1]		1.1%PC020	1.1%CS

[1] Formulations with identical polymer ratio but different total polymer content.

3.2.2. Evaluation of Mechanical Properties

The 1 × 3 cm film samples were analyzed with regard to their mechanical performance by using Texture Analyzer TA.XT. Plus (Stable Microsystems, Godalming, UK). Mechanical parameters, including σ_s, ε_s, E, and TR, were calculated for each of the analyzed composites at room temperature. Test parameters were set based on the previous scientific reports [24,25] and preliminary studies according to the tension test mode. The tensile grips were extended with pre-test, test, and post-test speed of 1 mm/s, with a cell loading of 5 kg, and the initial distance of the grips was 20 mm.

3.2.3. SEM Analysis

Cross-sections and average thickness of the films were evaluated with a scanning electron microscope (In-spect™S50, FEI Company, Hillsboro, OR, USA) at room temperature since the internal structure of the composites was regarded as the most important for the multilayers assessment. Film samples were placed on adhesive tapes fixed to the surface of a special stand and gold sprayed. Different magnifications (1000×–5000×) were utilized.

3.2.4. Swelling and Disintegration Test

The swelling behavior of the films in SSS and purified water was evaluated according to a previously developed and optimized method [23–25]. Briefly, 1 cm^2 films were put in the baskets dedicated for USP dissolution tests [44] and placed in 25 mL beakers with the swelling medium. The beakers were protected with aluminum foil and thermostated in a water bath at 37.0 ± 0.5 °C. At the set time intervals (15, 30, 60, 90, and 120 min), the baskets were removed from the beakers, carefully drained with cellulose wadding, and then weighted using the analytical balance. Mass fluctuations were expressed with the degree of swelling (α), according to the following equation:

$$\alpha\ (\%) = (W_s - W_0)/W_0 \times 100,$$

where W_s—weight of a film after swelling, W_0—initial weight of a film [45].

3.2.5. Thermal Analysis

Thermogravimetric analyses (TGA) and differential scanning calorimetry analyses (DSC) were performed with a Mettler Toledo Star TGA/DSC unit (Columbus, OH, USA). For TGA analysis, 3–5 mg samples of "pure" CS and PC or their physical mixtures in the weight ratios of 1:10, 1:1, and 10:1 (F20–F25) were placed in aluminum oxide crucibles and heated from 50 °C to 900 °C at a heating rate of 10 °C/min under argon. An empty pan was used as the reference. For the DSC test, 3–5 mg samples of CS, PC, their mixtures (F20–F25), and the films F1–F16 were placed in aluminum crucibles and heated from 0 °C to 480 °C at a heating rate of 10 °C/min under argon, and an empty pan was used as the reference here as well.

3.2.6. Turbidity Test

Turbidity measurements of CS/PC mixtures were performed by using a Hach Model 2100 N IS® Laboratory Turbidimeter (Loveland, CO, USA). Results were presented in nephelometric turbidity units (NTU), signifying the amount of scattered light reaching the detector. To prepare the PEC mixtures, 0.2% (*w*/*w*) CS in 1% (*w*/*w*) lactic acid and 0.2% (*w*/*w*) PC in water with addition of 0.5% (*w*/*w*) glycerol were utilized. With reference to the composition and the preparation technique, different polymer ratios and orders of polymer mixing were applied. Samples with the precipitated PECs were subsequently homogenized at 5000 rpm for 1 min to unify the particle size, then transferred to the sample cell for measurements. All results were performed within 15 min of PEC formation [8,23,46,47]. To recognize the potential impact of pH on the PEC formation process, the pH of the obtained dispersions was also recorded [24].

3.2.7. Measurements of Zeta Potential

Zeta potential values of CS/PC mixtures with different polymer ratio, PC type, and order of polymer mixing were determined with Zetasizer NanoZS90 (Malvern Instruments, Malvern, UK) [42].

3.2.8. FTIR Analysis

The attenuated total reflection FTIR (ATR–FTIR) spectra were made for single polyelectrolytes, physical mixtures, and both sides of films F1–F16. Spectra were recorded using Thermo Scientific Nicolet 6700 FTIR spectrophotometer (Thermo Scientific, Madison, WI, USA) equipped with an ATR accessory. Spectra were recorded against the background spectra and collected in the wavenumber range from 4000 cm^{-1} to 500 cm^{-1} by combining 32 scans with a resolution of 4 cm^{-1}.

3.2.9. Statistical Analysis

The quantitative variables were expressed as the mean ± SD using MS Excel software. The measurements of mechanical properties were considered significant at $p < 0.05$.

4. Conclusions

Multilayer films consisting of medium-molecular-weight CS and LM PC were developed and assessed as potential buccal drug delivery materials. Different parameters, including PC type, polymer ratio, and deposition order, were considered in the optimization process. Analysis of the mechanical properties and visual evaluation of the films enabled the selection of the most promising composites F1–F4, characterized by homogeneous appearance and optimal mechanical strength for maintaining physical integrity upon 2 h swelling. As SEM images illustrated, a layered structure of the systems was obtained. The CS/LM PC PECs were characterized by improved physicomechanical and thermal stability, and the ionic character of interpolymer bonds was confirmed inter alia by the FTIR analysis. LM PC was able to preserve its unique swelling behavior upon contact with calcium ions; nevertheless, the intensity of polymer–polymer interactions noticeably influenced the capability of water uptake. Considering the complex and multicompartment character of the highest-rated films, their potential in manufacturing buccal films cannot be ignored. The performed optimization was regarded as key for further investigations of multilayer films as antimicrobial agent carriers.

Supplementary Materials: The following supporting information can be downloaded at: https://www.mdpi.com/article/10.3390/ijms23158092/s1.

Author Contributions: Conceptualization, J.P. and K.W.; methodology, J.P., A.Z.W., P.M., A.B. and K.W.; software, J.P., A.Z.W., P.M. and A.B.; validation, J.P.; formal analysis, J.P.; investigation, J.P., A.Z.W., P.M. and A.B.; resources, J.P.; data curation, J.P., A.Z.W., P.M. and A.B.; writing—original draft preparation, J.P. and P.M.; writing—review and editing, A.Z.W. and K.W.; visualization, J.P. and A.Z.W.; supervision, K.W.; project administration, J.P. and K.W.; funding acquisition, J.P. All authors have read and agreed to the published version of the manuscript.

Funding: This research was funded by Medical University of Bialystok (project number: SUB/2/DN/22/002/2215).

Data Availability Statement: Except for the FTIR measurements available in the Supplementary Materials, data are contained within the article.

Acknowledgments: The authors thank Herbstreith & Fox and GmbH & Co. KG for providing pectin samples. Analysis (TGA, DSC, SEM, and ATR–FTIR) were performed in the Centre of Synthesis and Analysis BioNanoTechno of the University of Białystok. The equipment in the Centre was funded by the EU as a part of the Operational Program Development of Eastern Poland 2007–2013, projects: POPW.01.03.00-20-034/09-00 and POPW.01.03.00-20-004/11.

Conflicts of Interest: The authors declare no conflict of interest.

References

1. Potaś, J.; Winnicka, K. The Potential of polyelectrolyte multilayer films as drug delivery materials. *Int. J. Mol. Sci.* **2022**, *23*, 3496. [CrossRef] [PubMed]
2. Zhao, S.; Caruso, F.; Dähne, L.; Decher, G.; De Geest, B.G.; Fan, J.; Feliu, N.; Gogotsi, Y.; Hammond, P.T.; Hersam, M.C.; et al. The future of layer-by-layer assembly: A tribute to ACS nano associate editor Helmuth Möhwald. *ACS Nano* **2019**, *13*, 6151–6169. [CrossRef]
3. Criado-Gonzalez, M.; Mijangos, C.; Hernández, R. Polyelectrolyte multilayer films based on natural polymers: From fundamentals to bio-applications. *Polymers* **2021**, *13*, 2254. [CrossRef] [PubMed]
4. Gamzazade, A.I.; Nasibov, S.M. Formation and properties of polyelectrolyte complexes of chitosan hydrochloride and sodium dextransulfate. *Carbohydr. Polym.* **2002**, *50*, 339–343. [CrossRef]
5. Meka, V.S.; Sing, M.K.G.; Pichika, M.R.; Nali, S.R.; Kolapall, V.R.M.; Kesharwani, P. A comprehensive review on polyelectrolyte complexes. *Drug Discov. Today* **2017**, *22*, 1697–1706. [CrossRef] [PubMed]
6. Magoń, M.S. Layered polyelectrolyte complexes: Physics of formation and molecular properties. *J. Phys. Condens. Matter.* **2003**, *15*, 1781–1808.
7. Montenegro-Nicolini, M.; Morales, J.O. Overview and future potential of buccal mucoadhesive films as drug delivery systems for biologics. *AAPS PharmSciTech* **2017**, *18*, 3–14. [CrossRef]
8. Chollakup, R.; Smitthipong, W.; Eisenbach, C.; Tirrell, M. Phase behavior and coacervation of aqueous poly(acrylic acid)-poly(allylamine) solutions. *Macromolecules* **2010**, *43*, 2518–2528. [CrossRef]
9. Aranaz, I.; Mengibar, M.; Harris, R.; Panos, I.; Miralles, B.; Acosta, N.; Galed, G.; Heras, A. Functional characterization of chitin and chitosan. *Curr. Chem. Biol.* **2009**, *3*, 203–230.
10. Kumar, A.; Vimal, A.; Kumar, A. Why Chitosan? From properties to perspective of mucosal drug delivery. *Int. J. Biol. Macromol.* **2016**, *91*, 615–622. [CrossRef]
11. Hosseinnejad, M.; Jafari, S.M. Evaluation of different factors affecting antimicrobial properties of chitosan. *Int. J. Biol. Macromol.* **2016**, *85*, 467–475. [CrossRef] [PubMed]
12. Kim, I.Y.; Seo, S.J.; Moon, H.S.; Yoo, M.K.; Park, I.Y.; Kim, B.C.; Cho, C.S. Chitosan and its derivatives for tissue engineering applications. *Biotechnol. Adv.* **2008**, *26*, 1–21. [CrossRef]
13. Sundar-Raj, A.A.; Rubila, S.; Jayabalan, R.; Ranganathan, T.V. A review on pectin: Chemistry due to general properties of pectin and its pharmaceutical uses. *Sci. Rep.* **2012**, *1*, 550–553.
14. Cao, L.; Lu, W.; Mata, A.; Nishinari, K.; Fang, Y. Egg-box model-based gelation of alginate and pectin: A review. *Carbohydr. Polym.* **2020**, *242*, 116389. [CrossRef]
15. Nižić, L.; Potaś, J.; Winnicka, K.; Szekalska, M.; Erak, I.; Gretić, M.; Jug, M.; Hafner, A. Development, characterisation and nasal deposition of melatonin-loaded pectin/hypromellose microspheres. *Eur. J. Pharm. Sci.* **2020**, *141*, 105115. [CrossRef] [PubMed]
16. Jug, M.; Kosalec, I.; Maestrelli, F.; Mura, P. Development of low methoxy amidated pectin—Based mucoadhesive patches for buccal delivery of triclosan: Effect of cyclodextrin complexation. *Carbohydr. Polym.* **2012**, *90*, 1794–1803. [CrossRef] [PubMed]
17. Li, D.; Li, J.; Dong, H.; Li, X.; Zhang, J.; Ramaswamy, S.; Xu, F. Pectin in biomedical and drug delivery applications: A review. *Int. J. Biol; Macromol.* **2021**, *185*, 49–65. [CrossRef]

18. Murata, Y.; Maida, C.; Kofuji, K. Drug release profiles and disintegration properties of pectin films. *Materials* **2019**, *12*, 355. [CrossRef]
19. Akhgari, A.; Farahmand, F.; Afrasiabi Garekani, H.; Sadeghi, F.; Vandamme, T. The effect of pectin on swelling and permeability characteristics of free films containing Eudragit RL and/or RS as a coating formulation aimed for colonic drug delivery. *Daru* **2010**, *18*, 91–96.
20. Singh, S.; Jain, S.; Muthu, M.S.; Tiwari, S.; Tilak, R. Preparation and evaluation of buccal bioadhesive films containing clotrimazole. *AAPS Pharm. Sci. Tech.* **2008**, *9*, 660–667. [CrossRef]
21. Tejada, G.; Barrera, M.G.; Piccirilli, G.N.; Sortino, M.; Frattini, A.; Salomón, C.J.; Lamas, M.C.; Leonardi, D. Development and evaluation of buccal films based on chitosan for the potential treatment of oral candidiasis. *AAPS Pharm. Sci. Tech.* **2017**, *18*, 936–946. [CrossRef] [PubMed]
22. Kean, T.; Thanou, M. Biodegradation, biodistribution and toxicity of chitosan. *Adv. Drug Deliv. Rev.* **2010**, *62*, 3–11. [CrossRef] [PubMed]
23. Potaś, J.; Szymańska, E.; Basa, A.; Hafner, A.; Winnicka, K. Tragacanth gum/chitosan polyelectrolyte complexes-based hydrogels enriched with xanthan gum as promising materials for buccal application. *Materials* **2021**, *14*, 86. [CrossRef] [PubMed]
24. Potaś, J.; Szymańska, E.; Wróblewska, M.; Kurowska, I.; Maciejczyk, M.; Basa, A.; Wolska, E.; Wilczewska, A.Z.; Winnicka, K. Multilayer films based on chitosan/pectin polyelectrolyte complexes as novel platforms for buccal administration of clotrimazole. *Pharmaceutics* **2021**, *13*, 1588. [CrossRef] [PubMed]
25. Rajaram, D.M.; Laxman, S.D. Buccal mucoadhesive films: A review. *Sys. Rev. Pharm.* **2017**, *8*, 31–38. [CrossRef]
26. Nair, A.B.; Kumria, R.; Harsha, S.; Attimarad, M.; Al-Dhubiab, B.E.; Alhaider, I.A. In vitro techniques to evaluate buccal films. *J. Control. Release* **2013**, *166*, 10–21. [CrossRef] [PubMed]
27. Centkowska, K.; Ławrecka, E.; Sznitowska, M. Technology of orodispersible polymer films with micronized loratadine-influence of different drug loadings on film properties. *Pharmaceutics* **2020**, *12*, 250. [CrossRef]
28. Mishra, R.; Soni, K.; Mehta, T. Mucoadhesive vaginal film of fluconazole using cross-linked chitosan and pectin. *J. Therm. Anal. Calorim.* **2017**, *130*, 1683–1695. [CrossRef]
29. Tejada, G.; Piccirilli, G.N.; Sortino, M.; Salomón, C.J.; Lamas, M.C.; Leonardi, D. Formulation and in vitro efficacy of antifungal mucoadhesive polymeric matrices for the delivery of miconazole nitrate. *Mater. Sci. Eng. C* **2017**, *79*, 140–150. [CrossRef]
30. Saporito, F.; Sandri, G.; Rossi, S.; Bonferoni, M.C.; Riva, F.; Malavasi, L.; Caramella, C.; Ferrari, F. Freeze dried chitosan acetate dressings with glycosaminoglycans and tranexamic acid. *Carbohydr. Polym.* **2018**, *184*, 408–417. [CrossRef]
31. Sakloetsakun, D.; Preechagoon, D.; Bernkop-Schnürch, A.; Pongjanyakul, T. Chitosan-gum arabic polyelectrolyte complex films: Physicochemical, mechanical and mucoadhesive properties. *Pharm. Dev. Technol.* **2016**, *21*, 590–599. [CrossRef] [PubMed]
32. Abruzzo, F.; Bigucci, T.; Cerchiara, B.; Saladini, M.C.; Gallucci, F.; Cruciani, B.; Vitali, B. Luppi, Chitosan/alginate complexes for vaginal delivery of chlorhexidine digluconate. *Carbohydr. Polym.* **2013**, *91*, 651–658. [CrossRef] [PubMed]
33. Kononova, S.V.; Kruchinina, E.V.; Petrova, V.A.; Baklagina, Y.G.; Klechkovskaya, V.V.; Orekhov, A.S.; Vlasova, E.N.; Popova, E.N.; Gubanova, G.N.; Skorik, Y.A. Pervaporation membranes of a simplex type with polyelectrolyte layers of chitosan and sodium hyaluronate. *Carbohydr Polym.* **2019**, *209*, 10–19. [CrossRef] [PubMed]
34. Schlenoff, J.B.; Rmaile, A.H.; Bucur, C.B. Hydration contributions to association in polyelectrolyte multilayers and complexes: Visualizing hydrophobicity. *J. Am. Chem. Soc.* **2008**, *130*, 13589–13597. [CrossRef]
35. Fares, H.M.; Schlenoff, J.B. Diffusion of sites versus polymers in polyelectrolyte complexes and multilayers. *J. Am. Chem. Soc.* **2017**, *139*, 14656–14667. [CrossRef]
36. de Souza, R.F.B.; de Souza, F.C.B.; Moraes, A.M. Analysis of the performance of polysaccharide membranes in aqueous media as a tool to assist wound-dressing selection. *J. Appl. Polym. Sci.* **2017**, *134*, 45386–45397. [CrossRef]
37. Chabala, L.F.G.; Cuartas, C.E.E.; López, M.E.L. Release behavior and antibacterial activity of chitosan/alginate blends with Aloe vera and silver nanoparticles. *Mar. Drugs* **2017**, *15*, 328. [CrossRef]
38. Kassem, A.A.; Ismail, F.A.; Naggar, V.F.; Aboulmagd, E. Preparation and evaluation of periodontal films based on polyelectrolyte complex formation. *Pharm. Dev. Technol.* **2014**, *20*, 297–305. [CrossRef]
39. Ghaffari, A.; Navaee, K.; Oskoui, M.; Bayati, K.; Rafiee-Tehrani, M. Preparation and characterization of free mixed-film of pectin/chitosan/Eudragit((R)) RS intended for sigmoidal drug delivery. *Eur. J. Pharm. Biopharm.* **2007**, *67*, 175–186. [CrossRef]
40. Neufeld, L.; Bianco-Peled, H. Pectin-chitosan physical hydrogels as potential drug delivery vehicles. *Int. J. Biol. Macromol.* **2017**, *101*, 852–861. [CrossRef]
41. Mura, C.; Nácher, A.; Merino, V.; Merino-Sanjuan, M.; Carda, C.; Ruiz, A.; Manconi, M.; Loy, G.; Fadda, A.M.; Diez-Sales, O. N-Succinyl-chitosan systems for 5-aminosalicylic acid colon delivery: In vivo study with TNBS-induced colitis model in rats. *Int. J. Pharm.* **2011**, *416*, 145–154. [CrossRef] [PubMed]
42. Mahbubul, I.M. 3-stability and dispersion characterization of nanofluid. In *Micro and Nano Technologies, Preparation, Characterization, Properties and Application of Nanofluid*; Mahbubul, I.M., Ed.; William Andrew Publishing: Norwich, NY, USA, 2019; pp. 47–112.
43. Maciel, V.B.V.; Yoshida, C.M.P.; Franco, T.T. Chitosan/pectin polyelectrolyte complex as a pH indicator. *Carbohydr. Polym.* **2015**, *132*, 537–545. [CrossRef] [PubMed]
44. United States Pharmacopoeial Convention. *United States Pharmacopoeia and National Formulary (USP 41—NF 36)*; United States Pharmacopoeial Convention: Rockville, MD, USA, 2016.

45. Yüksel, N.; Dinç, E.; Onur, F.; Baykara, T. Influence of swelling degree on release of nicardipine hydrochloride from acrylic microspheres prepared by solvent evaporation method. *Pharm. Dev. Technol.* **1998**, *3*, 115–121. [CrossRef] [PubMed]
46. Strand, A.; Vähäsalo, L.; Ketola, A.; Salminen, K.; Retulainen, E.; Sundberg, A. In-situ analysis of polyelectrolyte complexes by flow cytometry. *Cellulose* **2018**, *25*, 3781–3795. [CrossRef]
47. Jeganathan, B.; Prakya, V.; Deshmukh, A. Preparation and evaluation of diclofenac sodium tablet coated with polyelectrolyte multilayer film using hypromellose acetate succinate and polymethacrylates for pH-dependent, modified release drug delivery. *AAPS PharmSciTech* **2016**, *17*, 578–587. [CrossRef]

International Journal of
Molecular Sciences

MDPI

Article

Effects of Neutralization on the Physicochemical, Mechanical, and Biological Properties of Ammonium-Hydroxide-Crosslinked Chitosan Scaffolds

Paola Hassibe Azueta-Aguayo [1,†], Martha Gabriela Chuc-Gamboa [1,†], Fernando Javier Aguilar-Pérez [1], Fernando Javier Aguilar-Ayala [1], Beatriz A. Rodas-Junco [2], Rossana Faride Vargas-Coronado [3] and Juan Valerio Cauich-Rodríguez [3,*]

1 Facultad de Odontología, Universidad Autónoma de Yucatán, Mérida 97000, Mexico
2 CONACYT—Facultad de Ingeniería Química, Universidad Autónoma de Yucatán, Mérida 97000, Mexico
3 Unidad de Materiales, Centro de Investigación Científica de Yucatán, Calle 43 No. 130 x 32 y 34, Colonia Chuburná de Hidalgo, Mérida 97205, Mexico
* Correspondence: jvcr@cicy.mx; Tel.: +52-999-942-8330
† These authors contributed equally to this work.

Abstract: It has been reported that chitosan scaffolds, due to their physicochemical properties, stimulate cell proliferation in different tissues of the human body. This study aimed to determine the physicochemical, mechanical, and biological properties of chitosan scaffolds crosslinked with ammonium hydroxide, with different pH values, to better understand cell behavior depending on the pH of the biomaterial. Scaffolds were either neutralized with sodium hydroxide solution, washed with distilled water until reaching a neutral pH, or kept at alkaline pH. Physicochemical characterization included scanning electron microscopy (SEM), elemental composition (EDX), Fourier-transform infrared (FTIR) spectroscopy, Raman spectroscopy, thermogravimetric analysis (TGA), and mechanical testing. In vitro cytotoxicity was assessed via dental-pulp stem cells' (DPSCs') biocompatibility. The results revealed that the neutralized scaffolds exhibited better cell proliferation and morphology. It was concluded that the chitosan scaffolds' high pH (due to residual ammonium hydroxide) decreases DPSCs' cell viability.

Keywords: biopolymer; pH effect; tissue engineering scaffolds; mechanical properties; biomaterial interactions with mesenchymal stem cells

1. Introduction

Tissue engineering is considered an important therapeutic tool in regenerative medicine; one of the crucial factors for tissue engineering is the scaffolds, which provide a platform for cells' adhesion while allowing their growth [1,2]. Scaffolds are usually made from different materials, including ceramics and polymers. Natural polymers such as chitosan, gelatin, fibrin, collagen, and alginate are widely used as scaffolding materials alone or in combination with other materials [1,3,4].

Tissue engineering of dental and oral tissues has been used in recent years as a reliable option to treat different pathologies. This includes alveolar reconstruction, which is considered a complex and challenging procedure for maxillofacial and periodontal surgeons [5]. The crucial objective of this therapy is increase alveolar bone mass in patients who have suffered from bone loss because of various conditions, such as periodontal disease, aging, osteoporosis, trauma, neoplastic pathology, or reconstructive surgery [6]. In consequence, the search for new biomaterials that promote cell proliferation to replace bone defects via scaffolds based on biopolymers is considered an area of interest in the dental field [5,6].

Chitosan (CHT) is a biopolymer derived from the deacetylation of chitin, which is one of the most abundant polysaccharides in nature [7,8]. Chitosan is bioabsorbable, biodegradable, and has been shown to be slowly degraded, mainly by enzymes such as chitosanases and lysozymes [9]. The ability to modify factors such as pH, morphology, and viscosity makes chitosan an ideal biomaterial to be used in the oral cavity [10–12]. Furthermore, chitosan derivatives promote fibroblast proliferation, suggesting that chitosan does not have toxic effects on this type of cell [13,14].

In vitro studies with chitosan-based scaffolds have shown that it is a biopolymer capable of osteoconduction, in addition to supporting adhesion and proliferation of osteoblasts and the formation of mineralized bone matrix; chitosan's highly versatile nature thus makes it attractive as a potential material for bone scaffolding [12,15].

However, one of its limitations is its low mechanical properties, which is why it is often physically or chemically crosslinked to provide better structural support [15,16].

Ammonium hydroxide (AH) has been used in tissue engineering to create scaffolds from gels with the proper porosity for cell proliferation [17]. Reyna et al. showed that physically crosslinking chitosan scaffolds with ammonium hydroxide improved the homogeneous morphology and porosity of the scaffolds [18].

Chitosan is a biopolymer that is soluble only in acidic media. Therefore, for the elaboration of the scaffolds, it is necessary to dissolve the chitosan in acids and, subsequently, eliminate the solvent from the scaffolds. For this, neutralization processes are carried out to regenerate the NH_2 groups in chitosan and make the surface hydrophilic and biocompatible with the cells [19]. Chitosan as a scaffold has been widely studied; however, there is little evidence on the effects of neutralization and the benefits for the physicochemical, mechanical, and biological properties of crosslinked chitosan scaffolds. Hence, the present study assessed the effects of neutralization on chitosan scaffolds crosslinked with AH

2. Results and Discussion

2.1. Preparation of CHT–AH Scaffolds

The scaffolds were obtained through physical crosslinking between CHT and ammonium hydroxide (Figure 1). Chitosan is insoluble in neutral and alkaline solutions but is soluble at a pKa of 6.5 due to its structure's protonation of free amino groups. Neutralization of chitosan solutions at a pH value above 6.2 results in the immediate formation of a hydrated gel-like precipitate [20]. The sponges obtained showed a porous structure due to the lyophilization process [21].

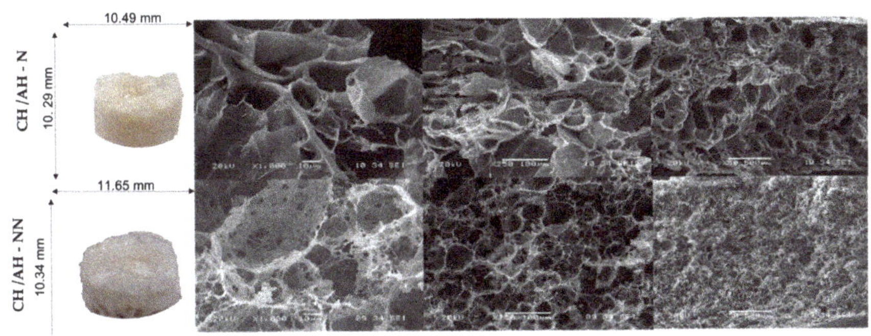

Figure 1. SEM surface morphology of neutralized (CHT/AH–N) and non-neutralized (CHT/AH–NN) CHT–AH scaffolds.

Szymon Mania et al. reported that porosity and pore size (≈100 μm) at the macroscopic and microscopic levels are important parameters for biomedical applications and adequate cell proliferation, since it is considered that porosity contributes to ensuring the adequate exchange of nutrients or gases and cell proliferation. They also reported that their foams were yellowish-white, like those obtained in our study. This coloration can be attributed to the raw materials from which they were obtained or the chemicals used for foam preparation [22]. Chitosan gels crosslinked with ammonium hydroxide have a white color. Because they are physically crosslinked, they are readily solubilized in an aqueous solution of acetic acid, as reported in the literature. However, they are not soluble in distilled water [4,23].

The scaffolds' elaboration was subjected to a freezing process prior to the lyophilization to improve the porosity. Freezing of a polymer solution causes a thermodynamic separation of the solution into polymer-rich and solvent-rich phases [24]. The polymer-poor phase's subsequent growth and coalescence form the scaffolds' pores when the solvent is removed during the lyophilization process.

Reyna et al. conducted a comparative study of two crosslinking methods (glutaraldehyde and ammonium hydroxide) in chitosan hydrogels containing collagen, observing that the chitosan scaffolds that were physically crosslinked exhibited a homogeneous morphology with greater porosity [18].

2.2. Physicochemical and Structural Characterization of CHT-AH Scaffolds

2.2.1. Surface Morphology by SEM

Figure 1 shows the morphology of neutralized and non-neutralized CHT-AH scaffolds at 50, 250, and 1000× magnifications. The SEM images showed highly interconnected, porous morphology, as reported in previous studies [18,25].

The structure of our scaffolds showed sufficient porosity and interconnected pores, so they could be considered suitable to allow the nutrition, proliferation, and migration of cells [26]. In addition, the morphology of our materials, for all samples, showed a relatively large pore size similar to that described by Rungsima et al., who also used a low-molecular-weight chitosan [3].

The more homogeneous morphology was found to correspond to the neutralized scaffolds with smaller voids/pores and heterogeneously distributed pores. A similar microstructure and surface morphology were observed on non-neutralized scaffolds, but they appeared less homogeneous.

We assumed that the crosslinking effect of ammonium hydroxide influenced the chitosan structure through hydrogen bond formation. Reyna et al. reported that the non-porous areas may be due to the presence of closed pores produced by the contraction of the foam structure [18]. In our samples, this could be due to the additional drying that the foams experienced after neutralization with NaOH and washing with distilled water, where small pores in the CHT-AH scaffolds could be observed. Jia Yan et al. reported similar morphology in their chitosan foams; they found discontinuous lamellae incompletely wrapped by the chitosan matrix, along with some small aggregates [18,27,28].

The pores that were observed in our foams could be attributed to interactions of hydrogen and amide bonds. This type of interaction improved the stability of the material and resulted in the formation of a three-dimensional foam, with porosity favorable for cell growth [29].

Surface topography is an important factor that alters cell adhesion and distribution on the biomaterial surface [30]. The porosity of scaffolds depends on the ratio of their composition, the types of crosslinking (i.e., physical or chemical), the crosslinking agents used, and the manufacturing method. Evidence is still lacking to clarify whether scaffolds with uniform pore size distribution are more efficient in tissue regeneration than those with variable pore size distribution. The distribution and morphology of the pores are designed according to their application. For example, in bone tissue engineering, a small pore size (<100 mm) has been reported to be associated with the formation of unmineralized fibrous or osteoid tissue. A pore size of up to 800 mm is appropriate to provide adequate space for cell growth. Moreover, previous studies have shown that pore size < 10 mm creates a larger surface area that stimulates greater ion exchange and bone protein adsorption [29].

Regarding pore size, a larger pore is preferred for cell growth and proliferation, since the pores will occlude later than smaller pores during progressive growth and, consequently, provide an open space for growth, supplying nutrients and oxygen [29].

In agreement with this, Nitar et al. [31] and Karimi [32] reported that chitosan scaffolds produced by freeze-gelation increased the growth and proliferation of pulposus cells of human intervertebral disks, attributing these characteristics to the formation of pores of various sizes, and concluded that the freeze-gelation process is a suitable method for fabricating various chitosan-based composite biomaterials for cell proliferation.

2.2.2. Elemental Composition by EDX

The results of the EDX analyses are shown in Table 1. The main difference between the neutralized and non-neutralized scaffolds was observed in the percentage of nitrogen obtained in the non-neutralized foams (7% at.), whose pH was 10.2, compared to the neutralized ones (12% at.) with a pH value of 7. This effect can be explained because incorporating the NaOH salt in chitosan modulates electrostatic and hydrophobic interactions and hydrogen bonds. It has been reported that the salts can cause three main effects: (a) increasing the pH, (b) inhibiting the immediate precipitation of the hydrated gel, and (c) imparting thermo-gelling characteristics at 37 °C. The addition of the neutralizing solution could cause reductions in electrostatic repulsion and increases in hydrogen bonds between chitosan chains [20].

Table 1. EDX elemental composition of neutralized and non-neutralized chitosan/AH.

Scaffolds	%C	%O	%N	%Na
Neutralized	48 ± 5	38 ± 2	11 ± 4	3 ± 1
Non-neutralized	51 ± 5	41 ± 1	7 ± 4	1 ± 1

2.2.3. Fourier-Transform Infrared (FTIR) Spectroscopy

Figure 2 shows the spectrum of the pristine chitosan, showing bands at 3358 cm^{-1} corresponding to the O–H groups, while amine group (N–H) symmetric strain vibration characteristics are observed at 3287 cm^{-1}. The bands at 2924 cm^{-1} and 2874 cm^{-1} are attributed to $-CH_2$ group stretching vibrations correlated with pyranose rings [33]. Bands located at 1646 cm^{-1} correspond to –C=O stretching of the amide group [33], while bands at 1580 cm^{-1} correspond to amide II [3], with the same intensity as the previous amide I band [34,35].

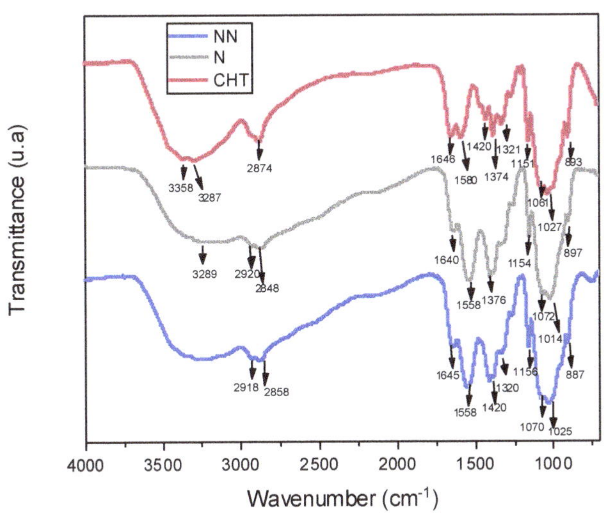

Figure 2. FTIR spectra of pristine CHT, neutralized AH-crosslinked CHT (N), and non-neutralized AH-crosslinked CHT (NN).

The infrared spectra of the non-neutralized ammonium-hydroxide-crosslinked CHT scaffolds (CHT-AH NN) exhibited bands at 3350–3280 cm^{-1} corresponding to stretching vibrations of OH and NH stretching, and at 2918 and 2858 cm^{-1} due to C–H bond stretching. NH_2 bending in amide I at 1645 cm^{-1} and N-H bending in amide II at 1558 cm^{-1} changed their intensity ratio, i.e., during crosslinking, amide I was reduced. The band located at 1424 cm^{-1} corresponded to C–H vibrations, whereas the peak at 1370 cm^{-1} was attributed to methyl vibrations in the acetamide group, and the 1320 cm^{-1} band was associated with amide III. The skeleton vibrations typical of the chitosan structure appeared at 1070–1025 cm^{-1} [30]

The existence of a large number of free COOH groups related to unreacted acetic acid can be characterized by a strong peak at 1730 cm^{-1} [36]. However, this was not observed in our samples. The characteristic bands of amide I, amide II, and amide III were found at 1637 cm^{-1}, 1548 cm^{-1}, and 1317 cm^{-1}, respectively [37,38]. Neutralization shifted the amino II band from 1580 cm^{-1} in the as-prepared CHT to 1558 cm^{-1} but did not change in intensity or peak position compared to the non-neutralized scaffold. The band located at 1158 cm^{-1} was assigned to the C–O–C of the glycosidic bond, and skeletal vibrations typical of the chitosan structure were observed at 1014 cm^{-1}. The peaks at 1377 cm^{-1} and 1420 cm^{-1} were assigned to the methyl symmetric deformation mode. Under acidic conditions, due to the excess of H+, a greater protonation of the amine and carbonyl groups on the oxygen is expected, with the consequent reduction in the bands. This is more evident as it approaches the dry state due to water's antagonistic "deprotonation" effect. The hydroxyl groups of chitosan most exposed to solvents tend to form hydrogen bonds with water molecules rather than OH–OH or NH–OH bonds within the polysaccharide. It can be affirmed that the link with ammonium hydroxide occurs only physically, giving rise to the interactions within the gels and composite foams [39].

2.2.4. Raman Spectroscopy

The characteristic bands of the non-neutralized chitosan scaffold (Figure 3) were observed at 2940 cm^{-1} (high intensity), 2900 cm^{-1}, 1645 cm^{-1} (medium intensity) with a shoulder at 1673 cm^{-1}, and low-intensity peaks at 1553 cm^{-1}, 1449 cm^{-1} (medium intensity), 1388 cm^{-1} (low intensity), 1277 cm^{-1}, 1031 cm^{-1}, 915 cm^{-1}, 860 cm^{-1}, and 557 cm^{-1}. In the neutralized scaffolds, the peak at 1645 cm^{-1} reduced its intensity, the peak at 1360 cm^{-1} was more intense, and another appeared at 1130 cm^{-1}.

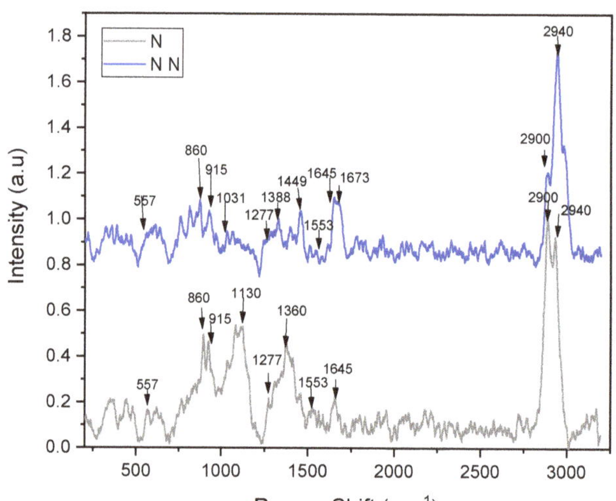

Figure 3. Raman spectra of neutralized (N) and non-neutralized (NN) AH-crosslinked CHT.

It has been reported that when using a 1064 nm laser, chitosan shows absorption bands at 2885 cm^{-1}, 1419 cm^{-1}, 1376 cm^{-1}, 1116 cm^{-1}, 900 cm^{-1}, 491 cm^{-1}, and 424 cm^{-1} [40]. Only some of these vibrations were observed with the 633 nm laser. Absorptions in the 2800–3000 cm^{-1} interval are derived from CH and CH_2 or even CH_3 bond stretching [41], whereas those in the 1000–1200 cm^{-1} interval indicate δ (C–C) and C–O–C band stretching [42]. The peaks at 1449 cm^{-1} and 1388 cm^{-1} can be assigned to the bending of CH groups in the pyranosidic skeleton. However, they have sometimes been associated with C–O–H stretching bonding, as reported by Schauenberg et al. [43].

The absorption of amide I observed at 1645 cm^{-1} and the band at 1553 cm^{-1} were of low intensity in Raman spectra in contrast to those observed in the infrared spectra, demonstrating that crosslinking with ammonium hydroxide was occurring. Interestingly, when the scaffold was neutralized, a split appeared in the signal at 2900 cm^{-1} and 2940 cm^{-1}. On the other hand, the absorptions at 1360 cm^{-1} and 1130 cm^{-1} were observed to be more intense.

2.2.5. Thermogravimetric Analysis (TGA)

The thermal stability of the sample was evaluated by determining the starting temperature of the degradation stage of the ammonium-hydroxide-crosslinked chitosan scaffolds and the maximum temperature during decomposition.

Figure 4 shows the TGA thermograms of the chitosan foams. In the neutralized NH_4OH-crosslinked chitosan scaffolds, the decomposition temperature (Td) was observed at 285 °C (Table 2). On the other hand, the decomposition temperature of the non-neutralized NH_4OH-crosslinked chitosan foam was observed at 316 °C (Table 2). These results suggest higher thermal stability in non-neutralized foams as the regenerated amino groups (from NH_4OH treatment) and acetate salts render them more stable—possibly because of the effects of the electrostatic interactions on the structure [35]. Studies performing thermogravimetric measurements under non-oxidizing conditions (e.g., nitrogen atmosphere) suggested that the chitosan sample contains approximately 10% of the adsorbed water, which evaporates at low temperatures (i.e., below 100 °C), which means that the water is physically adsorbed and/or is loosely bound to the chitosan molecules (stage 1). The decomposition accompanying the subsequent 10% weight loss starts above 100 °C and reaches the maximum rate at 168 °C (stage 2). Typical water with strong hydrogen bonds is released, possibly as a consequence of the evaporation of ammonium and acetic acid on the scaffold [44].

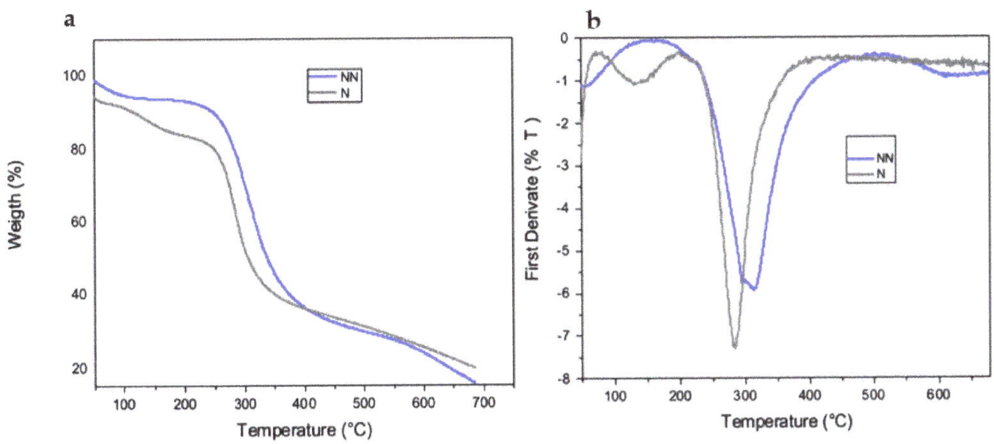

Figure 4. (**a**) Thermogravimetric analysis (TGA) and (**b**) DTGA thermograms of neutralized (N) and non-neutralized (NN) AH-crosslinked chitosan.

Table 2. Decomposition temperatures and weight loss of neutralized (N) and non-neutralized (NN) AH-crosslinked chitosan.

Scaffolds	T_d (°C)	T (°C) at 50% Weight Loss
Neutralized	285	278
Non-neutralized	316	263

The predominant stage of thermal degradation of low-molecular-weight chitosan occurs at a temperature range of 230–400 °C, with a 43% drop in the chitosan mass. This is caused by the depolymerization of the chitosan chains, the decomposition of the pyranose rings by dehydration and deamination and, finally, the opening reaction of the ring [16,45]. Additionally, the heat resistance is known to depend on the polysaccharide structure [46].

2.3. Compression Mechanical Test

The elastic modulus and compressive strength were determined by a mechanical compression test. The elastic modulus at compression (Ec) of the scaffolds was calculated in the elastic deformation zone of the foams, between 10 and 15% deformation in the strain–stress plot; the compressive strength (σ 10) of the foams was calculated at 10% deformation in all samples, as established in ASTM D1621 [47]. Table 3 reports the elastic modulus and compressive strength of the ammonium-hydroxide-crosslinked chitosan foams. The results indicate that the samples with the highest elastic modulus were the non-neutralized crosslinked chitosan scaffolds. The values obtained for the chitosan foams crosslinked with NH_4OH were similar to those obtained by Qin et al. in 2019, who reported low-molecular-weight chitosan foams with an elastic modulus of 19 kPa [44]. For bone tissue engineering, a requirement is that the scaffolds have similar mechanical properties to the bone under repair to ensure their mechanical integrity [30]. Although this study reports values that are inferior to those exhibited by either trabecular or cortical bone, it also provides evidence of the deleterious effect of neutralization.

Table 3. Scaffolds' mechanical properties under compression.

Scaffolds	Elastic Modulus (Ec) kPa	Compressive Strength (σ 10) kPa
Neutralized	8.8 ± 1.2	99.2 ± 15.7
Non-neutralized	19.1 ± 0.9 CV 4.7%	127.0 ± 10.9 CV 8.6
p-Values of t-test	0.011 *	0.176

* Statistically significant.

A scaffold for biomedical applications should provide a mechanical strength in the range of 50 to 350,000 kPa [3]. From the point of view of mechanical behavior, studies indicate that load transfer will be favorable when the modulus of elasticity between the implant and the bone tissue is similar. The modulus of elasticity of alveolar bone has been reported to be 20 GPa [48]. Therefore, in our study, the most suitable sample for the peri-implant bone tissue was the chitosan foam crosslinked with ammonium hydroxide with an elastic modulus of 19.1 kPa [46], considering that the DPSCs will produce the required bone.

Student's t-test was performed to compare the elastic modulus and compressive strength between the two materials, and significant differences were found (p = 0.011) in the comparison of the elastic modulus between the neutralized and non-neutralized crosslinked chitosan scaffolds. The analysis found no significant differences between the materials for the compressive strength at 10% strain (p = 0.176).

2.4. Evaluation of Cell Viability and Proliferation on the Scaffolds

Cytotoxic effects can prevent the in vivo integration of a biomaterial by modifying the natural assimilation process [49]. Therefore, it is necessary to evaluate biomaterials' cytotoxicity.

The viability of DPSCs grown on 2D or 3D (direct contact) chitosan scaffolds under varying neutralization regimens was assessed using the MTT assay. In the direct cytotoxicity test, it was observed that after 24 h the cells expressed adhesion and a fibroblastic morphology in the neutralized chitosan. However, there were physical changes in the chitosan structure, making the scaffolds translucent (Figure 5A(a,e)). At the end of the evaluation time (48 h), the cells adhered to the biomaterial, and more significant confluence was observed, showing that the cellular density increased over the examination period. In contrast, non-neutralized scaffolds did not allow cell adherence and proliferation (Figure 5A(c,g)), while the cell viability decreased by 90% (Figure 5B). We assumed this effect to be the result of excess ammonium hydroxide, which in large amounts is considered to be toxic to cells and was also trapped between the pores of the scaffold, creating a less viscous layer (or surface) to which the cells were unable to adhere. The presence of focal adhesions is directly or indirectly related to the nature and the amount of the surface adsorbed [19]. In our study, only neutralized scaffolds developed focal adhesions.

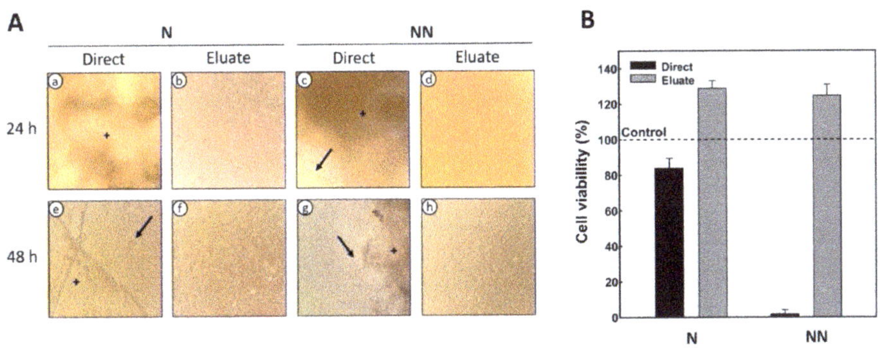

Figure 5. Microphotographs (inverted light) and viability of cultured DPSCs following direct or indirect contact with chitosan/AH. (**A**) Phase-contrast micrographs of DPSCs with direct or eluate contact at 24 and 48 h on neutral (N) and non-neutralized or alkaline (NN) scaffolds over 48 h. (**B**) MTT assay of DPSCs' viability on N and NN scaffolds after 48 h of culture. The arrows point to the cells attached to the culture surface; +: sponge fragment in the microscope field.

Indirect cytotoxicity tests, after 24 h, showed that cells on the neutralized chitosan scaffolds (N) had higher proliferation (Figure 5B) compared to those on the non-neutralized chitosan (NN) (Figure 5A(d)). After 48 h, higher cell confluency was observed for the indirect treatments on both scaffolds (Figure 5A(f,h)), and the viability was unchanged (Figure 5B). The cell viability of the scaffolds was higher than 80%, indicating their good cytocompatibility.

Intracellular (pHi) and pericellular (pHe) pH are considered to be influential factors in normal cell function, knowing that changes in their values affect signaling functions. Mammalian cell proliferation performs best at a permissive pH in the slightly alkaline range (7.0–7.2). [50]. Cell proliferation is known to be strongly affected by pH. However, many questions remain about whether pH alone can play a driving/signaling role in driving cell proliferation [50].

Noriega et al. reported that treatment with different concentrations of NaOH in chitosan scaffolds affected cell proliferation and differentiation [19]. The polyelectrolytic effect of chitosan in acid solutions is determined by the chelating capacity of the amino groups. In acidic solutions, amino groups are protonated to NH_3^+ with a pKa value of 6.1 to 6.4. When exposing the chitosan acetate surface in contact with a sodium hydroxide solution, the acetate ions associate with their counterion (Na^+), and the NH_3^+ loses a proton, forming a water molecule, which generates a microenvironment on a biomaterial surface with the characteristics of a polyelectrolyte. In addition, it has also been reported that higher viscosities are rendered at lower pH, while lower viscosities occur at higher pH

(0.1 M) [19]. In our scaffolds, the more viscous surface showed a detrimental effect on cell proliferation; this may have been due to the poor adhesion of the cells on the scaffold.

3. Materials and Methods

3.1. Materials

Chitosan (CHT)—with a low molecular weight of 183–364 g/mol^{-1} (calculated previously [18]) and a degree of deacetylation of 75–85% (batch number: STBF3282V provided by the supplier)—and ammonium hydroxide (AH, 28% NH_3 in H_2O ≥99.99% trace metals basis) were purchased from Merck (Saint Louis, MO, USA). Acetic acid was provided (AA, ≥99.5% purity) by J.T. Baker (Mexico). The 3-4,[5-dimethylthiazol-2-yl]-2,5-diphenyl tetrazolium bromide cell viability assay (MTT assay) was obtained from Merck (Saint Louis, MO, USA). Eagle's Minimum Essential Medium (MEM), fetal bovine serum, and penicillin/streptomycin were purchased from Sigma-Aldrich (St. Louis, MO, USA).

3.2. Methods

3.2.1. Scaffold Preparation

Chitosan–ammonium hydroxide scaffolds were prepared as reported in a previous work [18]. Briefly, the CHT solution was prepared by dissolving 1.5 g of CHT in 0.4 M acetic acid (20 mL). The mixture was stirred overnight with a magnetic stirrer at room temperature (25 °C) to ensure complete dissolution (the final pH of the solution was 4.5). Subsequently, the resulting solution was deposited in a 24-well cell culture plate, leaving spaces between wells. Afterward, 2.5 mL of 1N ammonium hydroxide (12.8 mL of NH_4OH per liter of water) was poured into the empty spaces of the culture dish. The 24-well cell culture plate containing the solutions was placed in a hermetically sealed container inside a fume hood. Physical crosslinking (i.e., gelation) was induced by ammonia diffusion for 48 h, as reported by Reyna et al. [18].

3.2.2. Neutralization Procedure of Chitosan–AH Scaffolds

Half of the obtained hydrogels were further neutralized with a sodium hydroxide (NaOH) solution (5% by weight) for 15–20 min. Subsequently, they were copiously rinsed with distilled water to remove ammonium acetate and excess NaOH; this procedure was repeated until reaching a neutral pH (~7.0 to ~7.2). The other half of the hydrogels were kept at the original pH (pH: 10.2) after crosslinking with AH. In order to obtain 3D scaffolds, hydrogels with (N) and without the neutralization treatment (NN) were refrigerated for 24 h, and then they were dried in a LABCONCO 4.5 FreeZone freeze-dryer at −50 °C and a pressure of 0.018 mBar for five days.

3.2.3. Characterization of Chitosan–AH Scaffolds

The morphology and pore size of the scaffolds were observed on a JEOL JSM-6360 LV scanning electron microscope (SEM) (Akishima, Tokyo, Japan) operating at 20 kV; each sample was coated with a thin layer of gold using a Denton Desk II Sputter Coater (Moorestown, NJ, USA). Energy-dispersive X-ray spectroscopy (EDX) (Oxford Instruments, INCA X-Sight 7582, High Wycombe, UK), coupled with the microscope, was used to determine the elemental composition. The chemical structures of the scaffolds were studied using an FTIR spectrophotometer (Nicolet Thermo-Scientific 8700, Waltham, MA, USA) with the attenuated total reflectance (ATR) technique. Spectra were collected over 4000 to 600 cm^{-1} with a zinc selenide crystal. The average number of scans was 100, with a resolution of 4.0 cm^{-1}, and corrected for H_2O and CO_2. Raman determination was performed using the InVia™ Raman Renishaw microscope (Wotton-under-Edge, Gloucestershire), with a laser of 633 nm, a power of 50% analyzed in the spectral interval between 200 and 3200 cm^{-1}, a power of 100%, grid of 1800, and an objective of 50× with an exposure time of 60 s. The thermogravimetric analysis was carried out using a TGA 8000™ from PerkinElmer (Waltham, MA, USA), in the temperature range 45–750 °C and under an inert atmosphere (i.e., nitrogen) at a heating rate of 10 °C/min.

3.2.4. Mechanical Analyses

Compression tests were performed according to ASTM D 1621-00. The specimens used (n = 5) were cylinders of 5.0 ± 0.3 mm in height and 6.0 ± 0.2 mm in diameter. The tests were performed on a Mini-Shimadzu AG-1 universal testing machine at room temperature (25 °C), using a 1 kN load cell, at a crosshead speed of 1 mm/min. The compressive strength (σ, MPa) was calculated using the following equation:

$$4F\sigma = \pi D2$$

where F is the maximum applied load (N) and D is the specimen's diameter (mm). The results obtained from the mechanical test (compression) were statistically analyzed using Student's t-test; p-values of <0.05 were considered statistically significant.

3.2.5. Cell Viability and Proliferation Studies

Direct Contact

DPSCs were seeded on 12-well plates at a density of 10^4 cells/well in MEM-alpha medium, 10% (v/v) fetal bovine serum, and 1% (v/v) antibiotic/antifungal solution. Plates were left for 24 h in the incubator at 37 °C, in a 5% CO_2 atmosphere, to adhere to the wells. After sterilization by UV-C light for 15 min, the materials to be tested were placed in the center of each well and on top of the adhered cells.

Indirect Contact

Indirect contact assays were conducted according to the practical guide for preparing samples and reference materials (ISO 10993-12) [51]. For this, 10 mg of chitosan was placed in contact with 10 mL of MEM-alpha for 24 h at 24 °C. After this, the extracts were centrifuged and filtered with a 0.22 µm Spritzer TPP® syringe filter. For the eluate tests, extracts replaced the culture medium in a 12-well plate with 104 cells/well. The seeded cells were incubated at 37 °C, 5% CO_2, and 95% RH for 48 h. DPSCs grown in the plates without extracts were used as 2D controls. All samples were checked daily using a Nikon Eclipse E600 inverted light phase-contrast microscope. For both techniques (i.e., direct and indirect), positive (i.e., cells with only the culture medium) and negative (i.e., cells with hydrogen peroxide) controls were included, which were also targeted with MTT solution.

4. Conclusions

Based on the results obtained, it can be concluded that there was a physical crosslinking of chitosan with ammonium hydroxide, which allowed the formation of chitosan gels in a hermetic environment. Later, the formation of foams after lyophilization was confirmed by FTIR, where slight changes in the amide ratio of the scaffolds were detected. The porosity of the scaffolds was corroborated by scanning electron microscopy. The compression test indicated that the neutralized NH_4OH-crosslinked chitosan scaffolds had lower elastic moduli than the non-neutralized scaffolds. The cell viability tests with DPSCs showed that the scaffolds produced from chitosan crosslinked with NH_4OH were non-toxic and biocompatible at alkaline and slightly alkaline pH, suggesting that this biomaterial has potential for tissue engineering applications in the biomedical field.

Author Contributions: Conceptualization, J.V.C.-R.; data curation, R.F.V.-C. and J.V.C.-R.; formal analysis, P.H.A.-A. and M.G.C.-G.; funding acquisition, F.J.A.-A.; investigation, M.G.C.-G., F.J.A.-P. and B.A.R.-J.; methodology, R.F.V.-C. and J.V.C.-R.; resources, J.V.C.-R.; supervision, F.J.A.-A.; writing—original draft, P.H.A.-A., M.G.C.-G., F.J.A.-P. and B.A.R.-J.; writing—review and editing, M.G.C.-G. and J.V.C.-R. All authors have read and agreed to the published version of the manuscript.

Funding: Funding from projects 1360 and 248378 from CONACYT is acknowledged, along with support from the Universidad Autónoma de Yucatán.

Institutional Review Board Statement: Not applicable.

Informed Consent Statement: Not applicable.

Data Availability Statement: The data presented in this study are available upon request from the corresponding author. The data are not publicly available due to its content being a thesis from a bachelor's degree in dentistry.

Conflicts of Interest: The authors declare no conflict of interest.

References

1. Preethi Soundarya, S.; Haritha Menon, A.; Viji Chandran, S.; Selvamurugan, N. Bone tissue engineering: Scaffold preparation using chitosan and other biomaterials with different design and fabrication techniques. *Int. J. Biol. Macromol.* **2018**, *119*, 1228–1239. [CrossRef] [PubMed]
2. Kheirjou, R.; Rad, J.S.; Khosroshahi, A.F.; Roshangar, L. The useful agent to have an ideal biological scaffold. *Cell Tissue Bank.* **2021**, *22*, 225–239. [CrossRef] [PubMed]
3. Chollakup, R.; Uttayarat, P.; Chworos, A.; Smitthipong, W. Noncovalent Sericin-Chitosan Scaffold: Physical Properties and Low Cytotoxicity Effect. *Int. J. Mol. Sci.* **2020**, *21*, 775. [CrossRef]
4. Oryan, A.; Alidadi, S.; Bigham-Sadegh, A.; Moshiri, A. Comparative study on the role of gelatin, chitosan and their combination as tissue engineered scaffolds on healing and regeneration of critical sized bone defects: An in vivo study. *J. Mater. Sci. Mater. Med.* **2016**, *27*, 155. [CrossRef]
5. Chen, F.M. Periodontal tissue engineering and regeneration. *Zhonghua kou qiang yi xue za zhi = Zhonghua kouqiang yixue zazhi = Chin. J. Stomatol.* **2017**, *52*, 610–614. [CrossRef]
6. Berbéri, A.; Fayyad-Kazan, M.; Ayoub, S.; Bou Assaf, R.; Sabbagh, J.; Ghassibe-Sabbagh, M.; Badran, B. Osteogenic potential of dental and oral derived stem cells in bone tissue engineering among animal models: An update. *Tissue Cell* **2021**, *71*, 101515. [CrossRef]
7. Kou, S.G.; Peters, L.M.; Mucalo, M.R. *Chitosan: A Review of Sources and Preparation Methods*; Elsevier B.V.: Amsterdam, The Netherlands, 2021; Volume 169, ISBN 0000000229340.
8. Philibert, T.; Lee, B.H.; Fabien, N. Current Status and New Perspectives on Chitin and Chitosan as Functional Biopolymers. *Appl. Biochem. Biotechnol.* **2017**, *181*, 1314–1337. [CrossRef] [PubMed]
9. Rodríguez-Vázquez, M.; Vega-Ruiz, B.; Ramos-Zúñiga, R.; Saldaña-Koppel, D.A.; Quiñones-Olvera, L.F. Chitosan and Its Potential Use as a Scaffold for Tissue Engineering in Regenerative Medicine. *Biomed Res. Int.* **2015**, *2015*, 821279. [CrossRef]
10. Wu, D.T.; Munguia-Lopez, J.G.; Cho, Y.W.; Ma, X.; Song, V.; Zhu, Z.; Tran, S.D. Polymeric scaffolds for dental, oral, and craniofacial regenerative medicine. *Molecules* **2021**, *26*, 7043. [CrossRef]
11. Fakhri, E.; Eslami, H.; Maroufi, P.; Pakdel, F.; Taghizadeh, S.; Ganbarov, K.; Yousefi, M.; Tanomand, A.; Yousefi, B.; Mahmoudi, S.; et al. Chitosan biomaterials application in dentistry. *Int. J. Biol. Macromol.* **2020**, *162*, 956–974. [CrossRef]
12. Singh, B.N.; Veeresh, V.; Mallick, S.P.; Jain, Y.; Sinha, S.; Rastogi, A.; Srivastava, P. Design and evaluation of chitosan/chondroitin sulfate/nano-bioglass based composite scaffold for bone tissue engineering. *Int. J. Biol. Macromol.* **2019**, *133*, 817–830. [CrossRef] [PubMed]
13. Patrulea, V.; Hirt-Burri, N.; Jeannerat, A.; Applegate, L.A.; Ostafe, V.; Jordan, O.; Borchard, G. Peptide-decorated chitosan derivatives enhance fibroblast adhesion and proliferation in wound healing. *Carbohydr. Polym.* **2016**, *142*, 114–123. [CrossRef] [PubMed]
14. Li, D.; Dai, F.; Li, H.; Wang, C.; Shi, X.; Cheng, Y.; Deng, H. Chitosan and collagen layer-by-layer assembly modified oriented nanofibers and their biological properties. *Carbohydr. Polym.* **2021**, *254*, 117438. [CrossRef] [PubMed]
15. Kozusko, S.D.; Riccio, C.; Goulart, M.; Bumgardner, J.; Jing, X.L.; Konofaos, P. Chitosan as a bone scaffold biomaterial. *J. Craniofac. Surg.* **2018**, *29*, 1788–1793. [CrossRef]
16. Turci, F.; Giulia, M.; Corazzari, I.; Nistic, R.; Franzoso, F.; Tabasso, S.; Magnacca, G. Advanced physico-chemical characterization of chitosan by means of TGA coupled on-line with FTIR and GCMS: Thermal degradation and water adsorption capacity. *Polym. Degrad. Stab.* **2015**, *112*, 1–9. [CrossRef]
17. Lauritano, D.; Limongelli, L.; Moreo, G.; Favia, G.; Carinci, F. Nanomaterials for periodontal tissue engineering: Chitosan-based scaffolds. A systematic review. *Nanomaterials* **2020**, *10*, 605. [CrossRef]
18. Reyna-Urrutia, V.A.; Mata-Haro, V.; Cauich-Rodriguez, J.V.; Herrera-Kao, W.A.; Cervantes-Uc, J.M. Effect of two crosslinking methods on the physicochemical and biological properties of the collagen-chitosan scaffolds. *Eur. Polym. J.* **2019**, *117*, 424–433. [CrossRef]
19. Noriega, S.E.; Subramanian, A. Consequences of Neutralization on the Proliferation and Cytoskeletal Organization of Chondrocytes on Chitosan-Based Matrices. *Int. J. Carbohydr. Chem.* **2011**, *2011*, 809743. [CrossRef]
20. Saravanan, S.; Vimalraj, S.; Thanikaivelan, P.; Banudevi, S.; Manivasagam, G. A review on injectable chitosan/beta glycerophosphate hydrogels for bone tissue regeneration. *Int. J. Biol. Macromol.* **2019**, *121*, 38–54. [CrossRef]
21. Wen, Z.; Zhang, L.; Chen, C.; Liu, Y.; Wu, C.; Dai, C. A construction of novel iron-foam-based calcium phosphate/chitosan coating biodegradable scaffold material. *Mater. Sci. Eng. C* **2013**, *33*, 1022–1031. [CrossRef]

22. Mania, S.; Partyka, K.; Pilch, J.; Augustin, E.; Cieślik, M.; Ryl, J.; Jinn, J.R.; Wang, Y.J.; Michałowska, A.; Tylingo, R. Obtaining and characterization of the PLA/chitosan foams with antimicrobial properties achieved by the emulsification combined with the dissolution of chitosan by CO2 saturation. *Molecules* **2019**, *24*, 4532. [CrossRef] [PubMed]
23. Maitra, J.; Shukla, V.K. Cross-linking in Hydrogels—A Review. *Am. J. Polym. Sci.* **2014**, *4*, 25–31. [CrossRef]
24. Silvestro, I.; Sergi, R.; Scotto d'Abusco, A.; Mariano, A.; Martinelli, A.; Piozzi, A.; Francolini, I. Chitosan scaffolds with enhanced mechanical strength and elastic response by combination of freeze gelation, photo-crosslinking and freeze-drying. *Carbohydr. Polym.* **2021**, *267*, 118156. [CrossRef] [PubMed]
25. Arasukumar, B.; Prabakaran, G.; Gunalan, B.; Moovendhan, M. Chemical composition, structural features, surface morphology and bioactivities of chitosan derivatives from lobster (*Thenus unimaculatus*) shells. *Int. J. Biol. Macromol.* **2019**, *135*, 1237–1245. [CrossRef] [PubMed]
26. Shavandi, A.; Bekhit, A.E.D.A.; Sun, Z.; Ali, M.A. Bio-scaffolds produced from irradiated squid pen and crab chitosan with hydroxyapatite/β-tricalcium phosphate for bone-tissue engineering. *Int. J. Biol. Macromol.* **2016**, *93*, 1446–1456. [CrossRef]
27. Yan, J.; Wu, T.; Ding, Z.; Li, X. Preparation and characterization of carbon nanotubes/chitosan composite foam with enhanced elastic property. *Carbohydr. Polym.* **2016**, *136*, 1288–1296. [CrossRef]
28. Thai, H.; Thuy Nguyen, C.; Thi Thach, L.; Thi Tran, M.; Duc Mai, H.; Thi Thu Nguyen, T.; Duc Le, G.; Van Can, M.; Dai Tran, L.; Long Bach, G.; et al. Characterization of chitosan/alginate/lovastatin nanoparticles and investigation of their toxic effects in vitro and in vivo. *Sci. Rep.* **2020**, *10*, 909. [CrossRef]
29. Abbasi, N.; Hamlet, S.; Love, R.M.; Nguyen, N. Journal of Science: Advanced Materials and Devices Porous scaffolds for bone regeneration. *J. Sci. Adv. Mater. Devices* **2020**, *5*, 1–9. [CrossRef]
30. Tamburaci, S.; Kimna, C.; Tihminlioglu, F. Bioactive diatomite and POSS silica cage reinforced chitosan/Na-carboxymethyl cellulose polyelectrolyte scaffolds for hard tissue regeneration. *Mater. Sci. Eng. C* **2019**, *100*, 196–208. [CrossRef]
31. Nwe, N.; Furuike, T.; Tamura, H. The Mechanical and Biological Properties of Chitosan Scaffolds for Tissue Regeneration Templates Are Significantly Enhanced by Chitosan from Gongronella butleri. *Materials* **2009**, *2*, 374–398. [CrossRef]
32. Karimi, Z.; Ghorbani, M.; Hashemibeni, B.; Bahramian, H. Evaluation of the proliferation and viability rates of nucleus pulposus cells of human intervertebral disk in fabricated chitosan-gelatin scaffolds by freeze drying and freeze gelation methods. *Adv. Biomed. Res.* **2015**, *4*, 251. [CrossRef] [PubMed]
33. Sethi, A.; Ahmad, M.; Huma, T.; Khalid, I.; Ahmad, I. Evaluation of Low Molecular Weight Cross Linked Chitosan Nanoparticles, to Enhance the Bioavailability of 5-Flourouracil. *Dose Response* **2021**, *19*, 15593258211025353. [CrossRef] [PubMed]
34. Chuc-Gamboa, M.G.; Vargas-Coronado, R.F.; Cervantes-Uc, J.M.; Cauich-Rodríguez, J.V.; Escobar-García, D.M.; Pozos-Guillén, A.; San Román del Barrio, J. The Effect of PEGDE Concentration and Temperature on Physicochemical and Biological Properties of Chitosan. *Polymers* **2019**, *11*, 1830. [CrossRef] [PubMed]
35. Menazea, A.A.; Eid, M.M.; Ahmed, M.K. Synthesis, characterization, and evaluation of antimicrobial activity of novel Chitosan/Tigecycline composite. *Int. J. Biol. Macromol.* **2020**, *147*, 194–199. [CrossRef]
36. Yang, Z.; Peng, H.; Wang, W.; Liu, T. Crystallization behavior of poly(ε-caprolactone)/layered double hydroxide nanocomposites. *J. Appl. Polym. Sci.* **2010**, *116*, 2658–2667. [CrossRef]
37. Przekora, A.; Benko, A.; Blazewicz, M.; Ginalska, G. Hybrid chitosan/β-1,3-glucan matrix of bone scaffold enhances osteoblast adhesion, spreading and proliferation via promotion of serum protein adsorption. *Biomed. Mater.* **2016**, *11*, 45001. [CrossRef]
38. Galli, C.; Parisi, L.; Elviri, L.; Bianchera, A.; Smerieri, A.; Lagonegro, P.; Lumetti, S.; Manfredi, E.; Bettini, R.; Macaluso, G.M. Chitosan scaffold modified with D-(+) raffinose and enriched with thiol-modified gelatin for improved osteoblast adhesion. *Biomed. Mater.* **2016**, *11*, 15004. [CrossRef]
39. Desbrie, J.; Brugnerotto, J.; Lizardi, J.; Goycoolea, F.M.; Argu, W. An infrared investigation in relation with chitin and chitosan characterization. *Polymer* **2001**, *42*, 3569–3580.
40. Zhang, K.; Peschel, D.; Helm, J.; Groth, T.; Fischer, S. FT Raman investigation of novel chitosan sulfates exhibiting osteogenic capacity. *Carbohydr. Polym.* **2011**, *83*, 60–65. [CrossRef]
41. Aryaei, A.; Jayatissa, A.H.; Jayasuriya, A.C. Mechanical and biological properties of chitosan/carbon nanotube nanocomposite films. *J. Biomed. Mater. Res. A* **2014**, *102*, 2704–2712. [CrossRef]
42. Zająąc, A.; Hanuza, J.; Wandas, M.; Dymińska, L. Determination of N-acetylation degree in chitosan using Raman spectroscopy. *Spectrochim. Acta-Part A Mol. Biomol. Spectrosc.* **2015**, *134*, 114–120. [CrossRef] [PubMed]
43. Machado, A.H.S.; Garcia, I.M.; de Souza da Motta, A.; Leitune, V.C.B.; Collares, F.M. Triclosan-loaded chitosan as antibacterial agent for adhesive resin. *J. Dent.* **2019**, *83*, 33–39. [CrossRef] [PubMed]
44. Qin, H.; Wang, K. Study on preparation and performance of PEG-based polyurethane foams modified by the chitosan with different molecular weight. *Int. J. Biol. Macromol.* **2019**, *140*, 877–885. [CrossRef] [PubMed]
45. Zawadzki, J.; Kaczmarek, H. Thermal treatment of chitosan in various conditions. *Carbohydr. Polym.* **2010**, *80*, 394–400. [CrossRef]
46. Robau-Porrua, A.; Pérez-Rodríguez, Y.; Soris-Rodríguez, L.M.; Pérez-Acosta, O.; González, J.E. The effect of diameter, length and elastic modulus of a dental implant on stress and strain levels in peri-implant bone: A 3D finite element analysis. *Biomed. Mater. Eng.* **2020**, *30*, 541–558. [CrossRef]
47. *D1621–1673*; Astm Standard Test Method for Compressive Properties of Rigid Cellular Plastics. ASTM Standard: West Conshohocken, PA, USA, 1991.

48. Brizuela, A.; Herrero-Climent, M.; Rios-Carrasco, E.; Rios-Santos, J.V.; Pérez, R.A.; Manero, J.M.; Mur, J.G. Influence of the elastic modulus on the osseointegration of dental implants. *Materials* **2019**, *12*, 980. [CrossRef]
49. Changotade, S.; Radu Bostan, G.; Consalus, A.; Poirier, F.; Peltzer, J.; Lataillade, J.-J.; Lutomski, D.; Rohman, G. Preliminary In Vitro Assessment of Stem Cell Compatibility with Cross-Linked Poly(ε-caprolactone urethane) Scaffolds Designed through High Internal Phase Emulsions. *Stem Cells Int.* **2015**, *2015*, 283796. [CrossRef]
50. Flinck, M.; Kramer, S.H.; Pedersen, S.F. Roles of pH in control of cell proliferation. *Acta Physiol.* **2018**, *223*, e13068. [CrossRef]
51. Sweidan, K.; Jaber, A.M.; Al-Jbour, N.; Obaidat, R.; Al-Remawi, M.; Badwan, A. Further investigation on the degree of deacetylation of chitosan determined by Potentiometric titration. *J. Excip. Food Chem.* **2011**, *2*, 16–25.

International Journal of
Molecular Sciences

MDPI

Article

Chitosan Membranes Containing Plant Extracts: Preparation, Characterization and Antimicrobial Properties

Luiza Madalina Gradinaru [1], Mihaela Barbalata-Mandru [1], Alin Alexandru Enache [2], Cristina Mihaela Rimbu [3], Georgiana Ileana Badea [4] and Magdalena Aflori [1,*]

1 "Petru Poni" Institute of Macromolecular Chemistry, 41A Grigore Ghica Voda Alley, 700487 Iasi, Romania; gradinaru.luiza@icmpp.ro (L.M.G.); mihaelamandru84@gmail.com (M.B.-M.)

2 S.C. Apel Laser S.R.L., 25 Vanatorilor Street, 077135 Ilfov, Romania; alin.enache@apellaser.ro

3 Department of Public Health, Faculty of Veterinary Medicine "Ion Ionescu de la Brad", University of Life Sciences, 8 Mihail Sadoveanu Alley, 707027 Iasi, Romania; crimbu@yahoo.com

4 National Institute of Research and Development for Biological Sciences, 296 Independentei Bd. District 6, 060031 Bucharest, Romania; truicageorgiana@yahoo.com

* Correspondence: maflori@icmpp.ro

Abstract: The main strategy of this study was to combine the traditional perspective of using medicinal extracts with polymeric scaffolds manufactured by an engineering approach to fabricate a potential dressing product with antimicrobial properties. Thus, chitosan-based membranes containing *S. officinalis* and *H. perforatum* extracts were developed and their suitability as novel dressing materials was investigated. The morphology of the chitosan-based films was assessed by scanning electron microscopy (SEM) and the chemical structure characterization was performed via Fourier transform infrared spectroscopy (FTIR). The addition of the plant extracts increased the sorption capacity of the studied fluids, mainly at the membrane with *S. officinalis* extract. The membranes with 4% chitosan embedded with both plant extracts maintained their integrity after being immersed for 14 days in incubation media, especially in PBS. The antibacterial activities were determined by the modified Kirby–Bauer disk diffusion method for Gram-positive (*S. aureus* ATCC 25923, *MRSA* ATCC 43300) and Gram-negative (*E. coli* ATCC 25922, *P. aeruginosa* ATCC 27853) microorganisms. The antibacterial property was enhanced by incorporating the plant extracts into chitosan films. The outcome of the study reveals that the obtained chitosan-based membranes are promising candidates to be used as a wound dressing due to their good physico-chemical and antimicrobial properties.

Keywords: chitosan membranes; antimicrobial resistance; *S. officinalis*; *H. perforatum*

1. Introduction

Antimicrobial resistance, one of the major challenges of the 21st century, is the leading cause of mortality worldwide, with the greatest burdens in low-resource regions, as indicated by a comprehensive study that evaluates and estimates the data available up to 2019 [1]. According to this report, the six leading pathogens associated with antimicrobial resistance deaths (*E. coli, S. aureus, K. pneumoniae, S. pneumoniae, A. baumannii* and *P. aeruginosa*) were responsible for 929,000 deaths attributable to antimicrobial resistance and 3.57 million deaths associated with this in 2019. Intervention strategies to address the challenge of antimicrobial resistance can be divided into several main categories, such as prevention through some community-based control programs focused on water, sanitation or hygiene, prevention and control of infections through vaccination, reduction in exposure to antibiotics, minimizing the use of antibiotics or development of new antibiotics [2–4]. Many researchers have recently focused on herbal-based treatment methods to treat a variety of human health problems due to the broad pharmacological importance of medicinal plants [5]. On the basis of the available scientific information, plant extracts and oils have been extensively studied as effective antimicrobial and anti-inflammatory therapeutics [6–9]. Hence, plant-derived antimicrobials have a wide range of activity depending on the species, topography and climate of the country of origin and may contain various categories of active principles, such as polyphenols, quinones, flavonoids, tannins, terpenoids, alkaloids, etc. [10].

Salvia officinalis, also known as sage, is an ornamental, culinary, aromatic and medicinal plant that is mostly used in the pharmaceutical, perfumery and food industries [11]. Although this plant is frequently used as a condiment, its medical uses have opened the way for a new and promising research line. Thus, the essential oil or extract has shown various biological activities, including antioxidant [12,13], antibacterial [14–16], antiviral [17,18] and antifungal [19,20] properties, and also antitumor or antimutagenic effects [21]. It is widely used in the treatment of various diseases, such as those of the nervous system, heart and blood circulation, respiratory system, digestive system, metabolic, and endocrine disease [22]. All of these bioactivities are due to the major phytochemicals, such as alkaloids, phenolic compounds, glycosidic derivatives, terpenes, etc. [23,24].

Hypericum perforatum is a herbaceous perennial plant known also as St. John's wort, which possesses a wide range of biological activities, such as antidepressant [25,26], antianxiety [27], anti-inflammatory [28], antimicrobial [29], wound healing [30] and antitumoral [31]. The biological activities are also due to the numerous phytochemicals that are included in the extracts, such as hypericin, pseudohypericin, rutin, quercetin, quercitrin, etc., leading to unique and combined medicinal effects [32]. Variations in these component contents in the plant are mainly related to the harvesting period, drying process and storage. Thus, due to its rich content of flavonoids, *H. perforatum* also presents antioxidant and free radical scavenger activity [33,34]. Other studies have shown its antimicrobial activity against a variety of bacterial and fungal strains [35,36]. Therefore, several phytochemicals found in the *H. perforatum* extract have been shown to support the desired pharmaceutical effect.

Chitosan is a natural carbohydrate polymer derived from the deacetylation of chitin through chemical or biological processes, which is found as a major component in the shells of crustaceans, such as crabs, shrimp and crawfish [37,38]. It is a cationic polysaccharide that exhibits a wide spectrum of antibacterial and antifungal activity against many bacteria, fungi and yeasts [39,40]. Moreover, due to its high biodegradability, biocompatibility, nontoxicity and similarity in structure to the extracellular matrix component, chitosan is widely used alone or blended with other natural or synthetic polymers in the food, pharmaceutical, medicine, agriculture, cosmetic industries, textiles, etc. [41]. Chitosan can be processed into different forms, such as nanoparticles, gel, membranes, nanofibers, beads and scaffolds, and it is widely applied in biomedical purposes, including drug delivery, regenerative therapy, tissue engineering, wound healing, cancer therapy, bioimaging or veterinary medicine [42,43]. The development of hydrogels based on chitosan is a useful tool for a variety of biomedical applications due to its high absorption capacity, ease of shaping, great capacity to produce a moist environment and ease of drug diffusion [44,45]. Several studies have found that chitosan alone has significant antimicrobial properties against various types of bacteria, including *S. aureus*, *L. plantarum*, *E. coli*, *Salmonella Enteritidis*, etc. [46,47]. Its antimicrobial activity is still unclear due to the complexity of its action on microorganisms that is not well understood, so there are several hypotheses regarding this aspect. One of the most accepted assumptions describes the electrostatic interactions between the positively charged amino groups of chitosan and the negatively charged microbial cell membrane, inducing changes in their permeability, leading to the inhibition of microbial growth [38,48,49]. Another assumption is the chelation of chitosan molecules with some metal ions, which damages the microorganism cell walls [46,50,51]. In addition, the effectiveness of the antimicrobial activity of chitosan is also dependent on its molecular weight. Therefore, low-molecular-weight chitosan is generally able to penetrate the cell wall, affecting DNA transcription [48]. Furthermore, the antimicrobial action of chitosan depends strongly on the type of microorganism. Thus, differences in the cell surface structure of the Gram-positive and Gram-negative bacteria lead to different sensitivity to chitosan. For example, more negatively charged cell surfaces of Gram-negative bacteria allow the binding of cationic chitosan to their phospholipids at low pH [47,48]. Taking into consideration these aspects, we used low-molecular-weight chitosan to prepare the membranes to increase both their extracellular and intracellular antimicrobial activity.

In recent years, researchers have become more and more interested in the synergistic effects of chitosan and various herbal extracts due to the broad pharmacological importance of medicinal plants and growing interest in alternative nanomedicine. Different scaffolds made of chitosan and various herbal extracts have been studied over the years, especially for wound dressing applications. Therefore, the possibility of incorporating natural extracts, such as *S. officinalis* extract, to enhance the antibacterial properties of chitosan has been explored, and a wide range of formulations was prepared. Thus, the *S. officinalis* extract was incorporated into chitosan films, hydrogels or nanoparticles in different forms: encapsulated in nano-form using niosome to enhance biodistribution and bioavailability [47], entrapped into an enzymatic biosensor based on multiwall carbon nanotubes–chitosan–laccase film for total polyphenolic content assessment [48], inserted into a chitosan-based hydrogel formulation for dermal applications [49] or entrapped in chitosan nanoparticles as a stable and protective vehicle to deliver rosmarinic acid [50]. *H. perforatum* extract or oil has also been used to prepare various chitosan-based matrices for wound healing [51–53], as a drug carrier [54] or in tissue engineering applications [55]. Thereby, the encapsulation of the plant extracts in chitosan-based carriers facilitates their delivery in a regulated manner, targeting therapeutic action, and the nanoscale dosage reduces side effects during medical treatment. Therefore, the aim of our study was to develop a novel scaffold by combining the antimicrobial properties of *S. officinalis* and *H. perforatum* extracts with the unique properties of low-molecular-weight chitosan and to investigate its suitability as a potential biomaterial used in tissue engineering or in wound healing applications. The main strategy was to develop a soft and stable chitosan-based scaffold embedded with plant extracts manufactured by an engineering approach. Moreover, we want to underline the potential application of these two extracts for biomedical formulations considering their natural benefits.

2. Results and Discussion

2.1. HPLC Characterization of Plant Extracts

Generally, the chemical composition of plant extracts varies depending on their geographic origin, botanical sources, storage time and conditions, extraction techniques and parts of the plants used during extraction (fruit, stems, leaves, roots or seeds), increasing the difficulty in comparing the modes of action of the main active compounds [8,23,52,53]. Considering these aspects, HPLC analysis was performed in order to identify and quantify the compounds found in the extracts based on the retention time and similarity of UV spectra with the standard substances.

The main constituents of *S. officinalis* extract and *H. perforatum* powder used in the experiments are presented in Table 1. *Salvia* and *Hypericum* species are known to be rich in phenolic acids and flavonoids. Thus, in the obtained *S. officinalis* extract was reported the quantitative determination of flavonoids, including catechin, naringenin, hesperidin, apigenin, pinocembrin, rutin, quercetin, myricetin, rhamnetin or kaempferol, in addition to the different phenolic acids, such as gallic acid, ellagic acid, caffeic acid, caftaric acid, p-coumaric acid, ferulic acid or rosmarinic acid. The presence of these compounds was detected in many species of *S. Officinalis* extracts [21,54–57] at different concentration levels. Numerous flavonoid compounds, such as isoquercitrin, rutin, epicatechin, kaempferol or myricetin, are found in the aboveground portions of the *H. perforatum* plant. Moreover, different phenolic acids (ellagic acid, caffeic acid, caftaric acid, p-coumaric acid or ferulic acid) appeared in both studied plants [57,58]. Actually, the synergistic effects of the variety of minor and major constituents of the studied plant extracts should be considered for their biological activity.

Table 1. HPLC analysis of *H. perforatum* powder and *S. officinalis* extract.

Nr. crt	Compound	*H. perforatum*		*S. officinalis*	
		t_R (min)	Conc (µg/mL)	t_R (min)	Conc (µg/mL)
1	Gallic acid	Nd	-	3.58 ± 0.01	subLOD
2	Catechin	Nd	-	7.62 ± 0.04	11.64 ± 1.57
3	Epicatechin	9.11 ± 0.03	90.61 ± 0.91	Nd	-
4	Ellagic acid	12.19 ± 0.003	71.92 ± 0.47	12.33 ± 0.01	38.22 ± 0.06
5	Hesperidin	14.33 ± 0.01	24.83 ± 1.34	14.52 ± 0.004	34.72 ± 1.17
6	Daidzein	Nd	-	Nd	-
7	Cinnamic acid	Nd	-	Nd	-
8	Apigenin	Nd	-	20.75 ± 0.01	21.88 ± 0.70
9	Genistein	21.20 ± 0.31	54.95 ± 0.59	Nd	-
10	Chrysin	Nd	-	27.29 ± 0.02	2.06 ± 0.29
11	Pinocembrin	27.68 ± 0.004	0.83 ± 0.09	27.62 ± 0.02	8.46 ± 0.32
12	Pinostrobin	Nd	-	Nd	-
13	Caftaric acid	6.36 ± 0.03	4.45 ± 0.31	6.62 ± 0.08	3.85 ± 0.12
14	Chlorogenic acid	7.99 ± 0.04	9.18 ± 1.69	Nd	-
15	Caffeic acid	9.11 ± 0.03	9.94 ± 0.05	8.85 ± 0.14	subLOD
16	Cynarine	9.91 ± 0.01	24.62 ± 0.01	10.00 ± 0.03	16.59 ± 0.05
17	P-Coumaric acid	11.63 ± 0.01	16.32 ± 0.04	11.64 ± 0.02	14.92 ± 0.16
18	Ferulic acid	12.62 ± 0.004	16.56 ± 0.21	12.81 ± 0.01	7.62 ± 0.37
19	Rosmarinic acid	Nd	-	14.93 ± 0.01	29.12 ± 0.10
20	Naringenin	Nd	-	21.30 ± 0.01	subLOD
21	Rutin	12.19 ± 0.003	105.97 ± 0.10	12.29 ± 0.01	1.79 ± 0.18
22	Isoquercetin	12.62 ± 0.004	183.95 ± 0.91	12.81 ± 0.01	124.18 ± 0.54
23	Myricetin	15.35 ± 0.03	9.13 ± 0.35	14.92 ± 0.01	34.36 ± 0.01
24	Quercetin	Nd	-	18.63 ± 0.01	9.83 ± 1.89
25	Kaempferol	21.25 ± 0.01	15.95 ± 0.12	21.61 ± 0.01	28.92 ± 0.29
26	Rhamnetin	24.41± 0.002	13.87 ± 0.23	25.06 ± 0.01	36.49 ± 0.41

2.2. ATR-FTIR Characterization of Chitosan and Chitosan Plant Extract Membranes

ATR-FTIR microscopy was used to determine the chemical binding and functional groups of the chitosan membranes and chitosan plant extract membranes. The recorded spectra of chitosan plant extract membranes were compared to those of chitosan membranes, as illustrated in Figure 1. Firstly, we analyzed the ATR-FTIR spectra of both CL3 and CL4 pure chitosan membranes, which presented a similar pattern. Thus, the pure chitosan membranes' spectra showed a broad absorption peak between 3600 cm^{-1} and 3100 cm^{-1} that can be attributed to O–H and N–H stretching vibrations. The peaks corresponding to asymmetric and symmetric stretching vibrations of C–H bonds in -CH_2 and -CH_3 groups appeared at 2920 and 2870 cm^{-1}, respectively. The absorption band between 1700 cm^{-1} and 1500 cm^{-1} was related to the vibrations of C=O bonds of amide groups (amide II) (1645 cm^{-1}) and to the vibrations of amine groups (1580 cm^{-1}). Bending vibrations of methyl and methylene groups were also visible at around 1416 cm^{-1} and 1377 cm^{-1}, respectively. The spectra between 1160 cm^{-1} and 1000 cm^{-1} are attributed to the vibrations of the CO groups. Thus, according to literature data [59], the small peak located around 1145 cm^{-1} could be attributed to the asymmetric vibrations of CO in the oxygen bridge, resulting from the deacetylation of chitosan. Then, the peaks around 1075–1030 cm^{-1} are attributed to the vibration of CO groups of the ring COH, COC and CH_2OH.

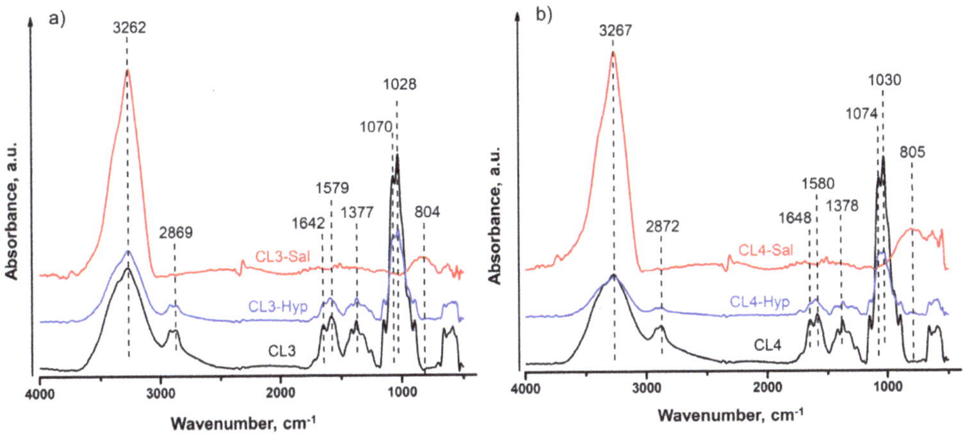

Figure 1. ATR-FTIR spectra of (**a**) 3 %wt and (**b**) 4 %wt pure chitosan and chitosan plant extract membranes.

Then, we investigated the IR spectra of chitosan membranes embedded with plant extracts, which clearly showed a different profile when compared to that of pure chitosan membranes. This is mainly due to the fact that the composition of the plant extracts is quite complex and dissimilar (Supporting Information Figure S1) and thus the absorption peaks characteristic of the chitosan structure overlap with those of the plant extract. Despite these inconveniences, some general conclusions could still be drawn. Therefore, analysis of the chitosan plant extract membranes showed that the profile of all the spectra changed after plant extract loading because of the intermolecular rearrangement and variations in the configuration of the main chain. Moreover, these changes are also due to the synergistic effect of phenolic compounds in the extracts and the interactions between them and the chitosan chain structure [60]. As a result, the broad absorption bands in the region between 3600 cm^{-1} and 3100 cm^{-1} corresponding to the -OH stretching vibrations have increased in intensity when compared to the spectrum of pure chitosan membranes. On this basis, it can be assumed that the chitosan molecules have established intra- and intermolecular hydrogen bonds through their -OH and -NH functional groups with -OH groups of the most prevalent components from the plant extracts, as proposed in the study of Bajic et al. [61].

2.3. SEM Evaluation of Chitosan and Chitosan Plant Extract Membranes

The SEM analysis was performed in order to visualize the surfaces of chitosan-based membranes before and after the addition of the plant extracts (Figure 2). This analysis can directly provide information on the morphology of the studied sample surfaces. Thus, the SEM micrographs revealed that all the samples presented ridge-and-valley structures with continuous and compact surfaces, without any visible cracks or holes. However, some variations in the surface morphology of the membranes were observed depending on the composition. The plant extract's incorporation changes the surface morphology, leading to rough surfaces after their addition as a result of the rise in height of the ridge-and-valley structures. This phenomenon may be due to the spatial organization and the interactions of the constituents during the preparation of the membranes.

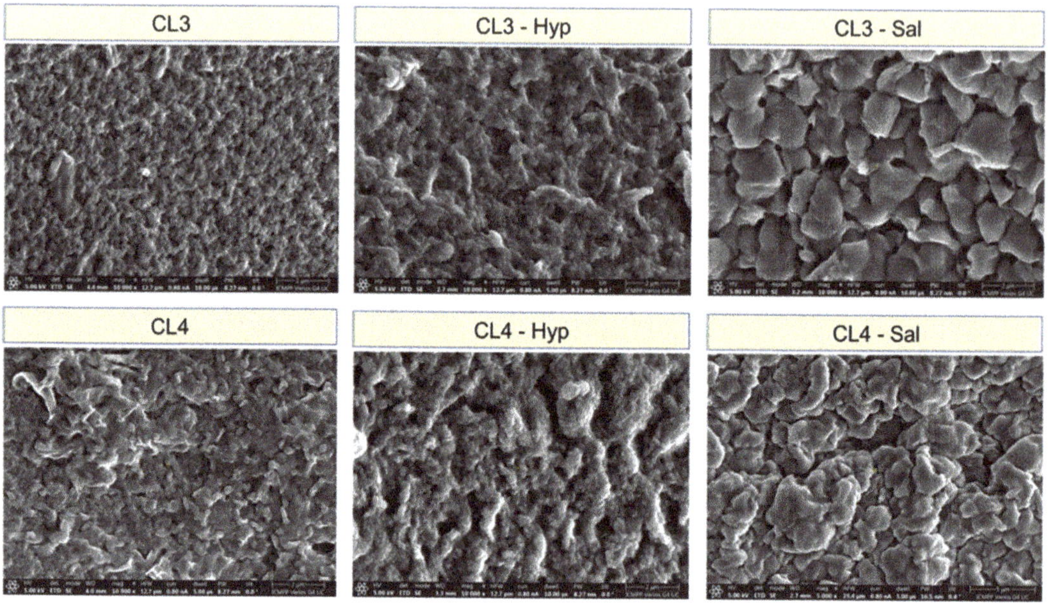

Figure 2. SEM micrograph of pure chitosan and chitosan plant extract membranes.

2.4. In Vitro Incubation

2.4.1. Swelling Capacity of Membranes

The water/fluid uptake capacity of the samples according to their plant extract composition was evaluated in Milli-Q water and PBS medium at 37 °C at specific intervals. The water uptake capacity/swelling of the investigated samples is presented in Figure 3. The swelling capacity of all the samples increased as time increased and reached equilibrium after around 30 min, maintaining their structural integrity for 300 min. After this time, the experiment was stopped because the samples began to deteriorate due to weight loss. The equilibrium swelling of hydrogels is the result of the balance of osmotic forces determined by the affinity for the solvent as well as by the elasticity of the network. Thus, the investigated membranes presented a sharp increase in water uptake in the first half hour and then began to level off. The sorption capacity of the samples with 4% chitosan (up to 200%) was higher than the samples with 3% chitosan (up to 100%). The PBS medium does not have a great influence on the sorption capacity when compared with pure water. The swelling degree profile in both media is almost the same. The samples without plant extract have the lowest swelling degree, while those with *S. officinalis* extract showed the highest swelling capacity. This trend was observed in all the samples. These results suggested that the addition of plant extracts led to an increase in the sorption capacity of fluids. This could be the result of fluids penetrating the spaces that plant extract had previously occupied inside the polymer network after it was released, leading to higher swelling ratios. The analogous results were presented by Drabczyk et al., who used *Aloe vera* embedded in hydrogel materials [62].

The swelling or water uptake capacity of a hydrogel is an important property, especially in wound healing when exudate absorption is required for healing [63]. Therefore, it is essential to develop some new materials with optimal sorption properties adequate for their use and to provide the optimal environment for wound healing.

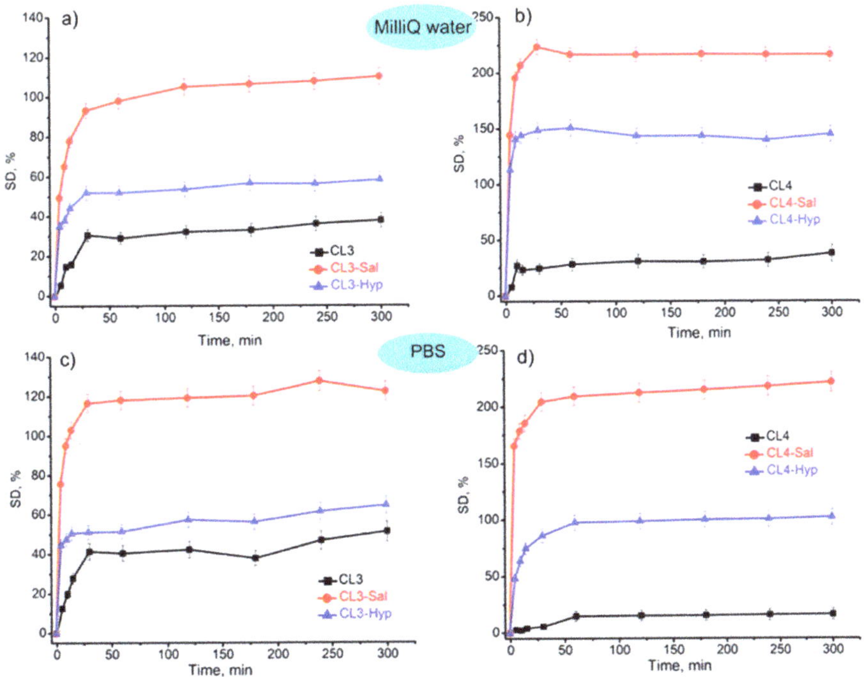

Figure 3. Swelling of membranes with or without plant extracts in (**a**,**b**) Milli-Q water and (**c**,**d**) PBS medium. The values are represented as mean ± standard deviation of triplicate experiments.

2.4.2. PH Metric Analysis

The influence of the membranes with or without plant extracts on the immersed fluids was also evaluated. This incubation study was conducted in order to observe the changes and perhaps the interactions that could appear when the membranes were soaked for 2 weeks in PBS and water. The results were illustrated in Figure 4. According to these results, it was observed that the pH decreases over time in both media. The samples immersed in Milli-Q water presented a decrease in the initial pH from 7.67 to 6.71 (CL3-Sal). The same trend was observed in PBS medium, when the pH decreased from 7.48 (initially) to 7.04 (CL4-Sal).

This decrease in pH is most probably due to the synergy of two factors. One of them is the release of the plant extracts in the incubated media since the initial pH of the plant extract solutions in ethanol was 6.79 and 6.48 for Sal and Hyp, respectively. Another could be the release of some molecules or ions from the membranes, mostly due to their partial degradation. Thus, the tested membranes had a small effect on the pH of the tested fluids.

Moreover, in Milli-Q water, the decrease in pH is more stable than in PBS medium, where fluctuation in the values was noted. This could be explained by the difference in the tested fluids. In the case of Milli-Q water, there were no free ions in the incubation medium that could interact with the tested membranes compared to PBS. The ions from the PBS medium could interact with the samples, leading to some variation in the values. Since the decrease in pH is around 7.0, we could draw the conclusion that the investigated membranes can be considered compatible with the selected fluids.

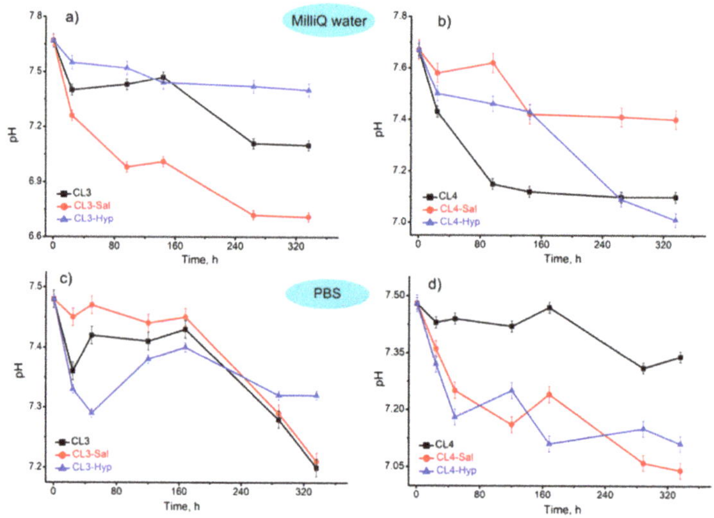

Figure 4. pH stability of membranes with or without plant extracts in (**a**,**b**) Milli-Q water and (**c**,**d**) PBS medium. The values are represented as mean ± standard deviation of triplicate experiments.

2.4.3. Mass Loss of Membranes

The loss in the weight of the investigated membranes with or without plant extracts was evaluated in Milli-Q water and PBS medium after 14 days of incubation. The mass loss was calculated using equation 2 and the results are illustrated in Figure 5. As observed, the degradation rate of the pure chitosan membranes (CL3 and CL4) was higher than the membranes with plant extracts. Thus, the mass loss of the CL3 and CL4 samples after incubation in water was around 84% and 74%, and in PBS was about 76% and 67%, respectively. The addition of the plant extracts led to the decrease in the membranes' mass loss up to 13% for the CL4-Hyp in PBS. This may be because the plant extracts could act as a crosslinker, which diminishes the hydrolytic degradation of the membranes. When comparing the mass loss in the two incubation media (Milli-Q water and PBS), it was observed that the membranes incubated in PBS were more stable, resulting in less mass loss. This is due to the presence of the ions in the PBS media, which led to a more packed structure with lower liquid absorption capacity [64].

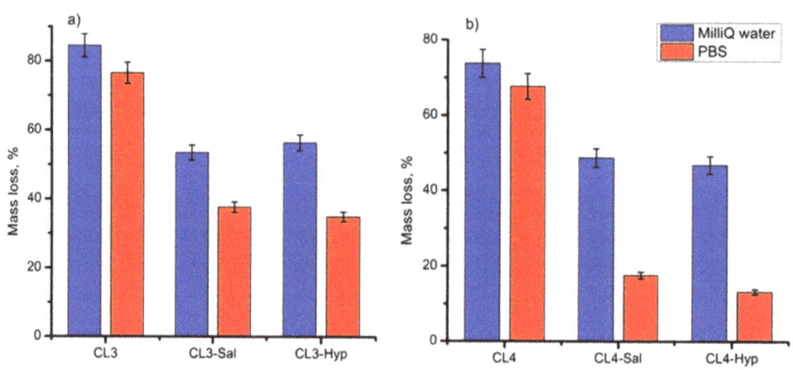

Figure 5. Mass loss after 14 days of incubation in Milli-Q water and PBS for (**a**) 3 %wt and (**b**) 4 %wt membranes with or without plant extracts. The values are represented as mean ± standard deviation of triplicate experiments.

In conclusion, the chitosan membranes with a concentration of 4% embedded with both plant extracts maintained their integrity after being immersed for 14 days in incubation media, especially in PBS. Thus, these membranes are suitable for applications that require higher stability, such as a wound dressing, which should remain in contact with the skin for a long period.

2.5. Antimicrobial Properties of Active Chitosan Membranes

As a key parameter, the antimicrobial property assessment is necessary for evaluating the eligibility and capability of these membranes. Mediterranean medicinal and aromatic herbs contain a wide range of bioactive compounds (polyphenols, terpenes, etc.) that are important constituents of herbal extracts. *S. officinalis* and *H. perforatum* extracts were chosen to be incorporated into the chitosan membranes due to their antimicrobial properties recognized in several works [14–16,29,35,36]. The antimicrobial capacity and potential health benefits of these bioactives could be associated with their synergistic effects.

The antimicrobial activity of the *H. perforatum* and *S. officinalis* extracts, as well as chitosan membranes loaded with both extracts, was tested in triplicate using the diffusimetric technique. The results were interpreted on the basis of the geometric mean of the values obtained by measuring the diameter of the inhibition zones. To highlight the accuracy of the obtained results, the standard deviation from the mean (mean ± SD) was also calculated (Table 2). Considering the differences in bacterial structure and behavior towards antimicrobial agents, the tests were performed against Gram-positive (*Staphylococcus aureus* ATCC 25923, *Methicillin-resistant Staphylococcus aureus* ATCC 43300) and Gram-negative (*Escherichia coli* ATCC 25922, *Pseudomonas aeruginosa* ATCC 27853) bacterial species.

Table 2. The mean value and standard deviation (SD) of the inhibition zones for the investigated samples obtained from the antimicrobial tests against different bacteria strains.

Samples	Initial Diameter	*S. aureus* ATCC 25923	*MRSA* ATCC 43300	*E. coli* ATCC 25922	*P. aeruginosa* ATCC 27853
	(Ø) (mm)	Mean ± SD Ø (mm)	Mean ± SD Ø (mm)	Mean ± SD Ø (mm)	Mean ± SD Ø (mm)
CL3-Hyp	7	8 ± 0.4	7 ± 0.34	10 ± 0.346	10.16 ± 1.06
CL4-Hyp	7	7.5 ± 0.60	9 ± 0.43	12.33 ± 0.61	8.2 ± 0.53
CL3-Sal	7	10.16 ± 0.35	11.36 ± 0.152	13.23 ± 0.55	12.2 ± 0.2
CL4-Sal	7	12.06 + 0.30	8.9 ± 0.53	9.23 ± 0.45	12.06 ± 0.30
CL3 + H_2O	7	7 ± 0	7 ± 0	7 ± 0	7 ± 0
CL4 + H_2O	7	7 ± 0	7 ± 0	7 ± 0	7 ± 0
CL3 + ethanol	7	7 ± 0	7.9 ± 0.41	7 ± 0	7 ± 0
CL4 + ethanol	7	7 ± 0	7 ± 0	9 ± 0.53	7 ± 0
H. perforatum Extract (10 ul) Control	7	12.13 ± 0.30	10 ± 0	13.06 ± 0.30	10.06 ± 0.23
S. officinalis Extract (10 ul) Control	7	12.3 ± 0.43	9.93 ± 0.32	12.9 ± 0.53	10.86 ± 0.47
Gentamicin (10 ug) Control	5	24 ± 0	0 ± 0	22 ± 0	20 ± 2

The antimicrobial activity of the chitosan membranes was influenced by the type of the plant extracts (Table 2, Figure 6). The results show that the extract of *S. officinalis* incorporated into the chitosan membranes has slightly stronger antimicrobial activity against Gram-positive bacteria than the chitosan membranes loaded with *H. perforatum* (Figure 6). In this regard, it was found that CL3-Hyp (8 mm) and CL4-Sal (12.06 mm) samples inhibited the multiplication of *S. aureus* species (Figure 7a). Against *MRSA* species (Figure 7b), the antimicrobial potential reversed so that the CL4-Hyp (9 mm) and CL3-Sal (11.36 mm) samples were most effective. This aspect is also maintained when Gram-negative species are tested, the extract of *S. officinalis* from chitosan membranes being more active than the extract of *H. perforatum* (Table 2). The exception is the CL4-Sal (9.23 mm) membrane, which was tested against *E. coli* (Figure 7c). Apparently, both chitosan membranes loaded with *S. officinalis* extract have almost identical inhibitory activity against *P. aeruginosa* (CL3-Sal, 12.2 mm and CL4-Sal, 12.06 mm), (Figure 7d).

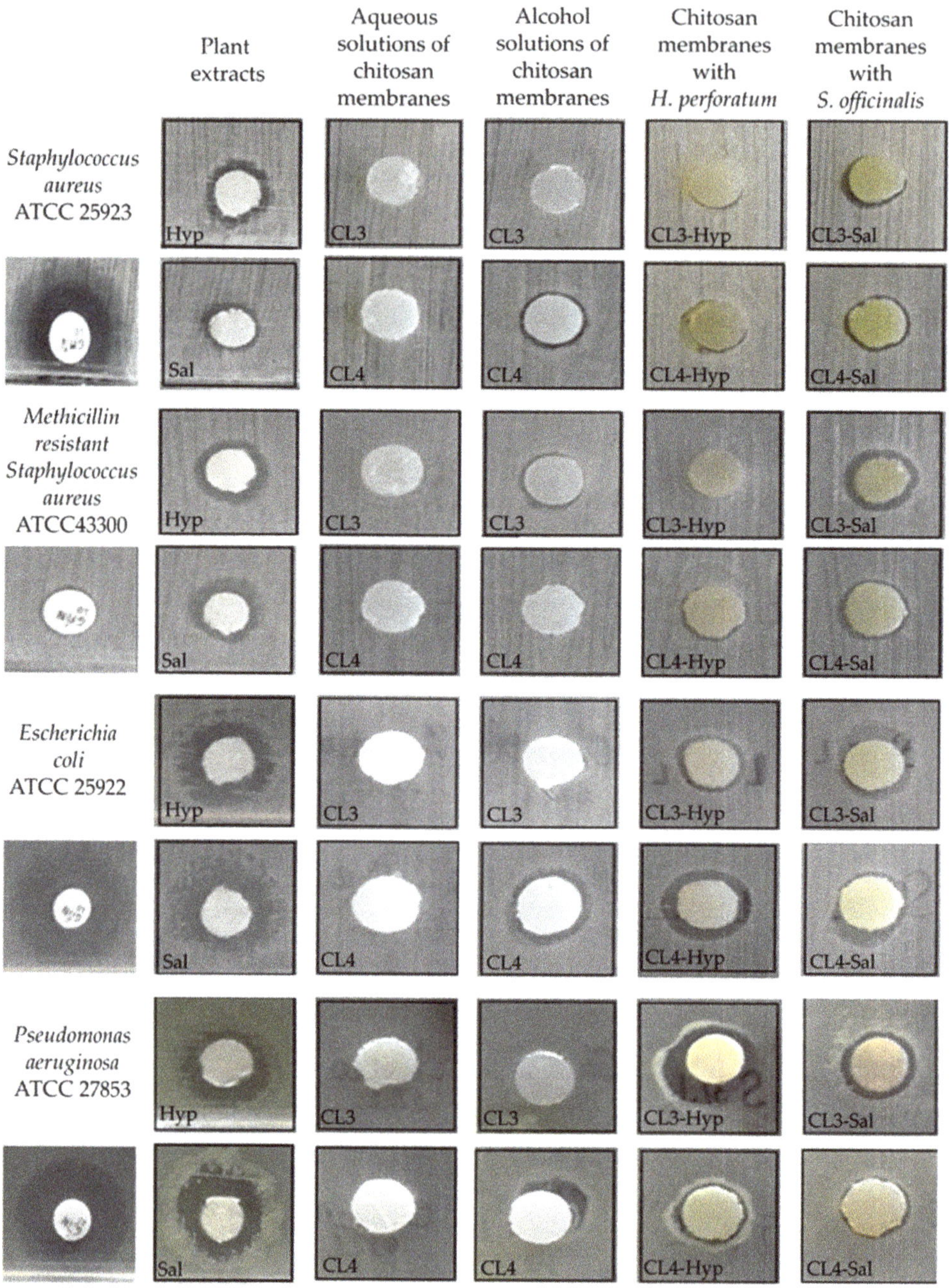

Figure 6. Illustration of the antimicrobial effect tested by diffusimetric method of the investigated samples using Gram-positive and Gram-negative bacteria.

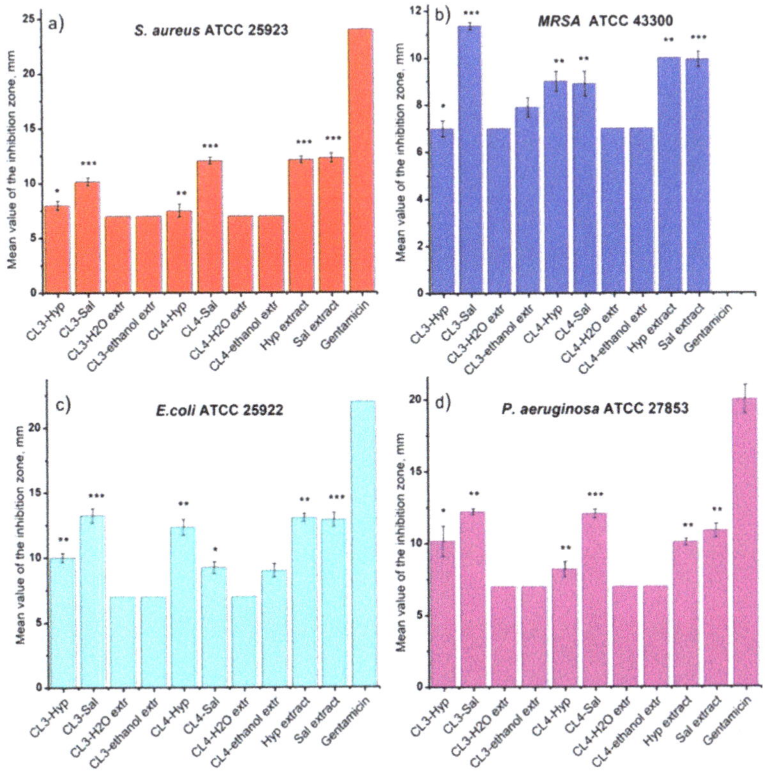

Figure 7. The antimicrobial activity of the investigated samples against Gram-positive (**a**) *S. aureus* and (**b**) *MRSA* and Gram-negative (**c**) *E. coli* and (**d**) *P. aeruginosa* bacteria. The values are represented as mean ± standard deviation of triplicate experiments. Statistical significance: *** $p < 0.001$; ** $0.001 < p < 0.01$; * $0.01 < p < 0.05$.

The tests of *S. officinalis* and *H. perforatum* extracts (10 μL) highlighted the antimicrobial potential of both plant extracts. The ranges of microbial inhibition obtained against *S. aureus* (*H. perforatum*—12.13 mm and *S. officinalis*—12.3 mm), *MRSA* (*H. perforatum*—10 mm and *S. officinalis*—9.93 mm), *E. coli* (*H. perforatum*—13.06 mm and *S. officinalis*—12.9 mm) and *P. aeruginosa* (*H. perforatum*—10.06 mm and *S. officinalis*—10.86 mm) are close in their values (Table 2 and Figures 6 and 7).

H. perforatum, the best-known species of the genus *Hypericum*, is widely used in traditional medicine for its sedative, analgesic, anthelmintic, anti-inflammatory, and antimicrobial properties [65,66]. The literature data frequently refer to the bactericidal potential of both extracts and essential oils of *T. perforatum*, which are mainly effective against Gram-positive bacteria [65,67,68], but not exclusively [69]. In our study, the pure extract of *H. perforatum* or embedded in chitosan membranes induced very similar antimicrobial activity against Gram-negative (*E. coli* and *P. aeruginosa*) and Gram-positive (*S. aureus*, *MRSA*) species. Such results were also mentioned by other authors [29]. It is known that the antimicrobial activity of plant extracts is the cumulative effect of polyphenols (flavonoids or hydrolysable tannins) containing one or more phenolic groups [70]. The most known active compounds of *H. perforatum* that exhibit antimicrobial activity include hypericin, pseudohypericin, naphthodianthrones, flavonoids, phloroglucinols, hyperforin and adhiperforin [66,71]. The main compounds identified in the extract from our study were isoquercitin, epicatechin, rutin and ellagic acid. These compounds are known to have in vitro antimicrobial properties [72,73], especially when they are combined with other identified molecules, attributing real biological properties to the *H. perforatum* extract.

The ethanolic extract of *S. officinalis* is known to have potent antifungal [74] and antibacterial properties against both Gram-positive and Gram-negative bacteria [75,76]. Its antimicrobial effects are closely related to the origin of the plants, the time of harvest and other factors. Sage essential oils are rich in thujone, camphor and 1,8-cineole [77], while camphor predominates in hydrolats [76]. They are considered the most important compounds for antibacterial activity [78]. The pure extract of *S. officinalis* embedded in chitosan membranes inhibited bacterial proliferation of both Gram-positive and Gram-negative species, with a slightly stronger antimicrobial effect than that of *H. perforatum*. These differences in the fluctuation of the antimicrobial activity were mainly due to the different composition and concentration of polyphenols.

In both extracts, isoquercetin was the major flavonoid, whose mechanism of antimicrobial action is based on lipid peroxidation in both Gram-negative and Gram-positive bacteria in response to oxidative stress, leading to membrane liquefaction and permeabilization, and ultimately apoptosis [73].

It is also known that positively charged amino groups of chitosan can electrostatically interact with negatively charged components on the surface of the microbial cell, leading to the damage of membranes' cell walls. Thus, the protonated groups of chitosan are bonded to the negatively charged teichoic acid of peptidoglycans, resulting in disruption of bacterial cell membranes, altering the cell permeability and barrier properties [38,46]. The death of the cells is due to the loss of essential cell materials, such as proteinaceous and other intracellular constituents [40]. This hypothesis is considered to be the widely accepted mechanism of antimicrobial activity of chitosan against Gram-positive and Gram-negative bacteria [47,48]. Furthermore, the chitosan with low molecular weight has the ability to penetrate through the cell wall structure into the cytoplasm of the bacteria, altering the DNA/RNA activity and ultimately inducing cell death [46]. This is because small chains exhibit easier mobility, attraction and ionic interactions than larger ones. In general, the lower the molecular weight, the higher will be the effectiveness in inhibiting microorganisms' growth and multiplication [79]. This is another proposed mechanism of antimicrobial action that is accepted in the literature.

The developed biomaterials based on chitosan embedded with both extracts of *H. perforatum* and *S. officinalis* can be used in many medical fields or textile technology due to their special antimicrobial properties. Considering the performances of the investigated chitosan-based membranes embedded with plant extracts and initiating a path toward the clinical implementation of such biomaterials, further investigations need to be carried out in future studies.

3. Materials and Methods

3.1. Materials

Low-molecular-weight chitosan with a degree of deacetylation of 75–85% and an average molecular weight of 50–190 kDa (based on viscosity), *Hypericum perforatum* extract (hypericin 0.3 mg/g), acetic acid (glacial, >99%) and ethanol (absolute for analysis) were purchased from Sigma-Aldrich (Taufkirchen, Germany). Sodium phosphate buffer solution (PBS), pH = 7.4, was prepared by following the standard protocol, using KH_2PO_4 (1.8 mM), Na_2HPO_4 (10 mM), NaCl (137 mM) and KCl (2.7 mM) dissolved in Milli-Q water. Milli-Q water (18.2 MW·cm) was produced by an Integrity+ Ultrapure water purification system (Adrona, Riga, Latvia). All other chemicals and reagents were of analytical grade and used as received without further purification.

3.2. Plant Materials and Extraction

The extraction of *S. officinalis* was described in a previous study [80]. Briefly, in a mortar, the dried plant buds or leaves were grounded. They were put through a sieve with mesh sizes between 20 and 30, and particle diameters ranging from 0.60–0.85 mm, in order to obtain reproducible extraction yields. The obtained product was stored in a sealed bag in a cool, dry location until they were used [81]. Aldrich® Soxhlet Extraction Equipment, Z556203 (St. Louis, MO, USA), was used for the extraction, which lasted 6 h. This extraction was carried out using 150 mL of ethanol and a known quantity (6 g) of dried materials. The temperature was maintained at 60 °C during the extraction. Then, the solvent was evaporated at 45 °C under reduced pressure in a rotary evaporator (Heidolph Instruments GmbH& CoKG, Schwabach, Germany) to obtain the crude ethanol extract. This was then stored at −20 °C.

3.3. Analysis of Polyphenols by HPLC-UV

In order to identify and characterize the phenolic compounds, a sample of 10 μL of the extracts was analyzed using a Shimadzu LC-20AT HPLC system equipped with a quaternary pump, solvent degasser, autosampler and a photodiode (PDA) detector. The separation of the compounds was performed on a 150 × 4.6 mm, 5 μm Kinetex C18 column, (Agilent Technologies, Santa Clara, CA, USA) using mobile phases A (water and phosphoric acid, pH 2.3) and B (acetonitrile) in a gradient mode (between 5 and 90% component B) for 45 min at a temperature of 35 °C, with a flow rate of 0.8 mL/min. The components of the mobile phase were filtered and degassed before starting the HPLC daily run, through Macherey-Nagel (MV) filters, 0.20 μm pore size and using an Elma-Elmasonic P sonication bath. Detection was performed by scanning from 190 to 800 nm and quantification was assessed at three specific wavelengths (280, 320 and 360 nm). The acquisition of data and the interpretation of the results occurred using the LC Solution software. Concentrations of standard compounds in extracts were determined from the peak areas using the equation for linear regression obtained from the calibration curves.

3.4. Preparation of Chitosan Membranes

In this study, we used low-molecular-weight chitosan for the preparation of the membranes due to its enhanced antimicrobial activity. Thus, the chitosan hydrogels were obtained as previously reported [82] by inducing the gelation of the chitosan solution with aqueous NaOH. Firstly, two chitosan solutions (3%, 4%wt.) were prepared by adding a stoichiometric amount of acetic acid in order to protonate all the amino groups of the chitosan molecules. The chitosan solutions were then placed in Petri dishes (D = 40 mm), immersed in a Berzelius glass containing 1 M NaOH aqueous solution and left a predetermined time to obtain the chitosan membranes. After the desired coagulation time, the Petri dishes were taken out from the NaOH solution. The uncoagulated chitosan solution was removed, and the chitosan membranes were placed into a Berzelius glass with deionized water in order to wash out the NaOH. The washing was carried out until the pH of the chitosan membranes supernatant was neutral.

3.5. Preparation of Chitosan-Based Membranes Containing Plant Extracts

The membranes containing plant extracts were prepared by absorption of plant solutions. Thus, 0.1% stock solutions of plant extracts were freshly prepared in ethanol. Each solution was sonicated for 10 min to ensure a homogenous solution. The chitosan membranes were cut into suitable sizes (Ø = 2 cm) according to the dimensions of the glass vials. Then, 1 mL from the plant extracts stock solution was added and the vials were shaken at 37 °C for 48 h until the solution was completely absorbed by the membranes. The resulting membranes were then freeze-dried using an ALPHA 1–2 LD Christ lyophilizer (Osterode, Germany). The samples were denoted as CLx, CLx-Sal and CLx-Hyp, respectively, where x represents the concentration of low-molecular-weight chitosan (3 and 4 %wt.)

3.6. Characterization of Chitosan-Based Membranes

Attenuated total reflection Fourier transform infrared (ATR-FTIR) spectra of the chitosan-based membranes with or without plant extracts were recorded using a Bruker LUMOS FTIR microscope spectrometer (Bruker Corporation, Karlsruhe, Germany) equipped with an ATR reflection module (Attenuated Total Reflection) and a diamond crystal. The previously prepared samples were cut into small pieces, placed directly on the ATR crystal and subjected to IR investigation. All the spectra were collected in the range 500–4000 cm^{-1}, and the measurements were made by averaging 64 scans at a resolution of 2 cm^{-1}. The spectra were recorded at room temperature.

The surface morphology of the chitosan-based membranes with or without plant extracts was studied by scanning electron microscopy (SEM). The samples were cut into small pieces, mounted on a stub, coated with a thin layer of platinum in a sputtering device and then examined on a Verios G4 UC scanning electron microscope (Thermo Scientific, Waltham, MA, USA) equipped with an energy-dispersive X-ray spectroscopy analyzer (Octane Elect Super SDD detector (AMETEK, Tokyo, Japan)).

3.7. *In Vitro Incubation*

3.7.1. Swelling Capacity of Membranes

The swelling ability of membranes was performed by conventional gravimetric procedure to calculate the water uptake capacity and equilibrium state of the samples with respect to time at body temperature (37 °C). Thus, the swelling degree (%SD) of chitosan-based membranes with or without plant extracts was determined in PBS and Milli-Q water. The membranes were cut in 1 × 1 cm dimensions, weighed and placed in closed glass containers each containing 10 mL of liquids. Then, the samples were incubated at 37 °C in a thermostated oven. At the predetermined time, the membranes were weighed after removing the water excess from the surface with filter paper. The swelling degree (%SD) of the membranes was calculated according to the following equation:

$$\%\,\mathrm{SD} = (\mathrm{W} - \mathrm{Wo})/\mathrm{Wo} * 100 \quad (1)$$

where W and Wo represent the weight of the membranes in the wet and dry states. The results were provided as mean ± standard deviation.

3.7.2. PH Metric Analysis

To evaluate the influence of the membranes on the incubation fluids in vitro, a pH metric study was performed. Our purpose was to observe the pH changes in the fluids used to incubate the investigated membranes over time. The change in the pH values is due to the interaction between the samples and the incubation fluids. Thus, membrane pieces were placed in closed glass containers containing phosphate buffer saline (PBS) and Milli-Q water and incubated at 37 °C. All samples were prepared in triplicate, and the pH was recorded after measurement stabilization, for two weeks, using a pH meter (Hanna Instruments, Woonsocket, RI, USA).

3.7.3. Mass Loss of Membranes

The mass loss was achieved by immersing the pre-weighed membrane pieces (1 × 1 cm) with or without plant extract in tightly closed glass containers containing 10 mL solutions of phosphate buffer saline (PBS) or Milli-Q water. The samples were kept at 50 rpm and 37 °C in a shaker incubator for 14 days. After this interval of time, the samples were rinsed with Milli-Q water and dried in an oven for 48 h. The degradation process was monitored by measuring the weight changes after 14 days of incubation. The percentage of weight loss was calculated using the following equation:

$$\%\,\mathrm{W} = (\mathrm{Wi} - \mathrm{Wt})/\mathrm{Wi} * 100 \quad (2)$$

where Wi and Wt are the weights before and after incubation in solutions.

3.8. *Evaluation of Antimicrobial Activity of Chitosan-Based Membranes*

The antibacterial properties of the pure plant extracts and embedded in chitosan membranes were evaluated using the modified Kirby–Bauer disk diffusion method [83].

The aqueous solutions (CL3 + H_2O, CL4 + H_2O) and the alcoholic solutions (CL3 + ethanol, CL4 + ethanol) served as controls for chitosan membranes without plant extracts. The extracts of *H. perforatum* (Hyp), *S. officinalis* (Sal) and a standardized antibiotic disk (gentamicin, 10 µg, Oxoid) served as controls for the evaluation of antimicrobial activity. Antimicrobial activity was tested against Gram-positive (*Staphylococcus aureus* ATCC 25923, *Methicillin-resistant Staphylococcus aureus* ATCC 43300) and Gram-negative (*Escherichia coli* ATCC 25922, *Pseudomonas aeruginosa* ATCC 27853) bacteria. Cell suspensions were prepared from the 24 h bacterial cultures at a density corresponding to a turbidity of 0.5 on the McFarland scale (1.5×10^8 bacterial cells/mL). Sterile Mueller–Hinton agar (Oxoid) was added to sterile Petri plates, melted and cooled to 45 °C, whereupon 1 mL of the bacterial suspension was added. After drying, usually for 5 min on the thermostat (37 °C), sterile filter paper disks (Ø 7 mm) were spread on the surface of the medium, onto which 10 µL of all test solutions was spotted. Plates were prepared and incubated at 37 °C for 24 h.

3.9. Statistical Analysis

The presented data correspond to the mean ± standard deviation of three experimental values. The statistical differences between data were evaluated by one-way analysis of variance (ANOVA), and $p < 0.05$ was considered to be statistically significant.

4. Conclusions

In this study, chitosan membranes with two different plant extracts were prepared and investigated in terms of antibacterial activity. Commercial *H. perforatum* and *S. officinalis* extracts were used for the preparation of membranes in order to determine their potential use as biomaterials for infection healing. The quantitative analysis of flavonoids and different phenolic acids of the used plant extracts was carried out by HPLC method. The ATR-FTIR spectra demonstrated the successful incorporation of both plant extracts in the chitosan matrix by the presence of the characteristic components.

The swelling degree of the investigated membranes was influenced by the addition of the plant extracts, mainly at the membrane with *S. officinalis* extract. The chitosan-based membranes with or without plant extracts had a slight impact on the pH of the incubation media (PBS and Milli-Q water).

A small decrease in the pH value was noticed over time in both cases. This can be attributed to the release of the plant extracts in the incubated solutions on the one hand and the release of some molecules or ions from the membranes on the other hand, mainly due to their partial degradation. The degradation rate of the pure chitosan membranes (CL3 and CL4) was higher than the membranes with plant extracts after 14 days of incubation. The addition of the plant extracts led to a decrease in the membranes' mass loss up to 13% for the CL4-Hyp in PBS.

The plant extracts conferred significant antibacterial effects to the chitosan-based membranes toward *S. aureus* ATCC 25923, *MRSA* ATCC 43300, *E. coli* ATCC 25922 and *P. aeruginosa* ATCC 27853. The extract of *S. officinalis* incorporated into the chitosan membranes has slightly stronger antimicrobial activity against Gram-positive bacteria than the chitosan membranes loaded with *H. perforatum*.

Additionally, synthetic and chemical antibacterial agents that are known to be harmful to the environment and human health were avoided. Instead, composite polymeric membranes were created using common biomedical materials, such as chitosan and herbal extracts that are completely eco-friendly. It is possible to combine the special properties of polymeric materials with entirely natural herbal solutions to lead an innovative and environmentally friendly approach with potential use in various pharmaceutical or medical applications. This could be induced by the morphological and antibacterial properties of the composites, selecting the proper compositions of polymer-extract types. Thus, all the results demonstrated that these membranes are suitable for applications requiring higher stability, such as a wound dressing that should remain in contact with the skin for a long period.

Supplementary Materials: The following supporting information can be downloaded at: https://www.mdpi.com/article/10.3390/ijms24108673/s1.

Author Contributions: Conceptualization, L.M.G. and M.A.; methodology, A.A.E., M.B.-M., C.M.R. and G.I.B.; investigation, M.B.-M., C.M.R. and G.I.B.; writing—original draft preparation, L.M.G. and M.A.; writing—review and editing, M.A. and L.M.G.; project administration, M.A. and A.A.E.; funding acquisition, M.A. All authors have read and agreed to the published version of the manuscript.

Funding: This research was funded by the Project "Partnerships for knowledge transfer in the field of polymer materials used in biomedical engineering" ID P_40_443, Contract no. 86/8.09.2016, SMIS 105689. co-financed by the European Regional Development Fund by the Competitiveness Operational Programme 2014–2020, Axis 1 Research, Technological Development and Innovation in support of economic competitiveness and business development, Action 1.2.3 Knowledge Transfer Partnerships.

Institutional Review Board Statement: Not applicable.

Informed Consent Statement: Not applicable.

Data Availability Statement: Not applicable.

Acknowledgments: The authors acknowledge the financial support of this research through the Project "Partnerships for knowledge transfer in the field of polymer materials used in biomedical engineering" ID P_40_443, Contract no. 86/8.09.2016, SMIS 105689, co-financed by the European Regional Development Fund by the Competitiveness Operational Programme 2014–2020, Axis 1 Research, Technological Development and Innovation in support of economic competitiveness and business development, Action 1.2.3 Knowledge Transfer Partnerships.

Conflicts of Interest: The authors declare no conflict of interest.

References

1. Murray, C.J.; Ikuta, K.S.; Sharara, F.; Swetschinski, L.; Robles Aguilar, G.; Gray, A.; Han, C.; Bisignano, C.; Rao, P.; Wool, E.; et al. Global burden of bacterial antimicrobial resistance in 2019: A systematic analysis. *Lancet* **2022**, *399*, 629–655. [CrossRef] [PubMed]
2. Swami, O.C. Strategies to Combat Antimicrobial Resistance. *J. Clin. Diagn. Res.* **2014**, *8*, 8–11. [CrossRef] [PubMed]
3. Annunziato, G. Strategies to overcome antimicrobial resistance (AMR) making use of non-essential target inhibitors: A review. *Int. J. Mol. Sci.* **2019**, *20*, 5844. [CrossRef] [PubMed]
4. Merkur, S.W.G. Tackling antimicrobial resistance globally. *Eurohealth* **2020**, *26*, 1–44.
5. Sen, T.; Samanta, S.K. Medicinal Plants, Human Health and Biodiversity: A Broad Review. In *Advances in Biochemical Engineering/Biotechnology*; Springer: Berlin/Heidelberg, Germany, 2014; Volume 147, pp. 59–110.
6. Hart, B.L. The evolution of herbal medicine: Behavioural perspectives. *Anim. Behav.* **2005**, *70*, 975–989. [CrossRef]
7. Limsuwan, S.; Subhadhirasakul, S.; Voravuthikunchai, S.P. Medicinal plants with significant activity against important pathogenic bacteria. *Pharm. Biol.* **2009**, *47*, 683–689. [CrossRef]
8. Heinrich, M.; Jiang, H.; Scotti, F.; Booker, A.; Walt, H.; Weckerle, C.; Maake, C. Medicinal plants from the Himalayan region for potential novel antimicrobial and anti-inflammatory skin treatments. *J. Pharm. Pharmacol.* **2021**, *73*, 956–967. [CrossRef]
9. Vaou, N.; Stavropoulou, E.; Voidarou, C.; Tsigalou, C.; Bezirtzoglou, E. Towards Advances in Medicinal Plant Antimicrobial Activity: A Review Study on Challenges and Future Perspectives. *Microorganisms* **2021**, *9*, 2041. [CrossRef]
10. Saxena, M.; Saxena, J.; Nema, R.; Sing, D.; Gupta, A. Phytochemistry of Medicinal Plants. *J. Pharmacogn. Phytochem.* **2013**, *1*, 168–182. [CrossRef]
11. Cameron, A. *Salvia Officinalis: Production, Cultivation and Uses*; Cameron, A., Ed.; Nova Science Publishers, Inc.: New York, NY, USA, 2021; ISBN 978-1-53619-531-6.
12. Walch, S. Antioxidant capacity and polyphenolic composition as quality indicators for aqueous infusions of *Salvia officinalis* L. (sage tea). *Front. Pharmacol.* **2011**, *2*, 79. [CrossRef]
13. Maleš, I.; Dragović-Uzelac, V.; Jerković, I.; Zorić, Z.; Pedisić, S.; Repajić, M.; Garofulić, I.E.; Dobrinčić, A. Non-Volatile and Volatile Bioactives of *Salvia officinalis* L., *Thymus serpyllum* L. and *Laurus nobilis* L. Extracts with Potential Use in the Development of Functional Beverages. *Antioxidants* **2022**, *11*, 1140. [CrossRef] [PubMed]
14. Mendes, F.S.F.; Garcia, L.M.; da Silva Moraes, T.; Casemiro, L.A.; de Alcântara, C.B.; Ambrósio, S.R.; Veneziani, R.C.S.; Miranda, M.L.D.; Martins, C.H.G. Antibacterial activity of *Salvia officinalis* L. against periodontopathogens: An in vitro study. *Anaerobe* **2020**, *63*, 102194. [CrossRef] [PubMed]
15. Rami, K.; Zheng Guo, L. Antimicrobial activity of essential oil of *Salvia officinalis* L. collected in Syria. *Afr. J. Biotechnol.* **2011**, *10*, 8397–8402. [CrossRef]
16. Ghezelbash, G.H.R.; Parishani, M.R.F.M. Antimicrobial activity of *Salvia officinalis* acetone extract against pathogenic isolates. *J. Herb. Drug* **2015**, *5*, 215–218.

17. Abou Baker, D.H.; Amarowicz, R.; Kandeil, A.; Ali, M.A.; Ibrahim, E.A. Antiviral activity of *Lavandula angustifolia* L. and *Salvia officinalis* L. essential oils against avian influenza H5N1 virus. *J. Agric. Food Res.* **2021**, *4*, 100135. [CrossRef] [PubMed]
18. Azad, M.M.; Al Mahmud, A.; Islam, M.S.; Gouhar, A.I. Common sage (*Salvia officinalis*) antiviral role: Potentiality of a Unani hand sanitizer in COVID-19 (corona virus) second wave control. *Asian J. Med. Biol. Res.* **2020**, *6*, 611–617. [CrossRef]
19. Rus, C.F.; Alexa, E.; Pop, G.; Sumalan, R.M.; Copolovici, D.M. Antifungal Activity and Chemical Composition of *Salvia officinalis* L. Essential Oil. *Res. J. Agric. Sci.* **2015**, *47*, 186–193.
20. Yilar, M.; Kadioglu, I.; Telci, I. Chemical composition and antifungal activity of *Salvia officinalis*, *S. criptantha*, *S. tomentosa* plant essentila oils and extracts. *Fresenius Environ. Bull.* **2018**, *27*, 1695–1706.
21. Ghorbani, A.; Esmaeilizadeh, M. Pharmacological properties of Salvia officinalis and its components. *J. Tradit. Complement. Med.* **2017**, *7*, 433–440. [CrossRef]
22. Miraj, S.; Kiani, S. A review study of therapeutic effects of *Salvia officinalis* L. *Der Pharm. Lett.* **2016**, *8*, 299–303.
23. Raal, A.; Orav, A.; Arak, E. Composition of the essential oil of *Salvia officinalis* L. from various European countries. *Nat. Prod. Res.* **2007**, *21*, 406–411. [CrossRef] [PubMed]
24. Jakovljević, M.; Jokić, S.; Molnar, M.; Jašić, M.; Babić, J.; Jukić, H.; Banjari, I. Bioactive Profile of Various *Salvia officinalis* L. Preparations. *Plants* **2019**, *8*, 55. [CrossRef] [PubMed]
25. Mennini, T.; Gobbi, M. The antidepressant mechanism of *Hypericum perforatum*. *Life Sci.* **2004**, *75*, 1021–1027. [CrossRef] [PubMed]
26. Wurglics, M.; Schubert-Zsilavecz, M. Hypericum perforatum: A "modern" herbal antidepressant—Pharmacokinetics of active ingredients. *Clin. Pharmacokinet.* **2006**, *45*, 449–468. [CrossRef] [PubMed]
27. Flausino, O.A.; Zangrossi, H.; Salgado, J.V.; Viana, M.B. Effects of acute and chronic treatment with *Hypericum perforatum* L. (LI 160) on different anxiety-related responses in rats. *Pharmacol. Biochem. Behav.* **2002**, *71*, 251–257. [CrossRef] [PubMed]
28. Huang, N.; Rizshsky, L.; Hauck, C.; Nikolau, B.J.; Murphy, P.A.; Birt, D.F. Identification of anti-inflammatory constituents in *Hypericum perforatum* and *Hypericum gentianoides* extracts using RAW 264.7 mouse macrophages. *Phytochemistry* **2011**, *72*, 2015–2023. [CrossRef] [PubMed]
29. Egamberdieva, D.; Wirth, S.; Behrendt, U.; Ahmad, P.; Berg, G. Antimicrobial Activity of Medicinal Plants Correlates with the Proportion of Antagonistic Endophytes. *Front. Microbiol.* **2017**, *8*, 199. [CrossRef]
30. Yasar, M.; Kaya, A.; Karaman, H.; Kavugudurmaz, M.; Polat, H.; Sagit, M.; Ozcan, I. Potential Curative Role of *Hypericum perforatum* in an Experimental Rat Model of Tympanic Membrane Perforation. *J. Int. Adv. Otol.* **2016**, *12*, 252–256. [CrossRef]
31. Menegazzi, M.; Masiello, P.; Novelli, M. Anti-Tumor Activity of *Hypericum perforatum* L. and *Hyperforin* through Modulation of Inflammatory Signaling, ROS Generation and Proton Dynamics. *Antioxidants* **2020**, *10*, 18. [CrossRef]
32. Coppock, R.W.; Dziwenka, M.S. John's Wort. In *Nutraceuticals*; Elsevier: Amsterdam, The Netherlands, 2016; pp. 619–631. ISBN 9780128021477.
33. Orčić, D.Z.; Mimica-Dukić, N.M.; Francišković, M.M.; Petrović, S.S.; Jovin, E.Đ. Antioxidant activity relationship of phenolic compounds in *Hypericum perforatum* L. *Chem. Cent. J.* **2011**, *5*, 34. [CrossRef]
34. Fathi, H.; Ebrahimzadeh, M.A. Antioxidant and free radical scavenging activities of *Hypericum perforatum* L. (st. John's wort). *Int. J. For. Soil Eros.* **2013**, *3*, 68–72.
35. Süntar, I.; Oyardı, O.; Akkol, E.K.; Ozçelik, B. Antimicrobial effect of the extracts from *Hypericum perforatum* against oral bacteria and biofilm formation. *Pharm. Biol.* **2016**, *54*, 1065–1070. [CrossRef] [PubMed]
36. Saddiqe, Z.; Naeem, I.; Maimoona, A. A review of the antibacterial activity of *Hypericum perforatum* L. *J. Ethnopharmacol.* **2010**, *131*, 511–521. [CrossRef] [PubMed]
37. Campos, C.A.; Gerschenson, L.N.; Flores, S.K. Development of Edible Films and Coatings with Antimicrobial Activity. *Food Bioprocess Technol.* **2010**, *4*, 849–875. [CrossRef]
38. Reshad, R.A.I.; Jishan, T.A.; Chowdhury, N.N. Chitosan and its Broad Applications: A Brief Review. *J. Clin. Exp. Investig.* **2021**, *12*, em00779. [CrossRef]
39. Kong, M.; Chen, X.G.; Xing, K.; Park, H.J. Antimicrobial properties of chitosan and mode of action: A state of the art review. *Int. J. Food Microbiol.* **2010**, *144*, 51–63. [CrossRef]
40. Rabea, E.I.; Badawy, M.E.T.; Stevens, C.V.; Smagghe, G.; Steurbaut, W. Chitosan as Antimicrobial Agent: Applications and Mode of Action. *Biomacromolecules* **2003**, *4*, 1457–1465. [CrossRef]
41. Zhao, D.; Yu, S.; Sun, B.; Gao, S.; Guo, S.; Zhao, K. Biomedical Applications of Chitosan and Its Derivative Nanoparticles. *Polymers* **2018**, *10*, 462. [CrossRef]
42. Xiong Chang, X.; Mujawar Mubarak, N.; Ali Mazari, S.; Sattar Jatoi, A.; Ahmad, A.; Khalid, M.; Walvekar, R.; Abdullah, E.C.; Karri, R.R.; Siddiqui, M.T.; et al. A review on the properties and applications of chitosan, cellulose and deep eutectic solvent in green chemistry. *J. Ind. Eng. Chem.* **2021**, *104*, 362–380. [CrossRef]
43. Morin-Crini, N.; Lichtfouse, E.; Torri, G.; Crini, G. Applications of chitosan in food, pharmaceuticals, medicine, cosmetics, agriculture, textiles, pulp and paper, biotechnology, and environmental chemistry. *Environ. Chem. Lett.* **2019**, *17*, 1667–1692. [CrossRef]
44. Ahmadi, F.; Oveisi, Z.; Samani, S.M.; Amoozgar, Z. Chitosan based hydrogels: Characteristics and pharmaceutical applications. *Res. Pharm. Sci.* **2015**, *10*, 1–16. [PubMed]

45. Domalik-Pyzik, P.; Chłopek, J.; Pielichowska, K. Chitosan-Based Hydrogels: Preparation, Properties, and Applications. In *Cellulose-Based Superabsorbent Hydrogels, Polymer and Polymeric Composites: A Reference Series*; Mondal, I.H., Ed.; Springer: Basel, Switzerland, 2019; pp. 1665–1693. ISBN 9783319778303. [CrossRef]
46. Yan, D.; Li, Y.; Liu, Y.; Li, N.; Zhang, X.; Yan, C. Antimicrobial Properties of Chitosan and Chitosan Derivatives in the Treatment of Enteric Infections. *Molecules* **2021**, *26*, 7136. [CrossRef] [PubMed]
47. Li, J.; Zhuang, S. Antibacterial activity of chitosan and its derivatives and their interaction mechanism with bacteria: Current state and perspectives. *Eur. Polym. J.* **2020**, *138*, 109984. [CrossRef]
48. Ke, C.-L.; Deng, F.-S.; Chuang, C.-Y.; Lin, C.-H. Antimicrobial Actions and Applications of Chitosan. *Polymers* **2021**, *13*, 904. [CrossRef] [PubMed]
49. Lima, M.; Gomes, L.C.; Teixeira-Santos, R.; Romeu, M.J.; Valcarcel, J.; Vázquez, J.A.; Cerqueira, M.A.; Pastrana, L.; Bourbon, A.I.; de Jong, E.D.; et al. Assessment of the Antibiofilm Performance of Chitosan-Based Surfaces in Marine Environments. *Int. J. Mol. Sci.* **2022**, *23*, 14647. [CrossRef] [PubMed]
50. Aflori, M.; Butnaru, M.; Tihauan, B.-M.; Doroftei, F. Eco-Friendly Method for Tailoring Biocompatible and Antimicrobial Surfaces of Poly-L-Lactic Acid. *Nanomaterials* **2019**, *9*, 428. [CrossRef] [PubMed]
51. Aflori, M. Chitosan-based Silver Nanoparticles Incorporated at the Surface of Plasma-treated PHB Films. *Chem. Lett.* **2017**, *46*, 65–67. [CrossRef]
52. Barra, A. Factors affecting chemical variability of essential oils: A review of recent developments. *Nat. Prod. Commun.* **2009**, *4*, 1147–1154. [CrossRef]
53. Borges, C.V.; Minatel, I.O.; Gomez-Gomez, H.A.; Lima, G.P.P. Medicinal Plants: Influence of Environmental Factors on the Content of Secondary Metabolites. In *Medicinal Plants and Environmental Challenges*; Ghorbanpour, M., Ajit, A., Eds.; Springer International Publishing: Berlin/Heidelberg, Germany, 2017; pp. 259–277. ISBN 9783319687179.
54. Vieira, S.F.; Ferreira, H.; Neves, N.M. Antioxidant and Anti-Inflammatory Activities of Cytocompatible *Salvia officinalis* Extracts: A Comparison between Traditional and Soxhlet Extraction. *Antioxidants* **2020**, *9*, 1157. [CrossRef]
55. Longaray Delamare, A.P.; Moschen-Pistorello, I.T.; Artico, L.; Atti-Serafini, L.; Echeverrigaray, S. Antibacterial activity of the essential oils of *Salvia officinalis* L. and *Salvia triloba* L. cultivated in South Brazil. *Food Chem.* **2007**, *100*, 603–608. [CrossRef]
56. Velickovic, D.; Randjelovic, N.; Ristic, M.; Smelcerovic, A.; Velickovic, A. Chemical composition and antimicrobial action of the ethanol extracts of *Salvia pratensis* L., *Salvia glutinosa* L. and *Salvia aethiopis* L. *J. Serbian Chem. Soc.* **2002**, *67*, 639–646. [CrossRef]
57. Temerdashev, Z.; Milevskaya, V.; Shpigun, O.; Prasad, S.; Vinitskaya, E.; Ryaboko, L. Stability of some biologically active substances in extracts and preparations based on st. John's wort (*Hypericum perforatum* L.) and sage (*Salvia officinalis* L.). *Ind. Crops Prod.* **2020**, *156*, 112879. [CrossRef]
58. Stamenković, J.; Radojković, I.; Đorđević, A.; Jovanović, O.; Stojanović, G. Optimization of HPLC method for the isolation of *Hypericum perforatum* L. methanol extract. *Biol. Nyssana* **2013**, *4*, 81–85.
59. Paluszkiewicz, C.; Stodolak, E.; Hasik, M.; Blazewicz, M. FT-IR study of montmorillonite–chitosan nanocomposite materials. *Spectrochim. Acta Part A Mol. Biomol. Spectrosc.* **2011**, *79*, 784–788. [CrossRef]
60. Abd-Elghany, A.A.; Mohamad, E.A.; El-Sakhawy, M.A.; Mansouri, S.; Ismail, S.H.; Elneklawi, M.S. Enhancement of mechanical properties of chitosan film by doping with sage extract-loaded niosomes. *Mater. Res. Express* **2022**, *9*, 035006. [CrossRef]
61. Bajić, M.; Ročnik, T.; Oberlintner, A.; Scognamiglio, F.; Novak, U.; Likozar, B. Natural plant extracts as active components in chitosan-based films: A comparative study. *Food Packag. Shelf Life* **2019**, *21*, 100365. [CrossRef]
62. Drabczyk, A.; Kudłacik-Kramarczyk, S.; Głąb, M.; Kędzierska, M.; Jaromin, A.; Mierzwiński, D.; Tyliszczak, B. Physicochemical Investigations of Chitosan-Based Hydrogels Containing Aloe Vera Designed for Biomedical Use. *Materials* **2020**, *13*, 3073. [CrossRef]
63. Norahan, M.H.; Pedroza-González, S.C.; Sánchez-Salazar, M.G.; Álvarez, M.M.; Trujillo de Santiago, G. Structural and biological engineering of 3D hydrogels for wound healing. *Bioact. Mater.* **2023**, *24*, 197–235. [CrossRef]
64. Bueno, C.Z.; Dias, A.M.A.; de Sousa, H.J.C.; Braga, M.E.M.; Moraes, Â.M. Control of the properties of porous chitosan–alginate membranes through the addition of different proportions of Pluronic F68. *Mater. Sci. Eng. C* **2014**, *44*, 117–125. [CrossRef]
65. Dall'Agnol, R.; Ferraz, A.; Bernardi, A.P.; Albring, D.; Nör, C.; Sarmento, L.; Lamb, L.; Hass, M.; von Poser, G.; Schapoval, E.E.S. Antimicrobial activity of some *Hypericum* species. *Phytomedicine* **2003**, *10*, 511–516. [CrossRef]
66. Ilieva, Y.; Marinov, T.; Trayanov, I.; Kaleva, M.; Zaharieva, M.M.; Yocheva, L.; Kokanova-Nedialkova, Z.; Najdenski, H.; Nedialkov, P. Outstanding Antibacterial Activity of *Hypericum rochelii*—Comparison of the Antimicrobial Effects of Extracts and Fractions from Four Hypericum Species Growing in Bulgaria with a Focus on Prenylated Phloroglucinols. *Life* **2023**, *13*, 274. [CrossRef] [PubMed]
67. Mazandaran, M.; Yassaghi, S.; Rezaei, M.B.; Mansourian, A.; Ghaemi, E.O. Ethnobotany and Antibacterial Activities of Two Endemic Species of Hypericum in North-East of Iran. *Asian J. Plant Sci.* **2007**, *6*, 354–358. [CrossRef]
68. Bagheri, R.; Bohlouli, S.; Maleki Dizaj, S.; Shahi, S.; Memar, M.Y.; Salatin, S. The Antimicrobial and Anti-Biofilm Effects of *Hypericum perforatum* Oil on Common Pathogens of Periodontitis: An In Vitro Study. *Clin. Pract.* **2022**, *12*, 1009–1019. [CrossRef] [PubMed]
69. Beran, M.; Horna, A.; Vorisek, V.; Berkova, E.; Korinkova, R.; Trousil, V.; Hrubanova, M. Antimicrobial polyhydroxybutyrate submicron fiber mat loaded with extract of *Hypericum perforatum*. *J. Plant Biotechnol.* **2022**, *49*, 257–270. [CrossRef]

70. Manso, T.; Lores, M.; De Miguel, T. Antimicrobial Activity of Polyphenols and Natural Polyphenolic Extracts on Clinical Isolates. *Antibiotics* **2021**, *11*, 46. [CrossRef] [PubMed]
71. Butterweck, V.; Schmidt, M.S. John's wort: Role of active compounds for its mechanism of action and efficacy. *Wien. Med. Wochenschr.* **2007**, *157*, 356–361. [CrossRef]
72. Arima, H.; Ashida, H.; Danno, G. Rutin-enhanced Antibacterial Activities of Flavonoids against *Bacillus cereus* and *Salmonella enteritidis*. *Biosci. Biotechnol. Biochem.* **2002**, *66*, 1009–1014. [CrossRef]
73. Yun, J.; Woo, E.-R.; Lee, D.G. Effect of isoquercitrin on membrane dynamics and apoptosis-like death in *Escherichia coli*. *Biochim. Biophys. Acta-Biomembr.* **2018**, *1860*, 357–363. [CrossRef]
74. Bassolé, I.H.N.; Juliani, H.R. Essential Oils in Combination and Their Antimicrobial Properties. *Molecules* **2012**, *17*, 3989–4006. [CrossRef]
75. Bouajaj, S.; Benyamna, A.; Bouamama, H.; Romane, A.; Falconieri, D.; Piras, A.; Marongiu, B. Antibacterial, allelopathic and antioxidant activities of essential oil of *Salvia officinalis* L. growing wild in the Atlas Mountains of Morocco. *Nat. Prod. Res.* **2013**, *27*, 1673–1676. [CrossRef]
76. Aćimović, M.; Pezo, L.; Čabarkapa, I.; Trudić, A.; Stanković Jeremić, J.; Varga, A.; Lončar, B.; Šovljanski, O.; Tešević, V. Variation of *Salvia officinalis* L. Essential Oil and Hydrolate Composition and Their Antimicrobial Activity. *Processes* **2022**, *10*, 1608. [CrossRef]
77. Gheorghita, D.; Robu, A.; Antoniac, A.; Antoniac, I.; Ditu, L.M.; Raiciu, A.-D.; Tomescu, J.; Grosu, E.; Saceleanu, A. In Vitro Antibacterial Activity of Some Plant Essential Oils against Four Different Microbial Strains. *Appl. Sci.* **2022**, *12*, 9482. [CrossRef]
78. Hammer, K.A.; Carson, C.F. Antibacterial and Antifungal Activities of Essential Oils. In *Lipids and Essential Oils as Antimicrobial Agents*; Wiley: Hoboken, NJ, USA, 2011; pp. 255–306. ISBN 9780470741788.
79. Goy, R.C.; De Britto, D.; Assis, O.B.G. A review of the antimicrobial activity of chitosan. *Polímeros* **2009**, *19*, 241–247. [CrossRef]
80. Barbălată-Mândru, M.; Serbezeanu, D.; Butnaru, M.; Rîmbu, C.M.; Enache, A.A.; Aflori, M. Poly(vinyl alcohol)/Plant Extracts Films: Preparation, Surface Characterization and Antibacterial Studies against Gram Positive and Gram Negative Bacteria. *Materials* **2022**, *15*, 2493. [CrossRef]
81. Wang, H.; Liu, Y.; Wei, S.; Yan, Z. Comparative seasonal variation and chemical composition of essential oils from the leaves and stems of *Schefflera heptaphylla* using microwave-assisted and conventional hydrodistillation. *Ind. Crops Prod.* **2012**, *36*, 229–237. [CrossRef]
82. Enache, A.A.; David, L.; Puaux, J.-P.; Banu, I.; Bozga, G. Kinetics of chitosan coagulation from aqueous solutions. *J. Appl. Polym. Sci.* **2018**, *135*, 46062. [CrossRef]
83. Weinstein, M. Performance Standards for Antimicrobial Disk Suspectibility Tests, Approved Standard-Eleventh Edition. *Clin. Lab. Stand. Inst.* **2018**, *38*, 2162–2914.

International Journal of *Molecular Sciences*

MDPI

Article

Hot Melt Extrusion as an Effective Process in the Development of Mucoadhesive Tablets Containing *Scutellariae baicalensis radix* Extract and Chitosan Dedicated to the Treatment of Oral Infections

Magdalena Paczkowska-Walendowska [1,*], Andrzej Miklaszewski [2], Daria Szymanowska [3], Krystyna Skalicka-Woźniak [4] and Judyta Cielecka-Piontek [1]

1 Department of Pharmacognosy, Poznan University of Medical Sciences, Rokietnicka 3, 60-806 Poznan, Poland
2 Institute of Materials Science and Engineering, Faculty of Materials Engineering and Technical Physics, Poznan University of Technology, Jana Pawła II 24, 61-138 Poznan, Poland
3 Department of Biotechnology and Food Microbiology, Poznan University of Life Sciences, 48 Wojska Polskiego Street, 60-627 Poznan, Poland
4 Department of Natural Products Chemistry, Medical University of Lublin, ul. Chodźki 1, 20-093 Lublin, Poland
* Correspondence: mpaczkowska@ump.edu.pl

Abstract: Hot Melt Extrusion (HME) technology was developed to obtain blends containing lyophilized *Scutellariae baicalensis* root extract and chitosan in order to improve the rheological properties of the obtained blends, including tableting and compressibility properties. (Hydroxypropyl)methyl cellulose (HPMC) in 3 different ratios was used as amorphous matrix formers. The systems were characterized using X-ray powder diffraction (PXRD), Fourier Transform Infrared Spectroscopy with Attenuated Total Reflectance (FTIR-ATR), and in vitro release, permeability, and microbiological activity studies. Then, the extrudates were used to prepare tablets in order to give them the appropriate pharmaceutical form. HPMC-based systems released baicalin more slowly, resulting in delayed peaks in the acceptor fluid. This behavior can be explained by the fact that HPMC swells significantly, and the dissolved substance must have diffused through the polymer network before being released. The best tabletability properties are provided by the formulation containing the extrudate with lyophilized extract HPMC 50:50 *w/w*. These tablets offer a valuable baicalin release profile while maintaining good mucoadhesive properties that condition the tablet's retention in the application site and the effectiveness of therapy.

Keywords: hot melt extrusion; *Scutellariae baicalensis radix*; chitosan; HPMC; controlled release

1. Introduction

Periodontal disease and other oral infections are a significant global burden on oral health, with severe periodontitis responsible for losing many teeth in the adult population worldwide. The WHO Global Oral Health Status Report (2022) estimated that oral diseases affect nearly 3.5 billion people worldwide. At the same time, severe periodontal disease is estimated to affect approximately 19% of the world's adult population, equivalent to more than 1 billion cases worldwide [1]. Over the past 50 years, both systemic and local administration have been crucial methods for drug delivery to treat oral infections. The use of systemic administration in treating oral infections has yielded some positive results and has been widely used for antibiotic therapy in the treatment of periodontitis [2]. However, systemic administration may result in issues including dysbacteriosis and inadequate biodistribution [3]. Because of these apparent disadvantages of systemic administration, there is a great need for local drug delivery systems to improve the prevention and treatment of periodontitis and other oral infections. Local drug delivery systems that are placed directly on the oral mucosa can provide a sufficiently high concentration of an active substance for a reasonably long period of time. Other significant advantages of local drug delivery systems include avoidance of gastrointestinal problems and first-pass metabolism due to direct application to a specific site; higher efficacy and fewer side effects due to controlled drug release; and improved patient compliance due to reduced dosing frequency and easier oral application and the ability to quickly remove the drug in the event of irritability [4]. Moreover, local drug delivery systems exert therapeutic effects mainly through the content of three types of active compounds, including antibacterial agents, modulators of inflammation, and alveolar bone and tissue repair agents for treating periodontitis.

One of the medicinal raw materials for which much has been described and evidence of medical use exists is *Scutellariae baicalensis radix* (Baikal Skullcap Root). The healing effect of *S. baicalensis* root is due to the presence of bioactive compounds, mainly flavones such as baicalin and wogonoside and their aglycones, baicalein and wogonin [5]. Thanks to the content of bioactive compounds, primarily baicalin, the plant material has anti-inflammatory properties by inhibiting the expression of proinflammatory mediators such as IL-1, IL-6, IL-8, and TNF in gingival tissues, antioxidant properties, and antibacterial properties against *Streptococcus mutans, Fusobacterium nucleatum, Aggregatibacter actinomycetemcomitans*, and *Porphyromonas gingivalis* for treating and preventing oral disease [6,7]. In addition, it has been shown that baicalein can increase the expression of osteogenic markers in human periodontal ligament cells, which is valuable in treating periodontitis [8].

Despite the many valuable health-promoting properties of *S. baicalensis* root, the solubility of baicalin is limited, which classifies it as IV BCS [9]. The low solubility limits the application of compounds also in local drug delivery systems. Therefore, one of the biggest challenges for pharmaceutical researchers has been increasing the solubility of the insoluble compound with pharmacological potency. Various methods have been developed, including particle size reduction, solubilization, and solid dispersion, with the latter, produced by hot melt extrusion (HME) becoming increasingly desirable [10,11]. Compared to traditional techniques, HME can offer numerous advantages, both economic benefits due to the shorter time to manufacture the final product and environmental benefits due to the elimination of solvents in the processing process [12]. From a pharmaceutical process point of view, HME involves pumping polymeric materials with a rotating screw at temperatures above their glass transition temperature to achieve molecular-level mixing of active compounds and thermoplastic binders, polymers, or both. The components are changed by this molecular mixing into an amorphous product with a homogeneous shape and density, which improves the dissolution profile of the poorly water-soluble substance [13].

As mentioned above, the low solubility of baicalin limits its application. Therefore, this work aimed to use HME technology to improve the physicochemical properties of baicalin as concluded in lyophilized *S. baicalensis* root extract. Nevertheless, the use of the extrudate itself in medicine is limited; hence, the optimization of the tableting process was done. Thus, the impact of the HME process on the rheological properties, such as tableting and compressibility properties, of the obtained extrudate-based blends was additionally assessed. Receiving appropriate baicalin release profiles as well as mucoadhesive functionality was indicated as being necessary for the development of this form of the drug.

2. Results and Discussion

In the first stage, a lyophilized extract of *Scutellariae baicalensis radix* was obtained according to the procedure described previously [14]. In an earlier study, the phytochemical and biological properties (including antioxidant and anti-inflammatory activity) of the obtained extract were confirmed. At the same time, the work aimed to use HME to improve the parameters for the release of active compounds from the obtained tablets and also to improve the tableting process itself. For a better understanding of all experimental work, all steps were collected in Figure S1 (Supplementary Materials).

As the first task, preparing three types of solid dispersions were possible using the hot melt extrusion technique. Importantly, in all cases, the torque measured during extrusion was similar (around 0.72 Nm), not causing any difficulty during processing. The process temperature (150 °C) did not decompose the active ingredient, i.e., baicalin (melting point 202–205 °C). This was also checked during the evaluation of the content of active compounds by the HPLC method (chromatogram of standards presented in Figure S2, Supplementary Materials), which was validated according to ICH guidelines and whose validation parameters are collected in Table S1 (Supplementary Materials). The phytochemical profile remained at the same level as in the original lyophilized extract: baicalin—2.61 mg per 100 mg of extract; baicalein—323.40 µg per 100 mg of extract; and wogonin—40.30 µg per 100 mg of extract. With the preserved phytochemical profile of bioactive compounds, there was no need to re-examine the biological properties because it is the content of active compounds that determines those effects.

Figure 1 shows the macroscopy pictures of HPMC-based extrudates. As seen, in the case of system-HPMC 75:25 *w*/*w* the inner structure appeared to be non-homogeneous, with a slightly rough surface and somehow "granular" inner structure with variable color. The internal structure changed as the amount of HPMC in the extrudate increased. So in the case of system HPMC 25:75 *w*/*w* outer structure appeared to be rather homogeneous and the surface relatively smooth.

(a) (b) (c)

Figure 1. Macroscopic pictures of hot melt extrudates (surfaces of blended and grounded) based on HPMC in three ratios: (**a**) system-HPMC 75:25 *w*/*w*; (**b**) system-HPMC 50:50 *w*/*w*; (**c**) system-HPMC 25:75 *w*/*w*.

The obtained extrudates were characterized by their structure (XRPD) and possible intermolecular chemical bond formation (FTIR-ATR).

The X-ray diffractograms (Figure 2) of the lyophilized extract and its system with chitosan show a large broadening of the diffraction peaks, which at low intensity indicates their amorphous structure, which was described previously [14]. HPMC is also amorphous in nature. So, hot melt extrudates are based on HPMC, which can be named an amorphous matrix former, transformed into an amorphous state, or molecularly dissolved in the carrier [15,16]. It is shown that the relationship in the 75/25 system shows the lowest degree of order (the lowest intensity and visibility of reflections). The increase in the amount of HPMC in the relationship with chitosan results in a clearer structural response of the extrudate system (an increase in the intensity of reflections and their visibility)—indirectly, this indicates a better reaction of the extract with chitosan to obtain an amorphous system. However, the addition of HPMC is necessary for processing reasons. The addition of the extract affects the position of reflections in the obtained extrudates. From the analysis carried out for the systems, the averaged result based on the chitosan base gives a reduction in the interplanar distances. However, no linear relationship was noted in all analyzed cases (Table 1).

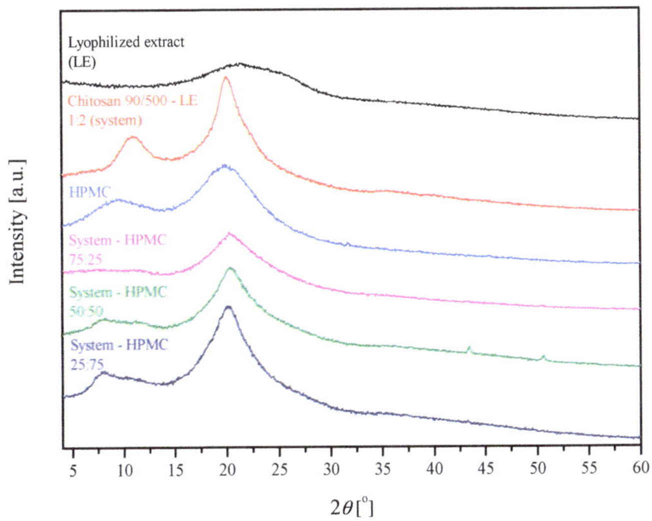

Figure 2. Diffractograms of extrudates.

Table 1. XRPD Signals' Positions.

Sample	Lyophilized Extract	Chitosan	HPMC	System-HPMC Extrudate 75:25	System-HPMC Extrudate 50:50	System-HPMC Extrudate 25:75
(1) Peak position [2θ]	-	11.14	9.77	-	-	-
(2) Peak position [2θ]	23.12	20.19	19.83	20.36	20.51	20.27
Matrix peak position displacement [2θ]	-	-	-	- (2) 0.17	- (2) 0.32	- (2) 0.08
Matrix peak position displacement [Å]	-	-	-	- (2) −0.03	- (2) −0.07	- (2) −0.02
(1) Peak position [2θ]	-	11.14	9.77	-	-	-
(2) Peak position [2θ]	23.12	20.19	19.83	20.36	20.51	20.27
Matrix peak position displacement [2θ]	-	-	-	- (2) 0.17	- (2) 0.32	- (2) 0.08

The obtained extrudates were characterized by their possibility to form intermolecular chemical bonds (FTIR-ATR) (Figure 3). Bands of *S. baicalensis* lyophilized extract at 3330 cm^{-1}, 1720 cm^{-1}, and 1660 cm^{-1} are characteristic for vibration of the O–H, –COOH, and C=O groups, while signals at 1600 cm^{-1} and 1580 cm^{-1} for the C=C vibration of the aromatic rings in the structure of flavones. The broad bands in the range 1200–900 cm^{-1} are characteristic of vibrations of C–O bonds of saccharides [14]. For the HPMC spectrum, a wide band was observed at 3300 cm^{-1}, associated with the presence of -OH groups. While the complex band between 1200 and 950 cm^{-1} is related to numerous C–O vibrations, including glycosidic C-O-C, C-OH, C-OCH_3, C-OCH_2CH_2OH [17]. In the case of extrudates, it can be observed that the bands at 3300 and 1600 cm^{-1} changed, broadened, and decreased in intensity, which means intermolecular hydrogen bonds between the extract and carrier, which has also been observed for solid dispersions of pure baicalin [18].

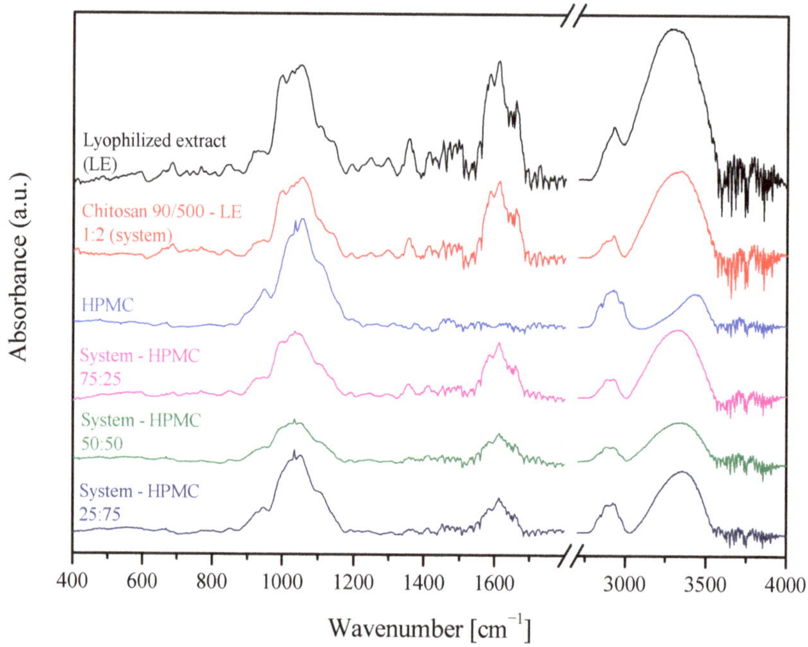

Figure 3. FTIR-ATR spectra of extrudates.

Figure 4 shows the release profiles of baicalin from ground hot melt extrudates based on HPMC in three different ratios. For comparison, the baicalin release from the lyophilized extract as well as the dissolution rate of pure baicalin are also shown. The dissolution rate of pure baicalin was very low; only 50% of pure baicalin was dissolved in 4 h due to its poor wettability and agglomeration. An increased dissolution rate of baicalin from the freeze-dried extract was observed, reaching 80% dissolved baicalin within 15 min, which is related to the change from crystalline to amorphous form. The HME process additionally improved the dissolution. Despite the slower solubility of baicalin, 80% over 90 min due to the presence of HPMC, the HME process improves wettability, reduces the size of baicalin dispersion, and prevents agglomeration of particles [19]. Differences in baicalin release from HPMC extrudates depend on the amount of carrier in the system. Firstly, differences in dissolution rates were statistically significant among the three extrudates (in all cases, f_1 was below 20 and f_2 was below 50). Secondly, with the increase in the amount of HPMC, baicalin dissolves to a lesser extent due to the hydration of the outer layer of the system, which causes the formation of a gel layer on its surface. This reduces the amount of water that enters the system's core, which can hinder the movement of the active compounds and cause them to dissolve slowly [20].

Additionally, permeability coefficients using the PAMPA test were established. While the permeation test is not critical when talking about local application, it is intended to more extensively check the material's properties after extrusion. The permeability coefficient for pure baicalin, calculated from equation no. 1 (n = 6), was $0.02 \pm 0.01 \times 10^{-6}$ cm/s, which is in line with previous research [21], and also confirms its low permeability, classifying baicalin as BCS IV [9]. Due to the improved solubility associated with the amorphization of baicalin, the permeability of the compound also increased. Permeability coefficients for extrudates were 0.96 ± 0.02, 0.74 ± 0.02 and $0.58 \pm 0.02 \times 10^{-6}$ cm/s, respectively, for extrudates 75:25, 50:50 and 25:75. The decrease in the permeation coefficient with the increase in the HPMC content in the system can also be explained by the formation of a gel layer, which makes it difficult for the active ingredients to reach the biological barrier.

However, thanks to the amorphization of the system, multiple increases in the permeation of baicalin can be observed, which in turn is associated with an increase in its dissolution rate from extrudates. Nevertheless, penetration above 1×10^{-6} cm/s was still not achieved, so it cannot be said that the system penetrates well. What is intended from the point of view of topical application within the oral cavity, a very well-constructed system has been achieved where baicalin appears at the application site in a higher dose due to the increase in release rate while not penetrating the systemic circulation, staying at the place of application.

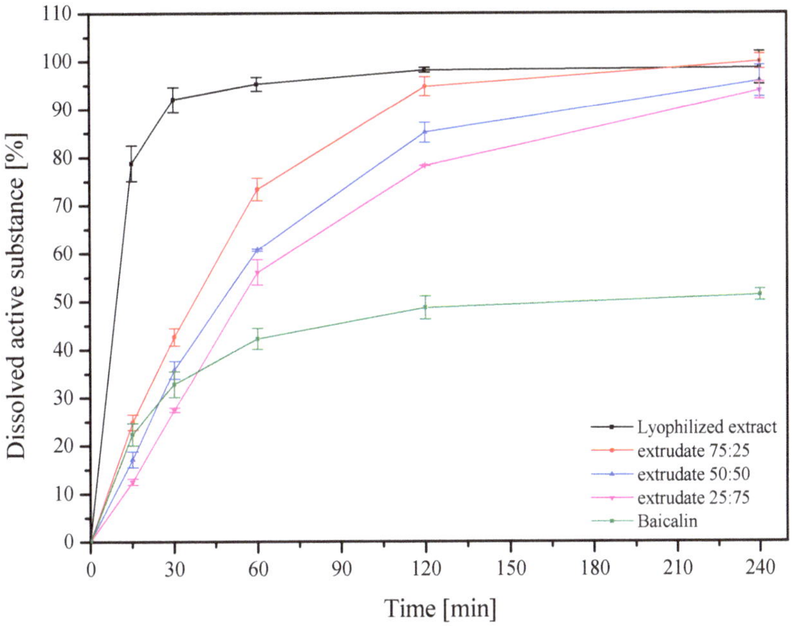

Figure 4. Dissolution profiles of baicalin from the lyophilized extract and extrudates (n = 6).

In the presented study, the antimicrobial activity of the prepared extrudates against microorganisms colonizing the oral cavity (e.g., *S. mutans*) and bacteria whose presence in the oral cavity causes the development of infection (e.g., *P. aeruginosa. S. aureus, E. aerogenes*) was investigated. The antimicrobial activity of binary systems was evaluated according to their inhibition zone diameter against six species of bacteria (Figure 5). Table 2 presents the results of the impact of those tested on the ability to increase micro- organisms in the medium. Apart from the obvious fact that the lyophilized extract has antibacterial activity, which was described earlier, it is worth noting that chitosan has an equally important effect on the activity of the extrudates. So, the results revealed the highest antimicrobial activity for extrudate 25:75. In contrast, the most significant increase in activity following the combination of *S. baicalensis radix* extract with chitosan was observed against *S. mutans*, the most sensitive strain and, at the same time, one of the pathogens causing periodontitis. In liquid cultures, three different concentrations of extrudates were used. The results in Table 2 shows that only a concentration of 100 mg/mL exhibited antimicrobial activity.

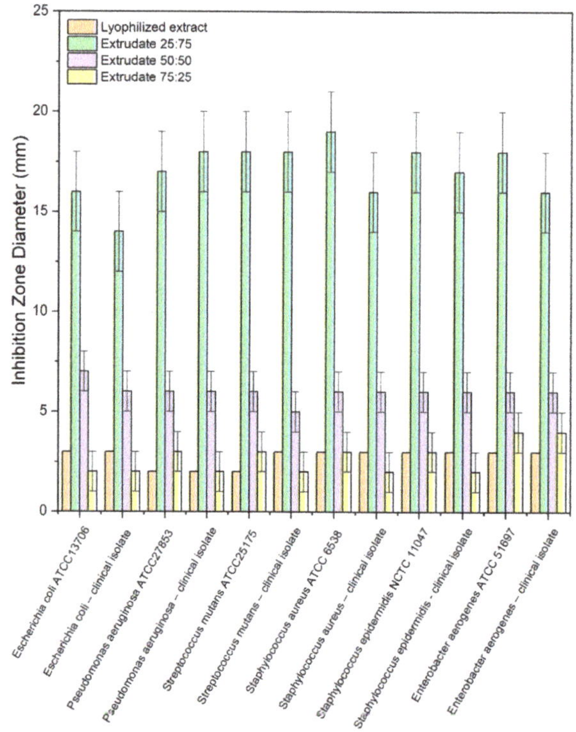

Figure 5. Influence of the lyophilized extract and its extrudates on indicator microorganisms (reference strains and clinical isolates) by the well-diffusion method.

Table 2. Influence of the raw material on indicator microorganisms (reference strains and clinical isolates) by the liquid culture method—concentration of the test sample 100 mg/mL.

	Lyophilized Extract	Extrudate 25:75	Extrudate 50:50	Extrudate 75:25
	Number of microorganisms [CFU]			
Escherichia coli ATCC13706	$6.9 \times 10^2 \rightarrow 3.9 \times 10^7$	$4.8 \times 10^2 \rightarrow$ nd	$2.0 \times 10^2 \rightarrow 3.9 \times 10^5$	$2.0 \times 10^2 \rightarrow 1.4 \times 10^7$
Escherichia coli—clinical isolate	$3.4 \times 10^2 \rightarrow 5.1 \times 10^7$	$2.0 \times 10^2 \rightarrow$ nd	$3.3 \times 10^2 \rightarrow 8.0 \times 10^4$	$2.9 \times 10^2 \rightarrow 3.3 \times 10^7$
Pseudomonas aeruginosa ATCC27853	$2.1 \times 10^2 \rightarrow 5.9 \times 10^7$	$2.7 \times 10^2 \rightarrow$ nd	$4.4 \times 10^2 \rightarrow 3.7 \times 10^5$	$2.0 \times 10^2 \rightarrow 5.8 \times 10^6$
Pseudomonas aeruginosa—clinical isolate	$2.5 \times 10^2 \rightarrow 6.0 \times 10^6$	$2.0 \times 10^2 \rightarrow$ nd	$5.8 \times 10^2 \rightarrow 3.6 \times 10^5$	$2.0 \times 10^2 \rightarrow 5.3 \times 10^7$
Streptococcus mutans ATCC25175	$6.9 \times 10^2 \rightarrow 3.9 \times 10^7$	$4.8 \times 10^2 \rightarrow$ nd	$2.0 \times 10^2 \rightarrow 3.9 \times 10^5$	$2.0 \times 10^2 \rightarrow 1.4 \times 10^7$
Streptococcus mutans—clinical isolate	$3.4 \times 10^2 \rightarrow 5.1 \times 10^7$	$2.0 \times 10^2 \rightarrow$ nd	$3.3 \times 10^2 \rightarrow 8.0 \times 10^5$	$2.9 \times 10^2 \rightarrow 3.3 \times 10^7$
Staphylococcus aureus ATCC 6538	$2.6 \times 10^2 \rightarrow 3.0 \times 10^7$	$3.6 \times 10^2 \rightarrow$ nd	$2.9 \times 10^2 \rightarrow 1.7 \times 10^4$	$1.9 \times 10^2 \rightarrow 5.9 \times 10^6$
Staphylococcus aureus—clinical isolate	$2.0 \times 10^2 \rightarrow 3.7 \times 10^7$	$1.9 \times 10^2 \rightarrow$ nd	$2.6 \times 10^2 \rightarrow 3.0 \times 10^5$	$3.4 \times 10^2 \rightarrow 3.0 \times 10^6$
Staphylococcus epidermidis NCTC 11047	$3.6 \times 10^2 \rightarrow 3.3 \times 10^6$	$2.0 \times 10^2 \rightarrow$ nd	$2.5 \times 10^2 \rightarrow 3.2 \times 10^4$	$2.1 \times 10^2 \rightarrow 3.6 \times 10^7$
Staphylococcus epidermidis—clinical isolate	$7.7 \times 10^2 \rightarrow 8.5 \times 10^4$	$2.9 \times 10^2 \rightarrow 3.3 \times 10^2$	$3.4 \times 10^2 \rightarrow 5.1 \times 10^4$	$2.9 \times 10^2 \rightarrow 3.0 \times 10^8$
Enterobacter aerogenes ATCC 51697	$2.6 \times 10^2 \rightarrow 3.0 \times 10^5$	$3.6 \times 10^2 \rightarrow 3.0 \times 10^2$	$2.9 \times 10^2 \rightarrow 1.7 \times 10^5$	$1.9 \times 10^2 \rightarrow 5.9 \times 10^6$
Enterobacter aerogenes—clinical isolate	$2.0 \times 10^2 \rightarrow 3.7 \times 10^6$	$1.9 \times 10^2 \rightarrow$ nd	$2.6 \times 10^2 \rightarrow 3.0 \times 10^4$	$3.4 \times 10^2 \rightarrow 3.0 \times 10^6$

nd—no detected.

In the next step, tablets containing all three types of extrudates were successfully prepared. A total of 6 formulations were prepared: three had extrudates (formulations F1, F3, and F5) and three contained the identical amounts of ingredients but in the form of powders, which were controls and comparative formulations (formulations F2, F4, and F6). HPMC was a carrier for hot melt extrusion in all formulations. Still, it also imparted mucoadhesive properties to the prepared systems.

Firstly, tablets containing formulations F1–F6 were initially characterized in terms of tabletability, compressibility, and compactability (Figure 6). The tabletability of the tablets decreased in the following order: F1~F3 > F5 > F6 > F4 > F2 (Figure 6a,b). Such an order is related to the composition of the individual components. In general, it can be said that extrudate-based tablets showed better tabletability properties than those containing powders. However, a smaller amount of HPMC in the tablet increases its tabletability. According to the overall trend of the compressibility profile, the porosity level or the solid fraction value decreases as the pressure load applied to the powder samples grows (Figure 6b). The compactibility profiles for all six formulations are generally comparable, with little benefit for powder-based tablets. Finally, a powder's compactability is defined as its ability to form coherent, strong tablets. Low-density tablets are obviously more porous because they have more pores, leading to poorer interparticle bonding. As a result, less power is needed to break down those tablets. The order of decreasing compactability appears to be as follows: F1 > F3 > F6 > F5 > F4 > F2 (Figure 6d,e). It can also be seen from Figure 6c that tablet tensile strength decreases exponentially with increasing porosity, which fits the Ryshkewitch equation (equations no. 5–6) where T_{S0} is the extrapolated tensile strength at zero porosity and T_{S0} is often used to compare bond strength. The T_{S0} of F2, F4, and F6 was, respectively, 1.5, 2.4, and 3.2 Mpa, the bonding strength of powders was poor. The T_{S0} of extrudates were enhanced to 4.7 and 4.2 for F1 and F3, and obtained a surprisingly low value for F6–1.7 Mpa. Bond strength increased, illustrating that the HME process can significantly improve bond strength. This may be due to better uniformity and closer contact during melting and extrusion. It can also be the result of the transformation from the crystalline to the amorphous form of the active compounds [19]. Based on the above parameters, the best tablet properties were obtained for formulations F1 (extrudates of system-HPMC 75:25) and F3 (extrudates of system-HPMC 50:50).

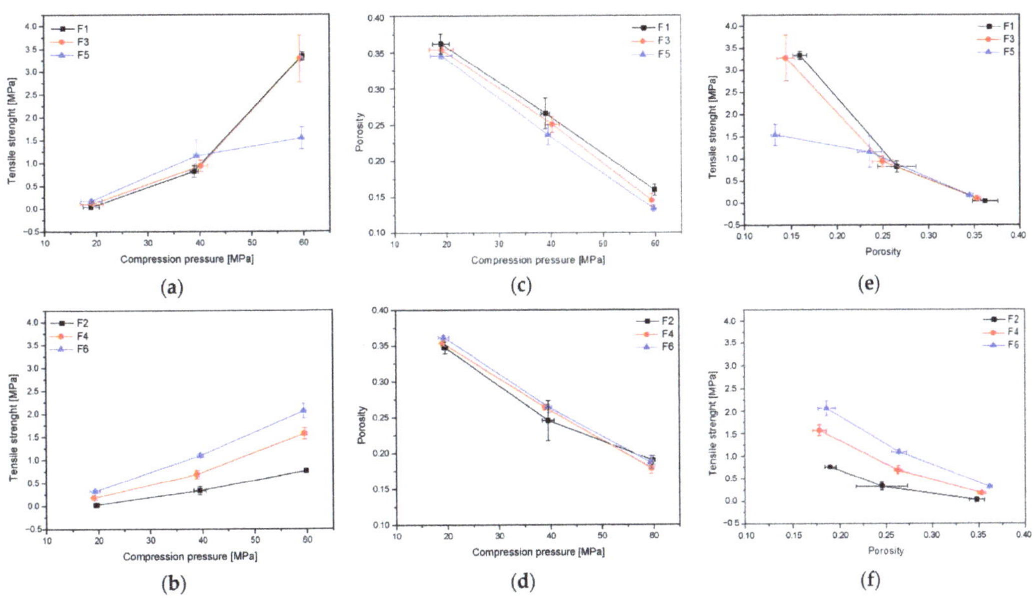

Figure 6. Tabletability (**a**,**b**), compressibility (**c**,**d**) and compactibility (**e**,**f**) of the extrudates and powder systems.

In the next step, the dissolution rate of baicalin from the F1–F6 formulation was assessed (Figure 7). As described above, an increased dissolution rate of baicalin from the lyophilized extract and extrudates was observed, which is related to the change from crystalline to amorphous form (Figure 4). The dissolution profiles of baicalin from extrudates and formulations prepared from them (extrudates 75:25 and F1, extrudates 50:50 and F3, and extrudates 25:75 and F5) differing in the pressure used to prepare the tablets were compared. In each case, it was noticed that the dissolution rate decreased with the increase in compression pressure, but the differences were not statistically significant (in all cases, f_1 was below 20 and f_2 was above 50). Importantly, baicalin release from extrudate-based formulations was relatively fast, even faster than release from powder-based tablets. In the case of powder-based tablets, a slow and controlled release was observed (Figure 7d–f). This substantial difference can be explained by the swelling behavior of these extrudates as well as HPMC, calculated from equation no. 7 (Figure 8). While powder-based systems rapidly absorb water upon contact with the release medium, extrudate-based tablets remained almost intact. The HPMC powder swelled significantly, and the active compounds had to pass through the polymer network, and a more extensive layer of gel formed around the powder tablets, which made it difficult for baicalin to dissolve and enter the release medium [22]. HME changes the behavior of HPMC, and the reprocessed carrier has less water absorption and sticky layer properties. In addition, changes in the release of baicalin can be observed depending on the amount of HPMC in the system. When the percentage of HPMC in the formulation increased, the baicalin release rate decreased simultaneously, both from extrudate- and powder-based tablets.

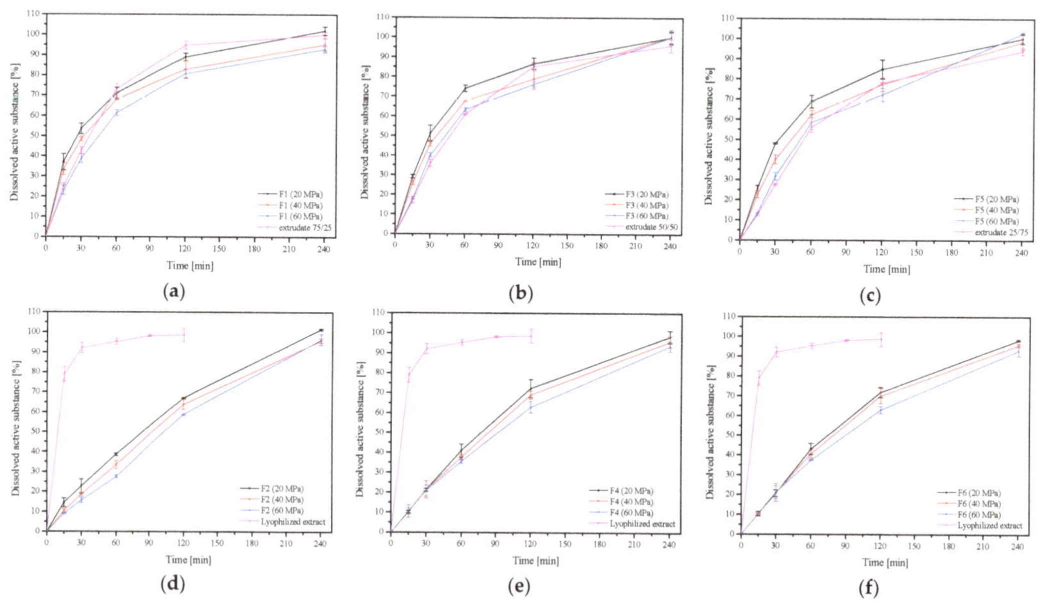

Figure 7. Dissolution profiles of baicalin from the extrudate-based tablets (**a–c**) and the powder-based tablets (**d–f**) (n = 6).

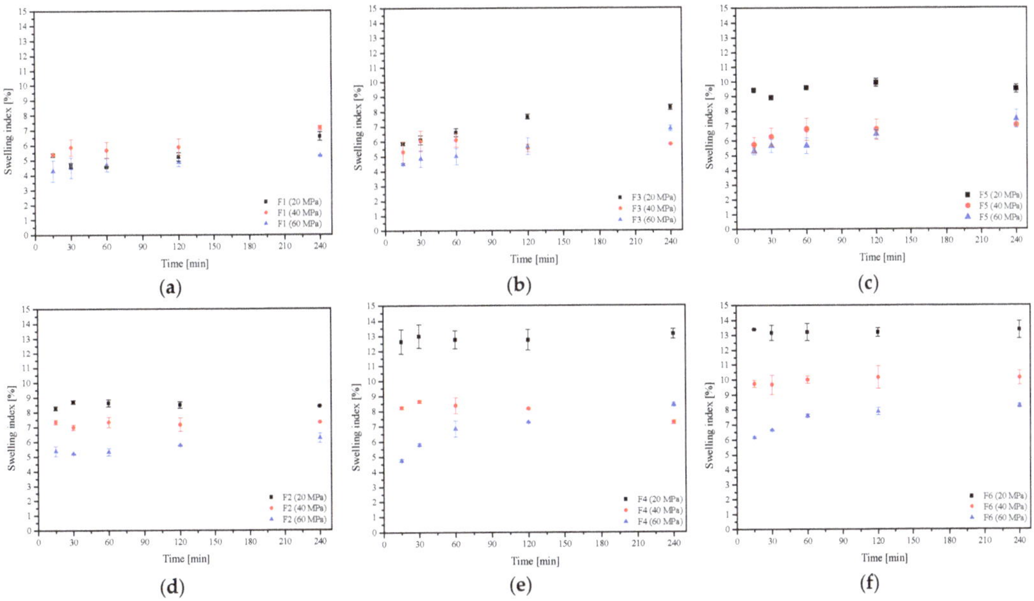

Figure 8. Swelling index of the extrudate-based tablets (**a–c**) and the powder-based tablets (**d–f**) ($n = 3$).

Differences in the release of baicalin from extrudate- and powder-based tablets are also visible in the kinetics of its release. Mathematical models describing the release kinetics of baicalin from formulations F1–F6 are collected in Table S2 (Supplementary Materials). As indicated above, the release of baicalin from powder-based tablets (formulations F2, F4, and F6) is considerably slowed down and controlled, and the release of baicalin follows zero-order kinetics. It means that the release rate of baicalin is constant over a period of time. Such controlled release systems are indicated and developed in therapeutic drug delivery systems, but if mucosal application within the oral cavity is desired, complete release of the substance should occur within 2 h because a longer stay of the tablet stuck to the mucous membrane may be uncomfortable for the patient. In this regard, baicalin release is preferred from an extrudate-based tablet (formulations F1, F3, and F5). For these formulations, as the most probable, the Higuchi model was shown, which best describes the release from the matrix system and suggests that the baicalin was primarily released by diffusion and that its release was from a homogeneous flat matrix that did not degrade [14]. Additionally, a good fit to Korsmeyer-Peppas with 'n' values in the range 0.45–0.89 indicated the release approximated the non-Fickian diffusion release mechanism [23]. The relative complexity of the prepared formulations may indicate that the active compound release is controlled by more than one mechanism; a coupling of polymer erosion, swelling, and dissolution, which were all involved in the release process, which is consistent with the previous literature data [24].

Finally, the mucoadhesive properties of formulations F1–F6 were evaluated by rheological measurements (Figure 9). The basis of the blends is a lyophilized extract with chitosan, to which HPMC was added as a carrier. The mucoadhesive properties of chitosan are widely known, and chitosan–mucin interact mainly electrostatically, supported by other types of interactions (e.g., hydrogen bonds and hydrophobic association) [25]. In this case, to demonstrate the mucoadhesive effect, an appropriate pH is necessary (pH < 6), so the pH of the oral cavity is on the limit [26]. However, HPMC is a non-ionic polymer; the medium's pH had no effect on how well it stuck to the mucosa. Thus, in the case of the described blends, HPMC is the primary mucoadhesive agent. It has a lot of hydroxyl groups, which allow it to form intermolecular bonds (including hydrogen interactions) with the components of mucus [27,28]. Formulations containing unprocessed HPMC, i.e., powder-based blends (formulations F2, F4, and F6) had more vital adhesion forces than their corresponding extrudate-based blends, possibly due to their elasticity, hydrogen bonding, molecular weight, and cross-linking. Internal forces are represented by viscosity, whereas the force needed to separate a polymer from a surface is known as adhesion force [28]. Moreover, the adhesive force of all blends decreased with a decrease in the HPMC content, which aligns with previous outcomes [29,30].

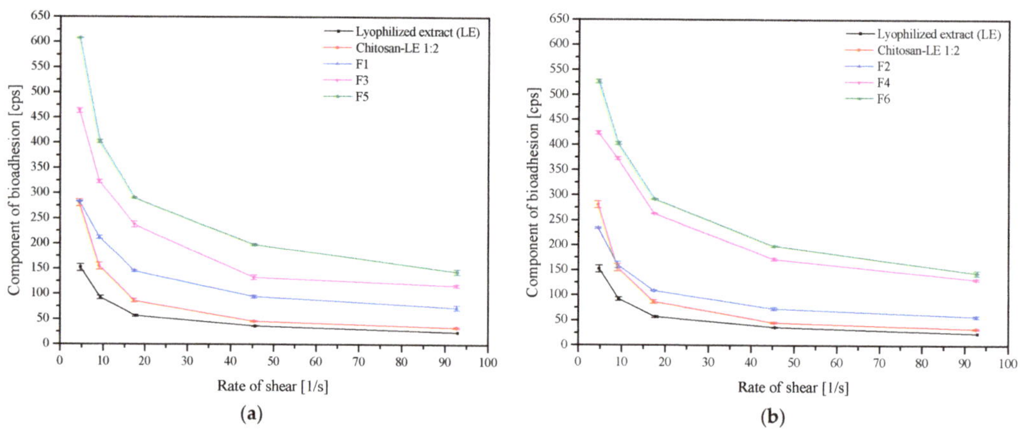

Figure 9. Component of bioadhesion of the extrudate-based tablets (**a**) and the powder-based tablets (**b**) ($n = 3$).

Tablets were additionally tested for their residence time to elaborate on their mucoadhesive behavior upon continuous contact with the medium-simulating saliva (Table 3). All of the formulations that were tested attached to the tissue right away, swelled progressively when they came into contact with the acceptor medium, and showed no evidence of disintegration at any point during the test. Despite the continuous movement of the cylindrical probe, the contact time of tablets F4 and F6 with the mucosal surface was preserved within 240 min of the test. In contrast, formulations F2 and F5 separated from the tissue after 220 min, F3 after 200 min, and F1 after 180 min (Table 3). This behavior of the tablets may be due to the higher viscosity and greater mucoadhesive strength of the unprocessed HPMC, as described above.

Table 3. The residence time of the extrudate- and powder-based tablets ($n = 3$).

Formulation	F1	F2	F3	F4	F5	F6
Residence time (min)	180 ± 5	220 ± 5	200 ± 5	>240	220 ± 5	>240

3. Materials and Methods

3.1. Plant Material

Plant raw material, *Scutellariae baicalensis radix*, was purchased from NANGA (Zlotow, Poland), the country of origin: China (Lot No. 243042021).

3.2. Chemicals and Reagents

Baicalin (≥95%, HPLC) was obtained from Sigma-Aldrich (Poznan, Poland). Excipients, such as chitosan with a degree of acetylation of 90% with a viscosity range of 500 mPas (marked as 90/500), was supplied from Heppe Medical Chitosan GmbH (Halle, Germany), (hydroxypropyl)methyl cellulose (HPMC) with an average Mn~90.000 (~15.000 cP), and magnesium stearate, were supplied by Sigma-Aldrich (Poznan, Poland). Microcrystalline cellulose (MCC) VIVAPUR 102 was supplied by JRS PHARMA (Rosenberg, Germany). Reagent for mucoadhesive tests: mucin from porcine stomach was obtained from Sigma-Aldrich (Poznan, Poland). HPLC grade acetonitrile and water were obtained from Merck. High-quality pure water and ultra-high-quality pure water were prepared using an Direct-Q 3 UV Merck Millipore purification system.

3.3. Preparation of Solid Dispersion Systems

3.3.1. Preparation of Extract System

5.0 g of the dried root of *Scutellariae baicalensis radix* was extracted four times with an ethanol–water mixture (8:2 v/v) for 60 min at 70 °C on an ultrasound-assisted water bath. The obtained extracts were collected and concentrated on a vacuum evaporator at a temperature 50°C to a volume of 20.0 mL (BÜCHI Rotavapor R-210) obtaining at that time DER 1:4. Then the extract was frozen and lyophilized (CHRIST 1–4 LSC, Osterode am Harz, Germany). The temperature on the freeze dryer shelf was heated and ranged from +15 °C to +20 °C; the temperature inside the product was estimated −4 °C; and the condensation temperature was set at −48 °C. The freeze-drying was conducted at reduced pressure (1.030 mbar) for 48 h. So obtained lyophilized extract was combined with chitosan 90:500 in a weight ratio of 2:1 and named as 'system' for further tests [14].

3.3.2. Hot Melt Extrusion (HME)

Extrusion was performed on a HAAKE MiniCTW micro-conical twin screw extruder (Thermo Scientific, Karlsruhe, Germany). The above-described system of lyophilized extract and chitosan in ratio 1:2 and carrier (HPMC) in three different ratios (Table 4) were mixed with a mortar and pestle and subsequently fed manually into the hopper of the extruder at barrel temperature of 150 °C and screw speed of 150 rpm. The extrudates were collected, ground softly manually with a pestle and mortar, passed through an 80 mesh sieve, and kept in a desiccator at room temperature for further analysis.

Table 4. Compositions of extrudates.

	Chitosan 90/500—Lyophilized Extract 1:2 (=System)	**HPMC**
	Ratio (w/w)	
Extrudate 75:25 m/m	75	25
Extrudate 50:50 m/m	50	50
Extrudate 25:75 m/m	25	75

3.3.3. Extrudate Characterization

Powder X-ray Diffraction (PXRD)

The crystallographic structure of the samples was analyzed by X-ray diffraction (XRD, Panalytical Empyrean, Almelo, The Netherlands) equipment with the copper anode (CuKα—1.54 Å) in a Brag-Brentano reflection mode configuration with 45 kV and 40 mA parameters. The measurement parameters were set up for 3–60° with a 45 s per step 0.05° in all cases.

Fourier Transform Infrared Spectroscopy with Attenuated Total Reflectance (FTIR-ATR)

The FTIR-ATR spectra were measured between 400 cm^{-1} and 4000 cm^{-1}, with a resolution set to 1 cm^{-1}, with a Shimadzu IRTracer-100 spectrometer equipped with a QATR-10 single bounce—diamond extended range—and LabSolution IR software.

3.3.4. Determinations of Active Components Content

The contents of the main active compounds (baicalin, baicalein, and wogonin) were determined by using the HPLC-Diode-Array Detection method described previously by

Paczkowska-Walendowska et al. [14]. Briefly, separations were performed on a Kinetex® C18 column, 5 μm particle size, 100 mm × 2.1 mm (Phenomenex, Poland). The mobile phase was composed of phosphoric acid 0.1% (A) and acetonitrile (B), with a gradient elution: 0–20 min, 10–40% B; 20–22 min, 10% B. The detection was performed at a wavelength (λ_{max}) of 280 nm. The flow rate of the mobile phase was set at 1.0 mL/min, and the column temperature was set at 30 °C. Injection volume was 10 μL. The test was repeated three times.

3.3.5. In Vitro Release Studies

An Agilent 708-DS apparatus was used for the dissolution studies. At 37 ± 0.5 °C, a typical paddle method was employed, with 50 rpm for stirring. Extrudates samples (~100 mg) were dissolved in 300 mL of an artificial saliva solution with the following ingredients: potassium chloride (1.20 g), sodium chloride (0.85 g), dipotassium hydrogen orthophosphate (0.35 g), magnesium chloride (0.05 g), calcium chloride (0.20 g), xylitol (20.0 g), and water up to 1L; the pH was adjusted to 6.8 by 1 M HCl. At certain intervals (15, 30, 60, 120, and 240 min), liquid samples were taken, and an equal volume of temperature-stabilized medium was substituted. A nylon membrane filter with a mesh size of 0.45 was used to filter the samples. The previously published HPLC method was used to ascertain the levels of baicalin in the filtered acceptor solutions. Sink conditions were preserved in the studies. The test was repeated for six samples of each substance.

The release profiles were compared by means of the model proposed by Moore and Flanner, which is based on two-factor values, f_1 and f_2.

3.3.6. Permeability Studies

The permeability of an active compound (baicalin) enclosed in systems through artificial biological membranes was investigated by using the PAMPA™ (parallel artificial membrane permeability assay) gastrointestinal tract (GIT) assay (Pion Inc., Billerica, MA, USA) according to the protocol supplied with the kit. Extrudate- and powder-based systems (concentration 10 mg/mL) were dissolved in donor solutions (artificial saliva solution at pH 6.8). The acceptor plates were loaded with acceptor Prisma buffer at pH 7.4. The plates were put together and incubated under the following conditions: temperature set at 37 °C for 15 min with continuous stirring at 50 rpm. Each experiment was repeated at least three times. The amount of permeated baicalin was determined using the HPLC method described above. The test was repeated six samples of each substance. The apparent permeability coefficients (P_{app}) were calculated from the following equation:

$$P_{app} = \frac{-ln\left(1 - \frac{C_A}{C_{equilibrium}}\right)}{S \times \left(\frac{1}{V_D} + \frac{1}{V_A}\right) \times t} \quad (1)$$

where V_D is the donor volume, V_A is the acceptor volume, $C_{equilibrium}$ is the equilibrium concentration $C_{equilibrium} = \frac{C_D \times V_D + C_A \times V_A}{V_D + V_A}$, C_D is the donor concentration, C_A is the acceptor concentration, S is the membrane area, and t is the incubation time (in seconds).

3.3.7. Microbiological Activity Assay

Well Diffusion Method

All microorganism strains were inoculated in Müeller-Hinton broth (pH 7.4) for approximately 16 h. The concentration of the suspensions was adjusted to 0.5 (optical density) by means of a spectrophotometer. Antimicrobial activity of the *S. baicalensis radix* lyophilized extract and extrudates were determined by the Agar well diffusion method against reference strains and clinical isolates of bacteria that colonize the oral cavity (*Escherichia coli, Pseudomonas aeruginosa, Streptococcus mutans, Staphylococcus aureus, Staphylococcus epidermidis,* and *Enterobacter aerogenes*). The 20 mL of sterilized Nutrient Agar was poured into sterile petri plates. Following solidification, 100 μL of standardized inoculate from each isolate was inoculated on Nutrient agar plates using sterilized spreaders. The wells were punched over the agar plates using a sterile gel puncher of 6 mm diameter. A measure of 100 μL of the lyophilized extract and extrudates was poured into separate wells. Samples were dissolved in 1% (*v*/*v*) dimethylsulphoxide (DMSO), which was used as a negative control. Plates were incubated at 37 °C for 24 h. Triplets of the experiment were maintained for each bacterial strain to ensure reliability. Following incubation, the diameter of the circular inhibitory zones formed around each well was measured in mm and recorded.

Liquid Culture Method

In the first stage of the research, strains of microorganisms were prepared. For this purpose, 0.1 g of bacterial lyophilisate was suspended in 10 mL of Müeller-Hinton liquid propagation medium. The samples were incubated at 37 °C for 18 h in order to activate and multiply the biomass. After incubation, the biomass was centrifuged from the substrate (14.000 rpm for 10 min). The supernatant was discarded, and the pellet was resuspended in 10 mL of 0.9% NaCl and centrifuged again. This procedure was performed three times. Then, the biomass was diluted in 0.9% NaCl, so that the concentration of microorganisms was 1.0×10^2 cfu/mL. At the same time, three solutions were prepared (the solvent was 0.9% NaCl) of the test samples at a concentration of 10, 50, and 100 mg/mL. Then, the dilutions prepared in this way were inoculated with the prepared suspension of microorganisms. The samples were mixed and incubated at 37 °C for 18 h. The number of microorganisms was analyzed before and after incubation using media intended for a given group of microorganisms.

3.4. Tableting Process

A laboratory scale, single-punch tableting equipment called the NP-RD10A Tablet Press was used to compressed tablets that were flat-faced and 8 mm in diameter (Natoli, Saint Charles, MO, USA). Utilizing a variety of compaction forces between 1000 and 3000 N, the compaction characteristics of tablets were evaluated (corresponding to compression pressures in a range from 20 to 60 MPa). When the desired compaction force was reached, the pressure was let go. Two types of formulations were prepared, containing extrudates or powder systems in appropriate proportions of ingredients. Table 5 lists the ingredients of the formulations.

Table 5. Compositions of formulations.

	Formulation 1 (F1)	Formulation 2 (F2)	Formulation 3 (F3)	Formulation 4 (F4)	Formulation 5 (F5)	Formulation 6 (F6)
			Content (mg) per 1 tablet			
Lyophilized extract—chitosan 90/500 2:1 *w*/*w* (=system)	-	75.0	-	50.0	-	25.0
HPMC 15.000 cP	-	25.0	-	50.0	-	75.0
System-HPMC 75:25 *w*/*w* extrudate	100.0	-	-	-	-	-
System-HPMC 50:50 *w*/*w* extrudate	-	-	100.0	-	-	-
System-HPMC 25:75 *w*/*w* extrudate	-	-	-	-	100.0	-
MCC	20.0	20.0	20.0	20.0	20.0	20.0
Stearate magnesium	1.2	1.2	1.2	1.2	1.2	1.2
SUM	121.2	121.2	121.2	121.2	121.2	121.2

3.4.1. Tablet Characterization

Immediately following the tablets compacting, the newly created tablets were weighed. A procedure outlined in Ph.Eur. 9th was used to control the uniformity of the tablet mass. A manual vernier caliper was also used to measure the diameter and thickness of 20 tablets that were chosen at random. Standard deviations and mean values were computed following all measurements (SD).

The tablet hardness was determined using the procedures outlined in Ph.Eur. 9th and was evaluated using the PTB-M manual tablet hardness testing device (Natoli, Saint Charles, MO, USA). Each hardness number is a mean with a standard deviation that represents the average of six measurements.

Tensile strength (σ) values were calculated on the basis of the breaking force (F) values (N), where d is the diameter of the tablet (mm) and h is the thickness of the tablet (mm) [31].

$$\sigma = \frac{2F}{\pi dh} \quad (2)$$

Solid fraction (SF) was calculated by the equation, where W_t is the weight of the tablet (mg), v is the tablet volume, and ρ_{true} is the powder true density (g/cm^3).

$$F = \frac{W_t}{\rho_{true} v} \quad (3)$$

The tablet porosity (ε) was calculated from the SF using the following equation:

$$\varepsilon = 1 - SF \quad (4)$$

Compactibility of the powders were analysed with the Ryshkewitch equation:

$$\varepsilon = \varepsilon_0 \times \exp(-b \times P) \quad (5)$$

$$T_S = T_0 \times \exp(-k \times \varepsilon) \quad (6)$$

where the porosity of powder when $p = 0$; b is a constant that is inversely proportional to the yield strength of the materials; T_S and T_0 are the tablet tensile strength and the limiting tablet tensile strength at zero porosity, respectively, and k is an empirical constant [19].

3.4.2. In Vitro Release Studies

In vitro release studies were performed according to the methodology described in Section 3.3.5. The test was repeated 6 times for each formulation.

The resulting active compound release profiles were fitted to the following mathematical models in order to study the release kinetics: [32]: zero-order equation: $F = k \times t$, first-order equation: $nF = k \times t$, Higuchi equation: $F = kt^{1/2}$, Korsmeyer-Peppas equation: $F = kt^n$, where F—the fraction of released drug, k—the constant associated with the release, and t—the time.

3.4.3. Swelling Index

Each tablet was individually weighted and placed in a 25 mL beaker that contained 10 mL of an artificial saliva solution at pH of 6.8 and at 37 ± 0.5 °C. Tablets were taken out, cleaned with filter paper, and reweighted at the preset intervals (15, 30, 60, 120, and 240 min). The swelling index was calculated by using the following formula:

$$SI = \frac{W_2 - W_1}{W_1} \quad (7)$$

where SI is the swelling index, W_1 is the initial weight of the tablet, W_2 is the weight of the tablet after the particular swelling time interval.

Each experiment was performed in triplicate.

3.4.4. In Vitro Assessment of Mucin-Biopolymer Bioadhesive Bond Strength

A viscometric method was used to quantify mucin-polymers' bioadhesive bond strength. The assessment was carried out according to Hassan and Gallo's procedure [33]. Each experiment was performed in triplicate.

3.4.5. Determination of the Residence Time

The residence time of tablets on regenerated cellulose membranes imitating porcine buccal mucosa was evaluated on an adjusted apparatus for the disintegration time test according to previous tests described by Paczkowska-Walendowska et al. [29]. Briefly, the medium was an artificial saliva solution at pH 6.8 maintained at 37 ± 0.5 °C. Each tablet was brought into contact with foil by putting on a finger force for 5 s. The time necessary to detach the formulation from the foil simulating mucosal tissue was measured within 4h of the performed test. Studies were carried out in triplicate.

3.5. *Statistical Analysis*

Software called Statistica 13.3 was used for the statistical analysis. The Shapiro-Wilk test was used to determine whether the results were normal. The ANOVA test, together with the post hoc Tukey's range test for multiple comparisons, was used to examine the variances between the mean values. At $p < 0.05$, differences between groups were deemed significant.

4. Conclusions

Extrudates containing *Scutellariae baicalensis radix* extract can be prepared at temperatures of about 150 °C, which does not decompose the active compounds. The proposed ground hot-melt extrudates based on HPMC show an interesting potential for improving the solubility of the poorly water-soluble active substance—baicalin. So, hot-melt extrusion is a good technique to improve the physicochemical properties of baicalin. Further, in order to obtain a suitable pharmaceutical form, the production process of mucoadhesive tablets containing extrudates was optimized. The prepared extrudates, differing in HPMC content, showed different tabletability, compressibility, and compactibility properties. As expected, the different content of the carrier influenced the release profile of baicalin from the tablets and the mucoadhesive properties. Higher HPMC content resulted in prolonged release of the substance, resulting from the diffusion of the substance through the polymer network. At the same time, the same carrier ensured that the tablets were kept in the affected area for a sufficiently long time. Importantly, the process did not reduce the biological, including microbiological, activity of the obtained extrudates.

Considering the complex matrix, both the tabletability/compactibility properties of the blends and the degree of release of the active substance, as well as mucoadhesive properties that give functionality to the developed tablets, should be considered. The best tabletability properties, a valuable baicalin release profile while maintaining sufficient mucoadhesive properties to condition the tablet's retention in the application site and the effectiveness of therapy, are provided by the F3 formulation, which contains the extrudate with lyophilized extract-HPMC 50:50 *w*/*w*.

Supplementary Materials: The following supporting information can be downloaded at: https://www.mdpi.com/article/10.3390/ijms24065834/s1.

Author Contributions: Conceptualization, M.P.-W. and J.C.-P.; methodology, M.P.-W., A.M. and D.S.; writing—original draft preparation, M.P.-W., A.M., D.S., K.S.-W. and J.C.-P.; writing—review and editing, M.P.-W. and J.C.-P.; supervision, J.C.-P.; project administration, M.P.-W. and J.C.-P.; funding acquisition, M.P.-W. and J.C.-P. All authors have read and agreed to the published version of the manuscript.

Funding: This research was funded by National Science Center (Poland), under Sonata grant (number 2020/39/D/NZ7/01824). The structural XRD research was financed by the Ministry of Science & Higher Education Poland.

Institutional Review Board Statement: Not applicable.

Informed Consent Statement: Not applicable.

Data Availability Statement: The data are contained within the article and Supplementary Materials.

Conflicts of Interest: The authors declare no conflict of interest.

References

1. Oral Health. Available online: https://www.who.int/news-room/fact-sheets/detail/oral-health (accessed on 5 August 2022).
2. Herrera, D.; Sanz, M.; Jepsen, S.; Needleman, I.; Roldán, S. A Systematic Review on the Effect of Systemic Antimicrobials as an Adjunct to Scaling and Root Planing in Periodontitis Patients. *J. Clin. Periodontol.* **2002**, *29* (Suppl. 3), 136–159; discussion 160–162. [CrossRef]
3. Hau, H.; Rohanizadeh, R.; Ghadiri, M.; Chrzanowski, W. A Mini-Review on Novel Intraperiodontal Pocket Drug Delivery Materials for the Treatment of Periodontal Diseases. *Drug. Deliv. Transl. Res.* **2014**, *4*, 295–301. [CrossRef]
4. Wei, Y.; Deng, Y.; Ma, S.; Ran, M.; Jia, Y.; Meng, J.; Han, F.; Gou, J.; Yin, T.; He, H.; et al. Local Drug Delivery Systems as Therapeutic Strategies against Periodontitis: A Systematic Review. *J. Control. Release* **2021**, *333*, 269–282. [CrossRef]
5. Zhao, Q.; Chen, X.-Y.; Martin, C. Scutellaria Baicalensis, the Golden Herb from the Garden of Chinese Medicinal Plants. *Sci. Bull.* **2016**, *61*, 1391–1398. [CrossRef] [PubMed]
6. Cui, L.; Feng, L.; Zhang, Z.H.; Jia, X.B. The Anti-Inflammation Effect of Baicalin on Experimental Colitis through Inhibiting TLR4/NF-KB Pathway Activation. *Int. Immunopharmacol.* **2014**, *23*, 294–303. [CrossRef]
7. Leung, K.C.-F.; Seneviratne, C.J.; Li, X.; Leung, P.C.; Lau, C.B.S.; Wong, C.-H.; Pang, K.Y.; Wong, C.W.; Wat, E.; Jin, L. Synergistic Antibacterial Effects of Nanoparticles Encapsulated with Scutellaria Baicalensis and Pure Chlorhexidine on Oral Bacterial Biofilms. *Nanomaterials* **2016**, *6*, 61. [CrossRef]
8. Ren, M.; Zhao, Y.; He, Z.; Lin, J.; Xu, C.; Liu, F.; Hu, R.; Deng, H.; Wang, Y. Baicalein Inhibits Inflammatory Response and Promotes Osteogenic Activity in Periodontal Ligament Cells Challenged with Lipopolysaccharides. *BMC Complement. Med. Ther.* **2021**, *21*, 43. [CrossRef] [PubMed]
9. Jakab, G.; Bogdán, D.; Mazák, K.; Deme, R.; Mucsi, Z.; Mándity, I.M.; Noszál, B.; Kállai-Szabó, N.; Antal, I. Physicochemical Profiling of Baicalin Along with the Development and Characterization of Cyclodextrin Inclusion Complexes. *AAPS PharmSciTech* **2019**, *20*, 314. [CrossRef] [PubMed]
10. Ashour, E.A.; Majumdar, S.; Alsheteli, A.; Alshehri, S.; Alsulays, B.; Feng, X.; Gryczke, A.; Kolter, K.; Langley, N.; Repka, M.A. Hot Melt Extrusion as an Approach to Improve Solubility, Permeability and Oral Absorption of a Psychoactive Natural Product, Piperine. *J. Pharm. Pharmacol.* **2016**, *68*, 989–998. [CrossRef]
11. Gao, N.; Guo, M.; Fu, Q.; He, Z. Application of Hot Melt Extrusion to Enhance the Dissolution and Oral Bioavailability of Oleanolic Acid. *Asian J.Pharm. Sci.* **2017**, *12*, 66–72. [CrossRef] [PubMed]
12. Patil, H.; Tiwari, R.V.; Repka, M.A. Hot-Melt Extrusion: From Theory to Application in Pharmaceutical Formulation. *AAPS PharmSciTech* **2015**, *17*, 20–42. [CrossRef] [PubMed]
13. Simões, M.F.; Pinto, R.M.A.; Simões, S. Hot-Melt Extrusion in the Pharmaceutical Industry: Toward Filing a New Drug Application. *Drug. Discov. Today* **2019**, *24*, 1749–1768. [CrossRef] [PubMed]
14. Paczkowska-Walendowska, M.; Cielecka-Piontek, J. Chitosan as a Functional Carrier for the Local Delivery Anti-Inflammatory Systems Containing Scutellariae Baicalensis Radix Extract. *Pharmaceutics* **2022**, *14*, 2148. [CrossRef]
15. Wu, H.; Liu, Y.; Ci, T.; Ke, X. Application of HPMC HME Polymer as Hot Melt Extrusion Carrier in Carbamazepine Solid Dispersion. *Drug. Dev. Ind. Pharm.* **2020**, *46*, 1911–1918. [CrossRef]
16. Krupa, A.; Cantin, O.; Strach, B.; Wyska, E.; Tabor, Z.; Siepmann, J.; Wróbel, A.; Jachowicz, R. In Vitro and in Vivo Behavior of Ground Tadalafil Hot-Melt Extrudates: How the Carrier Material Can Effectively Assure Rapid or Controlled Drug Release. *Int. J. Pharm.* **2017**, *528*, 498–510. [CrossRef]
17. Bajwa, G.S.; Sammon, C.; Timmins, P.; Melia, C.D. Molecular and Mechanical Properties of Hydroxypropyl Methylcellulose Solutions during the Sol:Gel Transition. *Polymer* **2009**, *50*, 4571–4576. [CrossRef]
18. Zhang, Y.; Luo, R.; Chen, Y.; Ke, X.; Hu, D.; Han, M. Application of Carrier and Plasticizer to Improve the Dissolution and Bioavailability of Poorly Water-Soluble Baicalein by Hot Melt Extrusion. *AAPS PharmSciTech* **2014**, *15*, 560–568. [CrossRef]
19. Feng, Z.; Li, M.; Wang, W. Improvement of Dissolution and Tabletability of Carbamazepine Solid Dispersions with High Drug Loading Prepared by Hot-Melt Extrusion. *Pharmazie* **2019**, *74*, 523–528. [CrossRef]
20. Li, C.L.; Martini, L.G.; Ford, J.L.; Roberts, M. The Use of Hypromellose in Oral Drug Delivery. *J. Pharm. Pharmacol.* **2005**, *57*, 533–546. [CrossRef]
21. Chanaj-Kaczmarek, J.; Osmałek, T.; Szymańska, E.; Winnicka, K.; Karpiński, T.M.; Dyba, M.; Bekalarska-Dębek, M.; Cielecka-Piontek, J. Development and Evaluation of Thermosensitive Hydrogels with Binary Mixture of Scutellariae Baicalensis Radix Extract and Chitosan for Periodontal Diseases Treatment. *Int. J. Mol. Sci.* **2021**, *22*, 11319. [CrossRef]
22. Joshi, S.C. Sol-Gel Behavior of Hydroxypropyl Methylcellulose (HPMC) in Ionic Media Including Drug Release. *Materials* **2011**, *4*, 1861–1905. [CrossRef]

23. Dash, S.; Murthy, P.N.; Nath, L.; Chowdhury, P. Kinetic Modeling on Drug Release from Controlled Drug Delivery Systems. *Acta Pol. Pharm.* **2010**, *67*, 217–223.
24. Fu, Y.; Kao, W.J. Drug Release Kinetics and Transport Mechanisms of Non-Degradable and Degradable Polymeric Delivery Systems. *Expert. Opin. Drug. Deliv.* **2010**, *7*, 429–444. [CrossRef]
25. Menchicchi, B.; Fuenzalida, J.P.; Bobbili, K.B.; Hensel, A.; Swamy, M.J.; Goycoolea, F.M. Structure of Chitosan Determines Its Interactions with Mucin. *Biomacromolecules* **2014**, *15*, 3550–3558. [CrossRef] [PubMed]
26. M. Ways, T.M.; Lau, W.M.; Khutoryanskiy, V.V. Chitosan and Its Derivatives for Application in Mucoadhesive Drug Delivery Systems. *Polymers* **2018**, *10*, 267. [CrossRef] [PubMed]
27. Agarwal, S.; Murthy, R.S.R. Effect of Different Polymer Concentration on Drug Release Rate and Physicochemical Properties of Mucoadhesive Gastroretentive Tablets. *Indian. J. Pharm. Sci.* **2015**, *77*, 705–714. [CrossRef]
28. Chen, X.; Yan, J.; Yu, S.; Wang, P. Formulation and In Vitro Release Kinetics of Mucoadhesive Blend Gels Containing Matrine for Buccal Administration. *AAPS PharmSciTech* **2018**, *19*, 470–480. [CrossRef]
29. Paczkowska-Walendowska, M.; Szymańska, E.; Winnicka, K.; Szwajgier, D.; Baranowska-Wójcik, E.; Ruchała, M.A.; Simon, M.; Cielecka-Piontek, J. Cyclodextrin as Functional Carrier in Development of Mucoadhesive Tablets Containing Polygoni Cuspidati Extract with Potential for Dental Applications. *Pharmaceutics* **2021**, *13*, 1916. [CrossRef] [PubMed]
30. Bandi, S.P.; Bhatnagar, S.; Venuganti, V.V.K. Advanced Materials for Drug Delivery across Mucosal Barriers. *Acta Biomater.* **2021**, *119*, 13–29. [CrossRef]
31. Tye, C.K.; Sun, C.C.; Amidon, G.E. Evaluation of the Effects of Tableting Speed on the Relationships between Compaction Pressure, Tablet Tensile Strength, and Tablet Solid Fraction. *J. Pharm. Sci.* **2005**, *94*, 465–472. [CrossRef]
32. Costa, P.; Sousa Lobo, J.M. Modeling and Comparison of Dissolution Profiles. *Eur. J. Pharm. Sci.* **2001**, *13*, 123–133. [CrossRef] [PubMed]
33. Hassan, E.E.; Gallo, J.M. A Simple Rheological Method for the in Vitro Assessment of Mucin-Polymer Bioadhesive Bond Strength. *Pharm. Res.* **1990**, *7*, 491–495. [CrossRef] [PubMed]

International Journal of *Molecular Sciences*

MDPI

Article

Chitosan Composites Containing Boron-Dipyrromethene Derivatives for Biomedical Applications

Aleksander Smolarkiewicz-Wyczachowski [1], Halina Kaczmarek [1,*], Jaroslaw Piskorz [2], Pawel Nowak [1] and Marta Ziegler-Borowska [1]

1 Faculty of Chemistry, Nicolaus Copernicus University in Torun, Gagarina 7, 87-100 Torun, Poland
2 Chair and Department of Inorganic and Analytical Chemistry, Poznan University of Medical Sciences, Rokietnicka 3, 60-806 Poznan, Poland
* Correspondence: halina@umk.pl

Abstract: The work is devoted to preparing and characterizing the properties of photosensitive composites, based on chitosan proposed for photodynamic therapy. Chitosan films with a 5% addition of two BODIPY dyes were prepared by solution casting. These dyes are dipyrromethene boron derivatives with N-alkyl phthalimide substituent, differing in the presence of iodine atoms in positions 2 and 6 of the BODIPY core. The spectral properties of the obtained materials have been studied by infrared and UV-vis absorption spectroscopy and fluorescence, both in solutions and in a solid state. Surface properties were investigated using the contact angle measurement. The morphology of the sample has been characterized by Scanning Electron and Atomic Force Microscopy. Particular attention was paid to studying the protein absorption and kinetics of the dye release from the chitosan. Adding BODIPY to the chitosan matrix leads to a slight increase in hydrophilicity, higher structure heterogeneity, and roughness, than pure chitosan. The presence of iodine atoms in the BODIPY structure caused the bathochromic effect, but the emission quantum yield decreased in the composites. It has been found that BODIPY-doped chitosan interacts better with human serum albumin and acidic α-glycoprotein than unmodified chitosan. The release rate of dyes from films immersed in methanol depends on the iodine present in the structure.

Keywords: chitosan; BODIPY; composites; fluorescence; protein adhesion; dye release

1. Introduction

Despite enormous progress in the diagnosis and treatment of cancer in recent decades, cancer is still difficult to cure. New methods used in cancer treatment include immuno-oncology, gene therapy, molecular targeted (personalized) therapy, and the photodynamic techniques [1–3]. The contemporary strategy for combating these dangerous diseases is based primarily on developing diagnostic methods, designing modern drugs and ways of precisely delivering them to the affected places.

A large group of currently used active medicinal substances consists of low-molecular organic compounds characterized by serious disadvantages—rapid metabolism, insufficient drug resistance, and lack of selectivity in numerous cases. It results in their rapid excretion from the body, inadequate distribution, and low therapeutic effectiveness. Therefore, an intensive search for carriers that could increase their potency is underway. Polymers, especially biopolymers, are promising materials that, when combined with bioactive substances, create conjugated systems capable of gradually releasing the drug at the target site, due to metabolic processes [4]. In addition, the polymer matrix can play a protective role in preventing the premature release or degradation [5,6].

Among the many macromolecular compounds currently used in biomedicine, chitosan (**CS**) is distinguished. It is a biopolymer consisting mainly of D-glucosamine linked by a β-(1→4) bond and a minor part of N-acetyl-D-glucosamine units that are the remainder of the chitin substrate. Due to various functional groups (OH, NH_2, CH_3CONH_2), chitosan has unique properties and is capable of many interactions (hydrogen, hydrophobic, and ionic bonding) and chemical reactions. Moreover, it is biocompatible, biodegradable, non-toxic, antimicrobial, and readily available. It allows for the understanding of its increasing application, not only in medicine (e.g., in surgery, tissue engineering, drug delivery systems, and wound healing), but in many other industries (packaging, cosmetics, agriculture, and veterinary) [7–9].

The discovery of the photodynamic reaction in the early 1900s contributed to the development of photodynamic anticancer therapy (PDT), which is now the subject of intense research [10–13]. This method involves giving the patient a photosensitizer, then exposing it to light and causing a chemical reaction that produces the reactive oxygen species, leading to apoptosis of the cancer cells. A particular advantage of PDT is its high selectivity and the absence of side effects, unlike in chemotherapy, radiotherapy, and surgery. Properly selected photosensitizing compounds, characterized by the specific absorption of radiation in the visible range and ability to fluorescence, as well as exposure conditions (presence of oxygen, dose), play a crucial role here [14]. A novelty in the PDT method is its combination with nanotechnology and nanomaterials. Such examples are polymeric nanocarriers loaded by a photosensitiser, hybrid nanoparticles, or a photoactive compounds encapsulated in a polymer shell that contribute to the proper administration and positive accumulation of the drug in tissues, with a higher efficiency of the formation of the reactive oxygen species and the increase the cancer cells death with no effect in the neighboring cells, as well as an improved stability [15–18].

However, the effectiveness of the PDT method still needs to be improved upon by an intense search for new effective photosensitizers and application methods. Over several decades, there have been many reports of such compounds with a potential use in PDT, including boron-dipyrromethene and its derivatives (called BODIPY dyes), reviewed recently by M. Poddar and R. Misra [19] and E. Antina et al. [20]. The structure of these compounds is based on the connection of two heterocyclic pyrrole rings through a difluoroboryl unit, resulting in a three-ring system with six resonating pairs of π electrons. The BODIPY core is flat and prone to chemical modification (e.g., substitution in λ, β or *meso* position), leading to changes in the photochemical properties.

There are several examples demonstrating the latest achievements in this field, e.g., BODIPY functionalized with lactose [21], phosphorylated BODIPY [22], diketopyrrolopyrrole-aza-BODIPY hybrids [23], N-BODIPY containing a para-nitrophenyl group at the *meso* position and a diamino-boron substituent with two tosyl pendant moieties [24], trifluoromethyl-substituted (i.e., with electron-withdrawing groups) and methoxy-substituted (with electron-donating groups) dyes [25].

The purpose of this study was to study the photosensitive composites based on chitosan (**CS**) containing two dipyrromethene boron difluoride derivatives (BODIPY dyes: **I** and **H**) with N-alkyl phthalimide substituent at the meso position. The dye marked **I**, contains two iodine atoms in positions 2 and 6 instead of hydrogen atoms (present in the **H** structure). In this way, it is possible to assess the influence of heavy atoms in BODIPY-type chromophores on the properties of chitosan modified with these dyes. It should be added that substituent in position 8 usually causes the desirable blue emission.

The spectroscopic and surface properties and the morphology of the obtained systems were characterized. Special attention was paid to the ability to interact with serum proteins by the composites and the release of dyes from chitosan films, which is relevant in biomedical applications.

The synthesis of applied modifying dyes: boron-dipyrromethene derivatives with substituted N-alkyl phthalimide groups has been previously described by Piskorz et al. [26]. The photodynamic antimicrobial activity of these compounds and their ability to generate singlet oxygen has also been proven in this work.

2. Results and Discussion

2.1. General Remarks and Visual Observation of Chitosan-BODIPY Composites

Two different BODIPY dyes were introduced into the chitosan matrix (Figure 1).

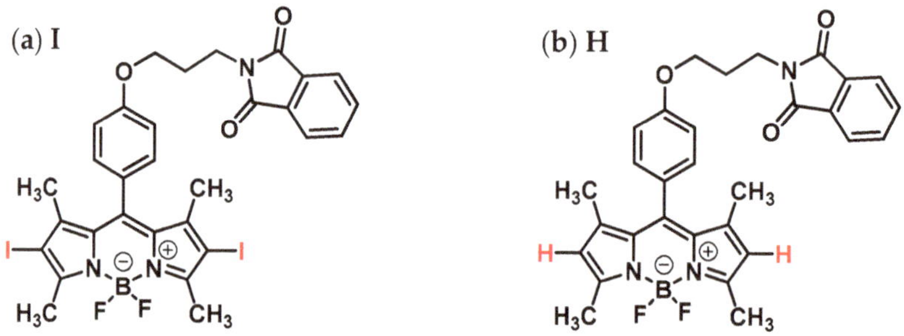

Figure 1. Chemical structure of BODIPY dyes used for chitosan modification: (**a**) 1,3,5,7-tetramethyl-8-{4-(3-phthalimide)propoxy]phenyl}-4,4-difluoro 4-bora-3a,4a-diaza-s-indacene (**I**); (**b**) 2,6-diiodo-1,3,5,7-tetramethyl-8-{4-(3-phthalimide)propoxy] phenyl}-4,4-difluoro-4-bora-3a,4a-diaza-s-indacene (**H**). Red atoms indicate the difference in the chemical structure of both compounds.

These dyes differ in the presence of two iodine or hydrogen atoms in positions 2 and 6 of the BODIPY cyclic structure.

The presence of the phthalimide substituents in the BODIPY structure can give the dye additional biological activities, including antimicrobial, anti-inflammatory, antinociceptive, anti-convulsant, and analgesic effects [26,27]. Moreover, heavy iodine atoms in the dye core increase the singlet oxygen production, which improves the therapeutic effect in treating tumors [28]. Introducing 5% wt BODIPY compounds into a chitosan solution and obtaining solid films by casting and evaporating solvents is uncomplicated. The composites obtained by this method were visually homogenous, transparent, and relatively flexible. The sample obtained with 5% **I** was deep purple, while the mixture with 5% **H** was intense orange (Figure 2). It can be added that the chitosan film itself was colorless.

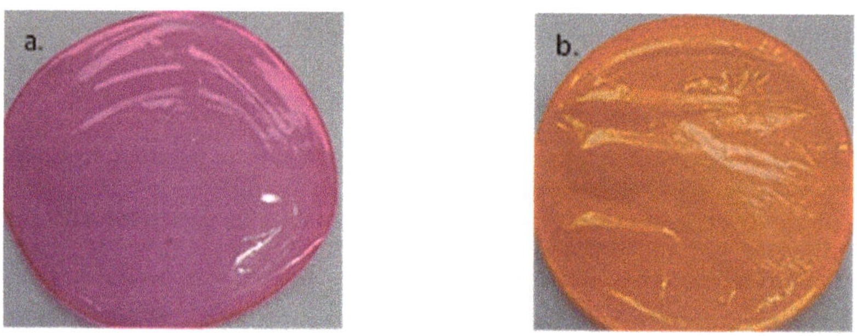

Figure 2. Photo of chitosan film (**a**) with 5% addition of **I** and (**b**) with 5% addition of **H** dyes.

2.2. Infrared Spectroscopy

FTIR-ATR spectra of applied **I** and **H** dyes are presented in Figure 3a,b. In both spectra, similar absorption bands appear, with slight shifts, as shown in Table 1. The band at 2928/2947 cm^{-1} is attributed to C-H stretching vibrations in $CH/CH_2/CH_3$ groups. The weak band of C-H in the aromatic rings is also seen at 3044 cm^{-1} and 3083 cm^{-1} in **I** and **H**, respectively.

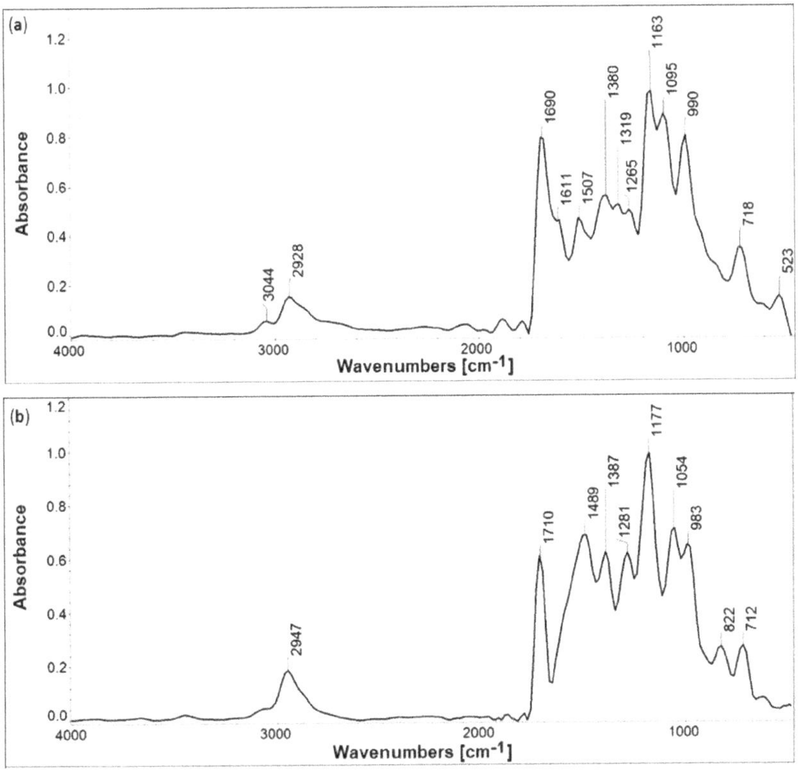

Figure 3. FTIR-ATR spectra of BODIPY: (**a**) **I** and (**b**) **H**.

Table 1. Maxima of the main absorption bands in **I** and **H** dyes, FTIR-ATR spectra, and their assignment.

IR Band Position, cm^{-1}		Assignment [29–32]
I	H	
3044	3083	C-H aromatic and heterocyclic, stretching
2928	2947	C-H aliphatic, stretching
1690	1710	C=O (amide I) stretching
1507	1489	N-H (amide II), C=C aromatic,. stretching
1380	1387	C-H deformation, B-N stretching
1163	1177	C-O stretching
1095	1054	C-O stretching
990	983	C-C stretching
-	822	C-H deformation
718	712	C-H, B-N deformation
523	-	C-I stretching

The carbonyl band occurs at 1690/1710 cm^{-1}, C=C in the aromatic rings at 1507/1489 cm^{-1}, C-O at the range of 1100–1400 cm^{-1}, and C-C at 990/983 cm^{-1}. The peak at 1380/1387 cm^{-1} may be due to the overlapping vibrations of the C-H (deformation) and B-N groups (stretching, in-plane) [29–31]. A band at 523 cm^{-1} in the **I** dye spectrum can be attributed to C-I stretching vibration [32]. The fingerprint region at 1000–1500 cm^{-1} is particularly rich in multiple bands, which is typical for such dyes. In this range, the main differences between both compounds appear.

Compared to pure, unmodified chitosan, FTIR spectra of chitosan-BODIPY composites show negligible differences (Figure 4a–c). In these three spectra, there are bands characteristic of polysaccharides, the main ones are: hydroxyl (with overlapping amine) at 3000–3600 cm^{-1} range, C-H at ~2880–2930 cm^{-1}, amide at ~1500–1700 cm^{-1}, C-O-C at ~1000–1200 cm^{-1}. The bands typical for the aromatic and heterocyclic rings in dyes are covered with intensive chitosan bands. However, the band arm at 578 cm^{-1} in **CS-I** spectrum indeed arises from the presence of the B-N group. The C=O amide I band, visible in CS at 1650 cm^{-1}, becomes the branch (at cm^{-1}) of the stronger amide II bands at 1566 cm^{-1} and 1593 cm^{-1} in the **CS-I** and **CS-H** spectrum, respectively. A detailed interpretation of the bands is presented in Table 2.

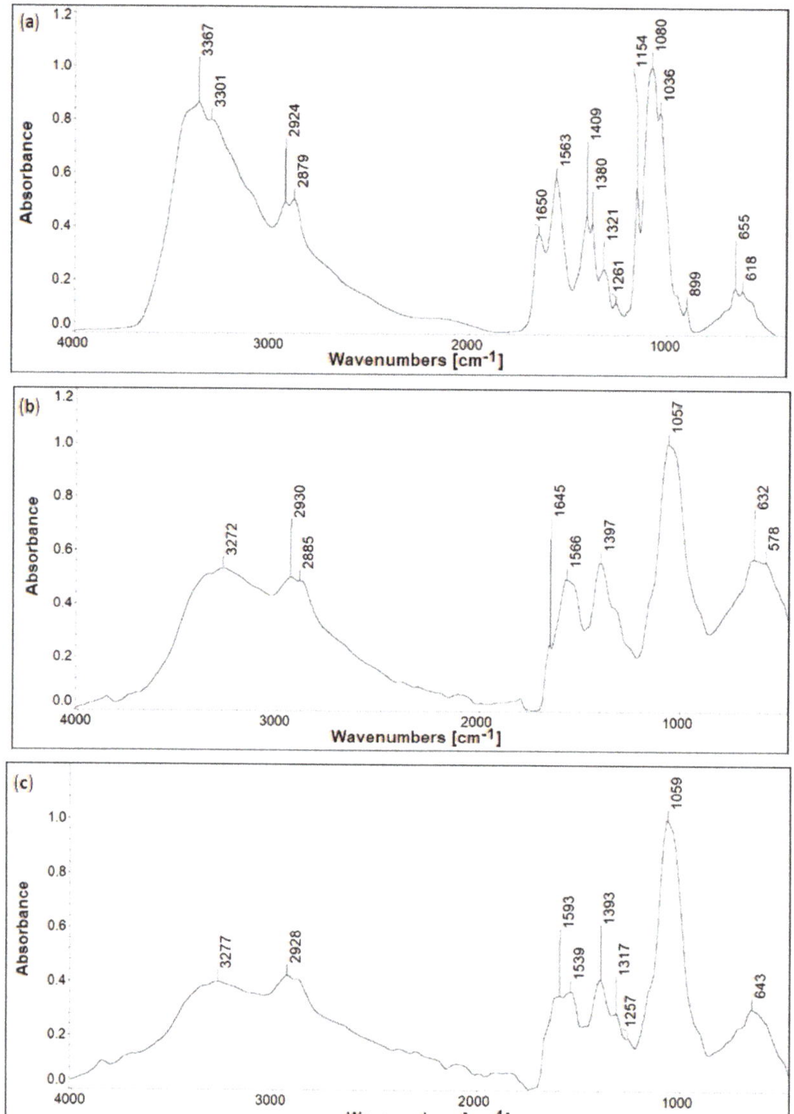

Figure 4. FTIR-ATR spectra of chitosan (**a**) and **CS-I** (**b**) and **CS-H** (**c**) composites.

Table 2. The main absorption bands (cm^{-1}) in the FTIR spectra of **CS**, **CS-I**, and **CS-H** composites and their assignment.

IR Band Position, cm^{-1}			Assignment [33–35]
CS	CS-I	CS-H	
3367	3272	3277	O-H/N-H stretching
2924, 2879	2930, 2885	2928, 2875	C-H stretching (symmetric and asymmetric)
1650	1645 (arm of amide II band)	1663 (arm of amide II band)	C=O (amide I) stretching
1563	1566	1539	N-H (amide II) stretching
1409	1397	1393	C-H/C-N deformation
1080	1057	1059	C-O-C deformation
899	barely visible	barely visible	C-H bending saccharide ring, out of the plane
655	632	643	C-H deformation
-	578	-	C-I stretching

Generally, the main detected differences appear in the range corresponding to the C=O and C-O vibrations and in the fingerprint region (600–1400 cm^{-1}). The disappearance of the weak band at 897 cm^{-1} in the spectra of the chitosan doped with dyes was also observed. These changes indicate the intermolecular interactions between chitosan and **I** or **H** dye. The dipole-dipole interactions are particularly probable due to the functional groups containing polarized chemical bonds: N-H, O-H, C=O in **CS**, C=O, B-F in **H** and **I** (polarity of C-I bond in **I** is negligible because of the slight differences between the electronegativity of carbon and iodine). Dipole-ion interactions between protonated amine groups (in **CS**) and anionic boron atoms (in dyes) cannot also be excluded (Scheme 1).

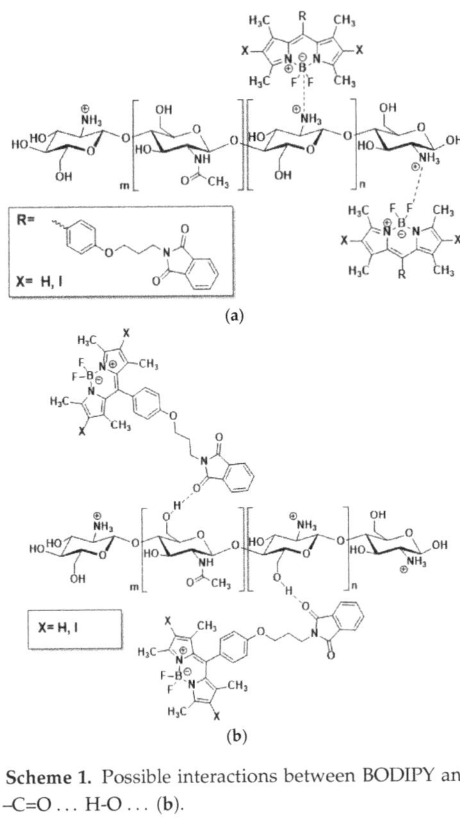

Scheme 1. Possible interactions between BODIPY and chitosan groups: -B^{-} … $^{+}NH_3^{-}$ (**a**) and –C=O … H-O … (**b**).

2.3. UV-Vis Absorption Spectroscopy

Electronic spectra of **I** and **H** dyes and CS-BODIPY indicate the strong bands in the visible range of 450–600 nm, resulting from the transitions of π electrons to the π^* excited state (Figure 5, Table 3). The absorption band in the composite spectra, both in solution and in the form of solid films, show a bathochromic effect in relation to the spectra of the dyes themselves, i.e., maxima are shifted towards longer wavelengths by 16 nm and 76 nm in **CS-I** and **CS-H**, respectively. Interestingly, no differences in maximum location are found in the spectra of films and solutions; thus, the solvents' effect seems negligible.

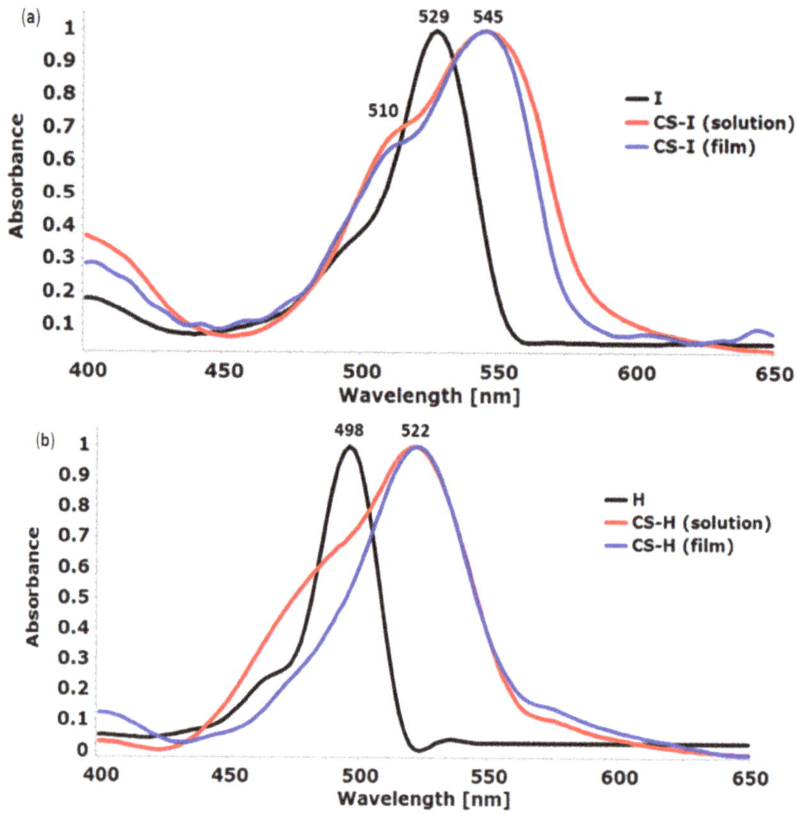

Figure 5. Normalized absorption spectra of BODIPY dyes (in acetone solutions) and chitosan composites (in solutions and solid films): **I** and **CS-I** (**a**); **H** and **CS-H** (**b**).

Table 3. Spectral characteristic of the studied samples: maxima of absorption (λ_{abs}) and emission (λ_{em}), stokes shifts ($\Delta\lambda$), and fluorescence quantum yield (Φ_F). Spectra of **I** and **H** dyes were recorded in acetone solution; **CS-I** and **CS-H** in aqueous acetic acid-acetone (3:1) solution.

Sample	λ_{abs}, nm	λ_{em}, nm	$\Delta\lambda$, nm	Φ_F, %
I solution	529	553	24	2.60
CS-I solution	545	571	26	1.95
CS-I film	545	603	58	0.99
H solution	498	513	15	19.4
CS-H solution	522	546/570	24/48	2.97
CS-H film	522	570/602	48/80	2.45

However, the significant broadening of these bands in **CS-I** and **CS-H** spectra is observed, compared to the related bands in **I** and **H** dye absorption spectra. The broadening is higher in the case of solutions than that in solid **CS-I** and **CS-H** spectra. It is caused by the interactions of solvents with dye-doped chitosan. As recently shown [36], the aromatic substituent in BODIPY also causes the broadening of the absorption band in aqueous solutions.

The asymmetric shape of absorption bands with the formation of the left-side branch, indicates the possibility of the dye aggregation. This asymmetry is very weakly marked in the spectrum of the **CS-H** film. The red shift of absorption bands in **I** and **CS-I**, compared to **H** and **CS-H**, is caused by the presence of heavy iodine atoms in the BODIPY structure.

2.4. Fluorescence

Fluorescence spectra for BODIPY dyes and CS-dye composites in solutions and solid state are shown in Figure 6. As can be seen, the emission maxima shift to longer wavelengths for the **CS-I** and **CS-H** solutions and films. Stokes shifts are particularly high in composite solid films—58 and 80 nm in **CS-I** and **CS-H**, respectively (Table 3).

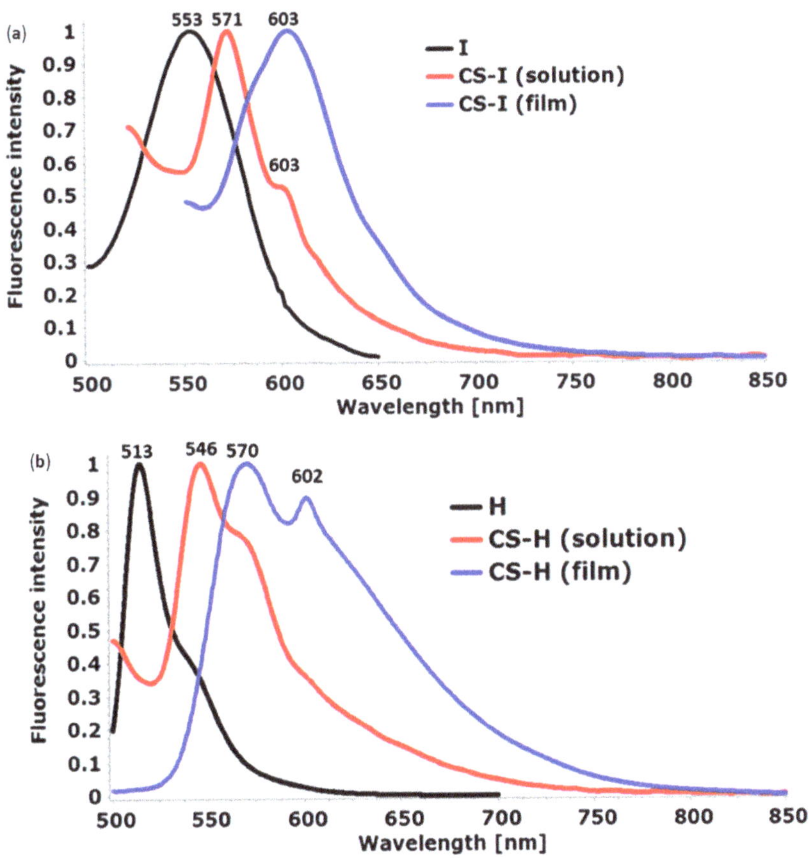

Figure 6. Normalized emission spectra of **I** (**a**) and **H** (**b**) dyes, as well as **CS-I** and **CS-H** in a solution or solid state.

As it is known from the literature, dyes of the BODIPY type are prone to aggregation [36–38]. The observed redshift suggests the formation of J-type aggregates in the chitosan matrix and solution. Due to the four methyl groups, the phenyl ring in the meso position does not rotate and lies in the same plane as the BODIPY, which would allow the formation of such aggregates. However, the phthalimide substituent is connected with the phenyl ring by a flexible propoxy spacer, enabling it to assume different conformations (with different degrees of twisting, relative to the BODIPY core). This makes it difficult to build aggregates. Thus, two opposing effects influence the structure of the chitosan-dye systems.

It was reported earlier that such a bathochromic shift of the band in the emission spectra, is profitable from the point of view of biomedical applications, as it leads to less absorption and, at the same time, a deeper penetration of light into tissues [39].

The intensity and quantum yield of fluorescence are relatively low. The sample without iodine is characterized by higher Φ_F values than the corresponding specimens containing iodine atoms, indicating the heavy atom's effect. The most intensive fluorescence (in acetone solution) exhibits **H** dye (Φ_F = 19.4%), whereas the **CS-I** film showed the highest quenching of radiation emission. As previously stated, fluorescence suppression is caused by efficient intersystem crossing (ISC), leading to a drop in the population of molecules in the singlet excited state [26,40]. However, the triplet state that arises from the ISC process promotes the formation of singlet oxygen or other reactive species [41]. Another reason for the decrease in the BODIPY fluorescence intensity in chitosan systems may be the above mentioned particle aggregation [36]. Moreover, fluorescence suppression was also attributed to water present in the solution mixture [36].

2.5. Contact Angle Measurements

To determine the surface free energy of chitosan film and its composites with BODIPY compounds, the contact angle was measured with two test liquids of different polarities: glycerin and diiodomethane (Table 4). It is mentioned that glycerin was chosen instead of H_2O as the polar liquid [42] since the water has recently been found to be unsuitable for testing biopolymer surfaces [43]. The contact angle of the water changes rapidly during the measurement, and the drop spreads over the biofilm surface [43].

Table 4. The mean values of the contact angle, surface free energy, and its polar and dispersion components for pristine chitosan and chitosan doped with dyes.

Sample	Contact Angle (θ,°)		Surface Free Energy (mJ/m²)		
	Test Liquid				
	Glycerin	Diiodomethane	SFE	SFE(D)	SFE(P)
CS	82.0	56.0	30.70	27.46	3.23
CS-I	72.9	61.1	30.71	22.07	8.64
CS-H	78.4	58.8	30.00	24.76	5.24

Based on the mean values of the contact angles and the calculated surface free energy (SFE), it can be stated that the surfaces of all obtained materials are hydrophobic (Table 4). This is also confirmed by the values of the SFE(D) dispersion component, which is much greater than the polar component SFE(P) in each studied specimen. Therefore, weak dispersion interactions, resulting from electron density fluctuations, dominate in the surface layers of the samples despite the presence of polar hydroxyl groups in the chitosan structure. Such surfaces are characterized by a poor wettability by water. Other authors obtained similar results for chitosan [44]. The hydrophobicity of this polysaccharide is explained by the presence of the hydrophobic nature of the CS backbone [45] or by the impurities, present even in small amounts [46]. It also depends on the deacetylation degree because the amino groups are hydrophilic, as opposed to the acetyl groups [9].

Chitosan films with the addition of **I** or **H** dye exhibited a slightly higher hydrophilicity than the un-doped chitosan. The location of ionic fragments of BODIPY and fluorine atoms at the top layer of the sample can explain it. In addition, both dyes have an 8-position substituent that contains two electronegative oxygen atoms and one nitrogen in the phthalimide ring, and an additional one oxygen atom in the propoxyphenyl linker. In the case of **CS-I**, iodine atoms also influence an increase in polarity (8.64 mJ/m^2). However, the slight difference between the electronegativity of iodine and hydrogen (2.20 and 2.66, respectively) does not explain the observed increase in the polarity of **CS-I**. The reason is probably because of the different conformation of the dyes in **CS-H** and **CS-I**, where a large I atom can cause additional steric hindrances (the covalent radius of **H** and **I** is 25 and 40 pm, respectively).

Results show that the surface free energy values of samples are within the range at which cell adhesion is possible (20–30 mJ/m^2) [47]. Thus, the potential application of the obtained materials as a drug carrier in the form of patches can be predicted.

2.6. Morphology (SEM and AFM)

SEM images of the chitosan cross-section (Figure 7a) indicate its compact homogeneous structure without any special details, in contrast to the samples with the addition of BODIPY dyes (Figure 7b,c). Apparent spherical inclusions with dimensions of approx. 0.2–0.5 μm can be observed in the **CS-H** sample (Figure 7c), where simultaneously, the roughness parameters take the highest values (Table 5). It can be explained by the precipitation of the BODIPY dye in a chitosan matrix during solvent evaporation and film molding. Sample **CS-I** presents a different image of the internal structure, which is also heterogeneous, but the scattered particles are much smaller than those in **CS-H** (Figure 7c).

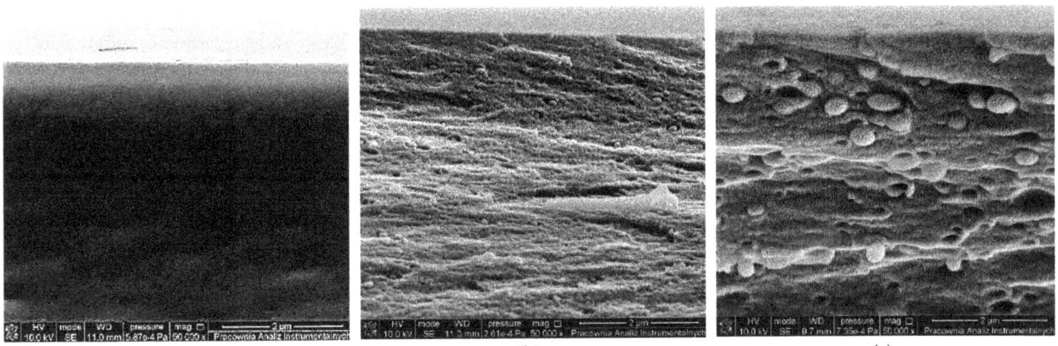

(a) (b) (c)

Figure 7. SEM images of a cross-section of chitosan (**a**), **CS-I** (**b**), and **CS-H** (**c**) films at 50,000× magnification.

Table 5. Roughness parameters for the chitosan film and chitosan doped with **I** and **H** dyes for the scan area 1 × 1 μm^2 and 10 × 10 μm^2.

Sample	Roughness Parameters [nm] for 1 × 1 μm^2 and 10 × 10 μm^2		
	R_q	R_a	R_{max}
CS	1.40/4.30	1.08/2.94	11.7/54.0
CS-I	1.66/1.98	1.31/1.51	15.1/41.8
CS-H	3.43/26.1	2.56/13.2	26.7/477

AFM allows for the observation of the surface topography and the assessment of the surface roughness, which, apart from SEP, also affects the adhesion to other materials, e.g., tissues. The surface topography of the samples is influenced by various factors, including the chemical structure of the compound, homogeneity of the composition, and the presence of inclusions or impurities. However, in the case of the tested samples, the differences in the surface structure are not very large (Figure 8). The surface of chitosan films shows a nodular structure but is relatively smooth, which is also observed by other authors [44,48]. The **CS-H** sample exhibits larger aggregates randomly dispersed in the chitosan matrix. This is also seen in the cross-section of this sample (Figure 8c, bottom).

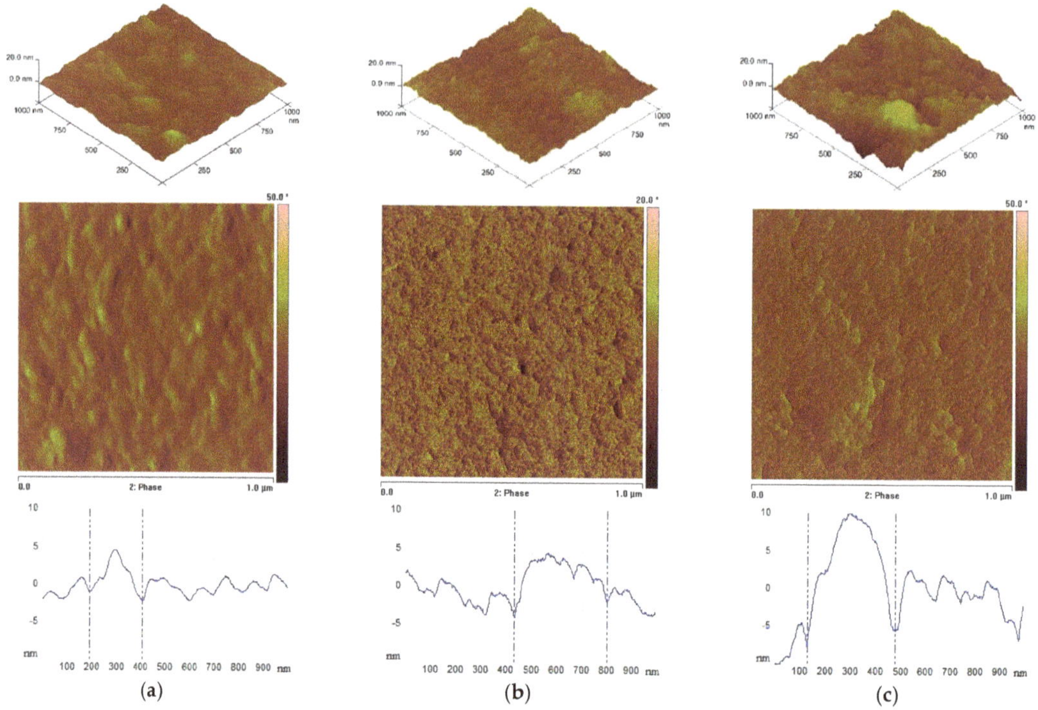

Figure 8. 3D-AFM (upper panel), phase AFM images (middle panel), and cross-sections (bottom panel) of chitosan (**a**), **CS-I** (**b**), and **CS-H** (**c**) films for a $1 \times 1\ \mu m^2$ scan area.

Furthermore, the determined roughness parameters are low for all studied samples (Table 5). R_q, R_a, and R_{max} values for **CS** and **CS-I** are similar and are only somewhat higher for **CS-H**. This trend is observed for smaller ($1 \times 1\ \mu m^2$) and larger ($10 \times 10\ \mu m^2$) scan areas. Thus, incorporated BODIPYs have a negligible effect on the surface geometry. It indicates that the precipitated dye particles are rather below than on the surface, as also shown by the SEM pictures. This can be an advantage because the biopolymer layer plays a protective role in such cases, e.g., against the destructive environmental or chemical factors. Moreover, it is a reason for the weakening of fluorescence in solid composites. Relatively high R_{max} values are caused by local structural imperfections or impurities in **CS**, and in the case of **CS-I** and **CS-H** systems—also by the presence of precipitated dye particles.

A low surface roughness indicates that the condition required for measuring the contact angle, i.e., $R_a < 0.5\ \mu m$ [49], is fulfilled.

2.7. Protein Adsorption

Non-toxic and biocompatible chitosan is a good material for drug delivery [50,51]. CS as a cationic biopolymer can be mucoadhesive through interaction with the mucosa of living organisms [52]. Due to the presence of protonated amino groups, ionic bonds can be formed with the sialic acid residue in the glycoprotein present in the mucosa. As a result of the adsorption of the biopolymer carrier, biologically active substances are released locally on the surface of the affected skin, and the treatment is effective. However, modified chitosan, e.g., in combination with a drug, may show other, sometimes unfavorable properties, a lower availability of positively charged amino groups, and thus a reduced adsorption capacity [51]. Therefore, studying the adsorption of proteins on biopolymer materials intended for therapeutic purposes is extremely important.

Two types of protein: human serum albumin (HSA) and α-acid glycoprotein (AGP), have been used to determine the interaction with BODIPY-modified chitosan films. The research results indicate that human serum albumin interacts better with chitosan films doped with dyes than with pure chitosan (Figure 9a). The increase in adsorbed HSA protein was observed up to 6h of incubation. However, after 24 h of the test, desorption of HSA from the composites was noted. This is convenient, due to the fact that the protein-bound drug should be released as the plasma concentration of the free fraction of the drug decreases. The reversible binding of plasma proteins by the resulting composites is therefore promising, in terms of pharmacological properties. The chitosan composite (**CS-I**) with the addition of dye **I** was the best for the adsorption of human serum albumin, among those three materials.

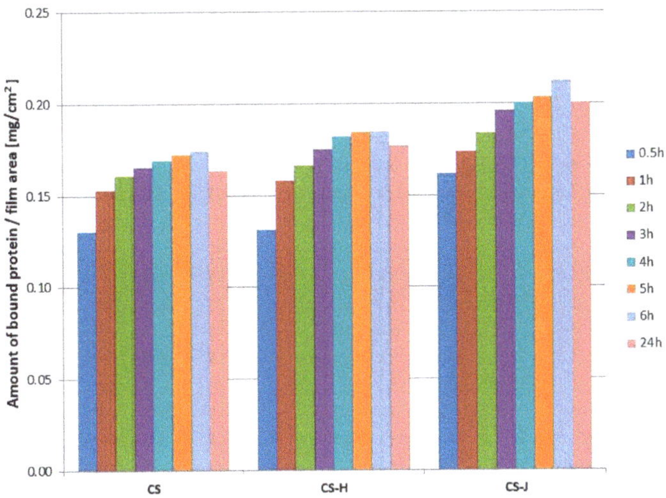

(a)

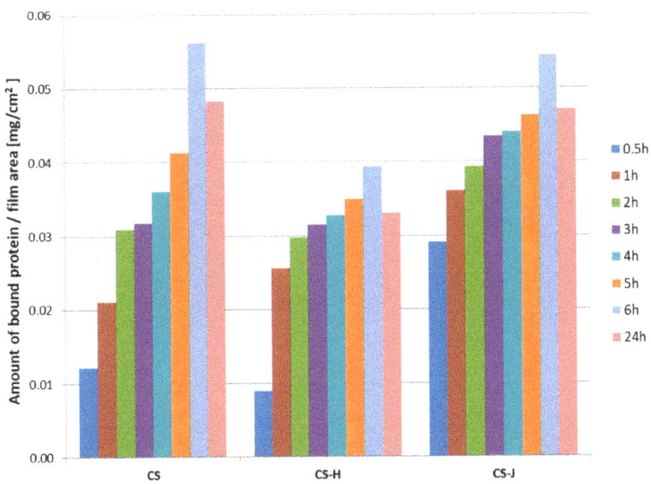

(b)

Figure 9. The amount of bound HSA (**a**) and AGP (**b**) proteins (in mg per 1 cm^2 of biofilm) of the tested samples: chitosan, chitosan doped with **I** and **H** dyes.

Sample **CS-I** also shows a greater ability to adsorb acidic α-glycoprotein (AGP) than **CS** and **CS-H** (Figure 9b) in the initial incubation period (up to 6 h). In this test, similarly to the HSA adsorption study, the desorption of AGP adsorbed on the materials was noticed after 6 h of incubation.

The chitosan used in this study has a relatively low molecular weight, which, as is known, allows not only for the external interaction with the target organism, but also has the ability to penetrate into tissues, resulting in intracellular interactions [9]. At the same time, in addition to the action of the photosensitive compound introduced, the biological effect of chitosan itself on infected skin areas can be enhanced. Attachment to proteins in the microbial cell leads to changes in their configuration and consequently to the disturbance of their vital functions. Moreover, as it has recently been reported, that pristine chitosan also exhibits anti-tumor activities [9], further enhancing the proposed systems' effectiveness.

2.8. Kinetics of the Dyes' Release

Another important feature of materials with potential medical applications, apart from the adsorption capacity, is the gradual release of the applied drug. Chitosan, thanks to its unique features (including biocompatibility, adhesion properties, cationic nature, and non-toxicity), has already been proposed as a drug carrier [53] and also as a photo-controlled delivery system [54].

The release kinetics of BODIPYs from the studied chitosan matrix is shown in Figure 10. As can be seen, the **H** dye is released much more slowly than the **I** dye. The concentration of this compound in the extract was noticeable after 5 min of the study, but the increase in its amount over time was relatively slow. Following 12 h of the study, its released amount constituted 35% of the mass of the **H** dye introduced into the sample. The **I** dye is released from the chitosan film much faster. Its release efficiency after 12 h is already 90%. The difference in the release rate of these two similar compounds is undoubtedly due to the presence of large iodine atoms in **CS-I**, which may constitute steric hindrances in intermolecular interactions. This probably makes it difficult for **I** dye molecules and **CS** macromolecules to move closer to each other, and therefore, a looser structure is formed. Thus, this dye is well dispersed in the chitosan matrix and more labile. Moreover, the lack of heavy atoms in the structure of the **H** dye causes the formation of larger dense clusters in the solid state (confirmed by SEM microscopy), from which the dye is released more slowly. Moreover, the particles of this dye can partially penetrate the chitosan random coils, from which they are eluted for a longer period of time by methanol.

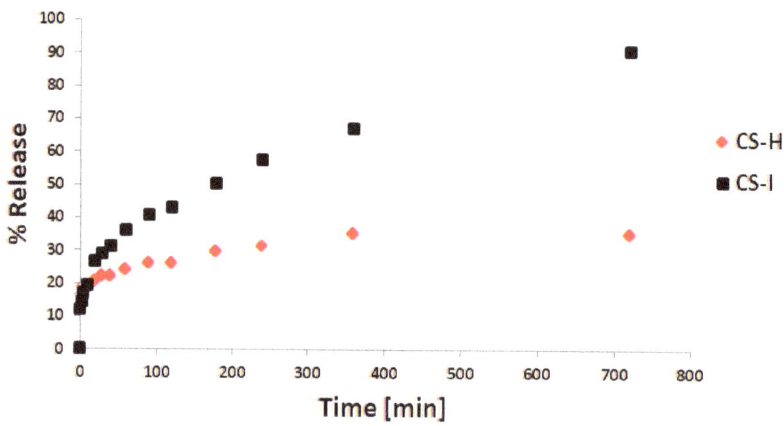

Figure 10. Kinetics of the release of dyes **I** and **H** from the chitosan film.

It is worth emphasizing that the released BODIPY dye can generate singlet oxygen under the influence of radiation with a wavelength of 615 nm [55]. The determined quantum yields of singlet oxygen generation Φ_{O2} are 0.48 and 0.06 for **I** and **H**, respectively. These results indicate a heavy atom effect, i.e., a significant increase in Φ_{O2} for a compound containing iodine atoms. Given the greater potential of compound **I** for photodynamic therapy and the fact that it releases from the chitosan composite much better than the less active compound **H**, it can be said that this chitosan-based material has potential as a new form of photosensitizer for this therapy.

2.9. Mechanical Properties

To determine the mechanical strength of the obtained films, tensile tests were carried out. These tests allowed us to record stress–strain curves and determine the breaking stress, ultimate elongation, and Young's modulus, which characterizes the material's elasticity. Young's modulus was calculated as the tangent of the angle in the initial, rectilinear course of the relationship where Hooke's law is fulfilled. The results from this analysis are presented in Table 6.

Table 6. Mechanical properties of the chitosan film and chitosan doped with **I** and **H** BODIPY.

Sample	Breaking Stress, MPa	Ultimate Elongation, %	Young's Modulus, MPa
CS	28.6 ± 0.8	2.9 ± 0.3	1169 ± 137
CS-I	22.5 ± 4.8	1.8 ± 0.2	423 ± 53
CS-H	8.6 ± 5.0	2.8 ± 0.8	767 ± 64

As can be seen, the addition of 5% BODIPY to the chitosan matrix deteriorates the mechanical properties, which is the result of a poor miscibility of the components. The drop in breaking stress is most significant in **CS-H**, but Young's modulus takes a relatively large value (767 MPa), indicating a good elasticity. In turn, the **CS-I** sample shows the largest decrease in Young's modulus, but the breaking stress changes are relatively minor, compared to this parameter in the chitosan alone. Low elongation values are typical for non-plasticized chitosan materials [56,57]. Changes in this parameter are relatively small in both dye-modified chitosan samples.

However, it can be stated that the obtained mechanical properties are sufficient for the proposed potential applications in the biomedical industry, e.g., in the form of patches or dressing materials containing photosensitizers.

3. Materials and Methods

3.1. Materials

Chitosan of a low molecular weight (50,000 g/mol) and deacetylation degree of 75–85% was supplied by Chemat (Konin, Poland); acetic acid (99.5–99.9%) by Avantor Performance Materials Poland; solvents (methanol, acetone) and salt hydrates—monosodium phosphate, $NaH_2PO_4{\cdot}2H_2O$, and disodium phosphate, $Na_2HPO_4{\cdot}12H_2O$, were purchased from Chempur (Piekary Slaskie, Poland). Glycerin, diiodomethane, and proteins—human serum albumin (HSA) and α-acid glycoprotein (AGP) were provided by Sigma-Aldrich. Chitosan and the other reagents (of high purity) were used without further purification.

Two BODIPY chromophores (**I** and **H**), synthesized in a two-step process, described previously in detail [26], were supplied by J. Piskorz from Poznan University of Medical Sciences in Poland. The chemical structures of these dyes have been confirmed by mass spectrometry (HRMS) and ^{1}H and ^{13}C NMR spectroscopy. The chemical formulae, systematic names, and abbreviations are shown in Figure 1.

3.2. Preparation of the Chitosan Composites with BODIPY Dyes

One g of chitosan was dissolved in a 250 mL flask containing 100 mL of 1% acetic acid. Then, 50 mg of the dye (**I** or **H**) previously dissolved in three drops of acetone was added. The mixture was stirred until the compound was completely dissolved. Homogeneous solutions were poured onto Petri dishes and placed on a leveled table. A chitosan film alone without adding a dye was also prepared as a reference. Following the evaporation of the solvent, the films were dried in the dark. The film thickness was **CS**-0.026 mm, **CS-I**-0.012 mm, and **CS-H**-0.032 mm. The abbreviations used in study are **CS-I** and **CS-H**.

3.3. Research Methodology

3.3.1. FTIR and UV-Vis Absorption Spectroscopy

Infrared spectra were obtained using Spectrum Two (Perkin Elmer, Waltham, MA, USA) with an ATR attachment containing a diamond crystal (angle of incidence 45°, one reflection). The range was 4000–400 cm^{-1}, the scanning speed was 0.2 cm/s, and the number of scans was 32.

The UV-Vis spectra were recorded with a Shimadzu UV-1800 double–beam spectrophotometer (Kioto, Japan) in the visible range of 400–800 nm in quartz cuvettes with an optical length of 1 cm. To determine the wavelength at which the maximum absorption by the dyes used occurred, they were dissolved in acetone, diluting the obtained solutions so that their absorbance was close to 1. The final concentration was 3.125×10^{-3} M for both dyes. The spectra of chitosan composites in the solid thin films were also measured.

3.3.2. Fluorescence and Quantum Yield Determination

The emission spectra of the samples in the solutions (**I**, **H**—in acetone, **CS-I**, **CS-H**—in 1% CH_3COOH + acetone) and in the solid films (**CS-I**, **CS-H**) were recorded with the use of the Jasco FP-800 spectrofluorimeter (Tokyo, Japan) in the visible range of 400–900 nm. The dye concentrations in the solutions were chosen so that the emission intensity was within the scope of 0.1–0.2. The concentration obtained by diluting the solutions was 1.5625×10^{-4} M and 3.125×10^{-4} M for **I** and **H**, respectively.

Quantum fluorescence yields were determined using the Jasco FP-800 spectrofluorimeter equipped with an integrating sphere. The same acetone solutions (of absorbance in the range of 0.1–0.2) and thin solid films were used. The measurement was performed in quartz cuvettes with an optical path length of 3 mm. The excitation wavelength of 300 nm was used. The quantum yield study using the integrating sphere is based on the absolute method. It compares the number of emitted photons with the number of photons absorbed by the sample, which allows for omitting the use of a standard with known values of quantum fluorescence yields [58].

3.3.3. Contact Angle Measurement

The measurement of the static contact angle (Θ, with accuracy $\pm 2°$) was performed using the DSA goniometer from KRUSS GmbH (Hamburg, Germany) using the sessile drop method at constant room temperature. Films with a carefully selected, smooth surface and two test liquids of different polarities: glycerin and diiodomethane, were used for these studies. Drops with three µL volume were applied to the surface of the samples using a syringe with a needle diameter of 1.5 mm. The contact angle was determined immediately after the drop was applied. Several measurements were made for each specimen. The surface free energy (SFE) was calculated using Young's equation and the Owens–Wendt method, which assumes that the total surface free energy (SFE) is the sum of two components: polar SFE(P) and dispersive SFE(D) [59]. For analysis of the drop shape, observed with a digital camera, and SFE calculation, the software supplied by the producer was used.

3.3.4. Morphology Analysis by SEM and AFM

The surface morphology of the obtained samples was examined with the scanning electron microscope model 1430 VP by LEO Electron Microscopy Ltd., (Cambridge, UK). A gold layer of about 18 nm thickness was sputtered on the specimens. The images of the internal structure (cross-sections) on brittle fractures of the samples (in liquid nitrogen) were taken. The surface topography and roughness were also examined using a digital instrument multimode nanoscope atomic force microscope. The following roughness parameters were determined from the surface profile: the root mean square roughness—R_q, arithmetic mean—R_a, and maximum height within the definition area—R_{max}.

3.3.5. Protein Adsorption

The interaction of the proteins: human serum albumin (HSA) and α-acid glycoprotein (AGP) with the chitosan membrane and dye-doped chitosan membranes was investigated using the Jasco FP-800 spectrofluorimeter. A solution of human serum albumin (HSA) with a concentration of 6 μM and α-acid glycoprotein (AGP) with a concentration of 10 μM in phosphate buffer pH = 7.4 and a concentration of 0.05 M was prepared. Square film samples with dimensions of 2 × 2 cm^2 were immersed in 2 mL of protein solutions and then incubated at 36 °C in a VWR International Thermomixer at 600 rpm. The fluorescence spectra in the wavelength range of 285–400 nm for HSA and 300–400 nm for AGP were recorded at various time intervals. The measurements were repeated three times.

A standard curve was prepared for each of the proteins used to calculate the amount of adsorbed proteins on the obtained composites. The human blood albumin serum standard solution concentration in the phosphate buffer was 9 μM. The HSA solutions with a concentration of 1 to 9 μM were prepared. A calibration curve was obtained by performing spectrofluorimetric measurements at a wavelength of 333 nm, i.e., for the maximum emission intensity. Similarly, a calibration curve of a standard solution of α-acid glycoprotein serum in phosphate buffer (10 μM) was obtained by performing spectrofluorimetric measurements at a wavelength of 329 nm (at maximum emission).

3.3.6. Kinetics of the Dyes' Release

A piece of the square-shaped film (1 × 1 cm^2) containing 5 wt. % dye (**I** or **H**) and 20 cm^3 of methanol were introduced into Eppendorf test tubes. The filled tubes were placed in the thermomixer and shaken at 1000 rpm. Following the shaking, 4 mL of extract were collected to obtain the UV-Vis absorption spectra. A calibration curve was previously designated for methanolic solutions of different dye concentrations (from 0 to 5 mg/L) to quantify the dye released. The standard deviation for the release after 12 h was 1.86 and 1.94 for **H** and **I** dye, respectively.

3.3.7. Mechanical Properties

Mechanical properties of pure chitosan films and chitosan doped with the BODIPY compounds were studied using the Shimadzu EZ-Test E2-LX machine (Shimadzu, Kioto, Japan) based on the ISO standard (ISO 527-2:2012). The samples of each type of film were prepared and cut into a straight strips. Thus, the prepared specimens were clamped between pneumatic holders. The tensile tests were performed at an elongation rate of 20 mm/min. The measurement was repeated at least five times for each sample, and the results were averaged.

4. Conclusions

The photosensitive composites of BODIPY with chitosan were prepared by physical mixing and characterized by spectroscopic and scanning electron microscopy methods. Films have flat, smooth surfaces, while the internal structure is heterogeneous. FTIR spectroscopy proves the intermolecular interactions between the dyes and biopolymer, which can lead to aggregation, not only in a solid state. The introduction of BODIPY to chitosan causes a slight increase in the hydrophilicity, which positively affects their gradual release in time and human protein adsorption.

BODIPY-modified chitosan becomes fluorescent and exhibits a bathochromic shift of the absorption and emission bands relative to the original dyes, which is advantageous for potential medical applications [39]. The Stokes shifts are much higher in the films than those in the solutions. An additional therapeutic effect may be exerted by the phthalimide substituent in the meso position, which, due to its large volume, hinders the formation of aggregates. The previously confirmed antibacterial effect of the BODIPY is an additional benefit of this system [26].

Iodine introduced into the BODIPY structure causes a decrease in the fluorescence quantum yield, and this trend also concerns the **CS-H** and **CS-I** compositions (films and solutions).

The fluorescence, good adsorption, gradual dye release, and mechanical properties of the obtained materials allow us to propose them for biological sensing and imaging or for the production of patches that deliver a photosensitizer to the affected skin in PDT. However, before introducing these materials into medical practice, it would be necessary to perform biocompatibility tests in vitro and in vivo.

By gaining the ability to fluoresce, the CS-BODIPY system does not lose the valuable physicochemical properties of the macromolecular compounds. Thin, durable films can be obtained, which facilitates the application on the skin surface. It should be added that the dyes themselves exist in powder form and do not have film-forming properties. The simplicity of obtaining chitosan-BODIPY films suggests potential benefits in the implementation in practice.

Author Contributions: Conceptualization, M.Z.-B.; methodology, M.Z.-B. and A.S.-W.; formal analysis, A.S.-W., J.P. and P.N.; investigation, A.S.-W., P.N. and J.P.; data curation, A.S.-W.; project administration, M.Z.-B.; writing—original draft preparation, H.K.; writing—review and editing, M.Z.-B., H.K. and A.S.-W.; supervision, M.Z.-B. All authors have read and agreed to the published version of the manuscript.

Funding: Tested BODIPY dyes were obtained in the range of grant No. 2016/21/D/NZ7/01540 from the National Science Centre, Poland.

Institutional Review Board Statement: Not applicable.

Informed Consent Statement: Not applicable.

Data Availability Statement: Not applicable.

Conflicts of Interest: The authors declare no conflict of interest.

References

1. Debela, D.T.; Muzazu, S.G.; Heraro, K.D.; Ndalama, M.T.; Mesele, B.W.; Haile, D.C.; Kitui, S.K.; Manyazewal, T. New approaches and procedures for cancer treatment: Current perspectives. *SAGE Open Med.* **2021**, *9*. [CrossRef] [PubMed]
2. Zugazagoitia, J.; Guedes, C.; Ponce, S.; Ferrer, I.; Molina-Pinelo, S.; Paz-Ares, L. Current Challenges in Cancer Treatment. *Clin. Ther.* **2016**, *38*, 1551–1566. [CrossRef] [PubMed]
3. Orzetti, S.; Tommasi, F.; Bertola, A.; Bortolin, G.; Caccin, E.; Cecco, S.; Ferrarin, E.; Giacomin, E.; Baldo, P. Genetic Therapy and Molecular Targeted Therapy in Oncology: Safety, Pharmacovigilance, and Perspectives for Research and Clinical Practice. *Int. J. Mol. Sci.* **2022**, *23*, 3012. [CrossRef] [PubMed]
4. Sung, Y.K.; Kim, S.W. Recent advances in polymeric drug delivery systems. *Biomater. Res.* **2020**, *24*, 12. [CrossRef]
5. Kaczmarek, H.; Rybczyński, P.; Maćczak, P.; Smolarkiewicz-Wyczachowski, A.; Ziegler-Borowska, M. Chitosan as a Protective Matrix for the Squaraine Dye. *Materials* **2021**, *14*, 1171. [CrossRef]
6. Kaczmarek, H.; Tafelska-Kaczmarek, A.; Roszek, K.; Czarnecka, J.; Jędrzejewska, B.; Zblewska, K. Fluorescent Chitosan Modified with Heterocyclic Aromatic Dyes. *Materials* **2021**, *14*, 6429. [CrossRef]
7. Calero, J.L.; Berríos, Z.; Suarez, O.M. Biodegradable Chitosan Matrix Composite Reinforced with Titanium Dioxide for Biocidal Applications. In *Renewable and Sustainable Composites*; Pereira, A., Fernandes, F., Eds.; IntechOpen: London, UK, 2019.
8. Muxika, A.; Etxabide, A.; Uranga, J.; Guerrero, P.; de la Caba, K. Chitosan as a bioactive polymer: Processing, properties and applications. *Int. J. Biol. Macromol.* **2017**, *105*, 1358–1368. [CrossRef]
9. Kou, S.; Peters, L.; Mucalo, M. Chitosan: A review of molecular structure, bioactivities and interactions with the human body and micro-organisms. *Carbohydr. Polym.* **2020**, *282*, 119132. [CrossRef]
10. Yanovsky, R.L.; Bartenstein, D.W.; Rogers, G.R.; Isakoff, S.T.; Chen, S.T. Photodynamic therapy for solid tumors: A review of the literature. *Photodermatol Photoimmunol Photomed.* **2019**, *35*, 295–303. [CrossRef]
11. Correia, J.H.; Rodrigues, J.A.; Pimenta, S.; Dong, T.; Yang, Z. Photodynamic Therapy Review: Principles, Photosensitizers, Applications, and Future Directions. *Pharmaceutics* **2021**, *13*, 1332. [CrossRef]
12. Algorri, J.F.; Ochoa, M.; Roldán-Varona, P.; Rodríguez-Cobo, L.; López-Higuera, J.M. Photodynamic Therapy: A Compendium of Latest Reviews. *Cancers* **2021**, *13*, 4447. [CrossRef] [PubMed]

13. Gunaydin, G.; Gedik, M.E.; Ayan, S. Photodynamic Therapy for the Treatment and Diagnosis of Cancer–A Review of the Current Clinical Status. *Front. Chem.* **2021**, *9*, 686303. [CrossRef] [PubMed]
14. Crous, A.; Abrahamse, H. Photodynamic therapy of lung cancer, where are we? *Front. Pharmacol.* **2022**, *13*, 932098. [CrossRef] [PubMed]
15. Mokwena, M.G.; Kruger, C.A.; Ivan, M.T.; Heidi, A. A review of nanoparticle photosensitizer drug delivery uptake systems for photodynamic treatment of lung cancer. *Photodiagn. Photodyn. Ther.* **2018**, *22*, 147–154. [CrossRef] [PubMed]
16. Lu, B.; Huang, X.; Mo, J.; Zhao, W. Drug delivery using nanoparticles for cancer stem-like cell targeting. *Front. Pharmacol.* **2016**, *7*, 84. [CrossRef]
17. Jiang, S.; Zhu, R.; He, X.; Wang, J.; Wang, M.; Qian, Y.; Wang, S. Enhanced photocytotoxicity of curcumin delivered by solid lipid nanoparticles. *Int. J. Nanomed.* **2017**, *12*, 167–178. [CrossRef]
18. Chen, H.; Sun, X.; Wang, G.D.; Nagata, K.; Hao, Z.; Wang, A.; Li, Z.; Xie, J.; Shen, B. LiGa 5 O 8: Cr-Based theranostic nanoparticles for imaging-guided X-ray induced photodynamic therapy of deep-seated tumors. *Mater. Horiz.* **2017**, *6*, 1092–1101. [CrossRef]
19. Poddar, M.; Misra, R. Recent advances of BODIPY based derivatives for optoelectronic applications. *Coord. Chem. Rev.* **2020**, *421*, 213462. [CrossRef]
20. Antina, E.; Bumagina, N.; Marfin, Y.; Guseva, G.; Nikitina, L.; Sbytov, D.; Telegin, F. BODIPY Conjugates as Functional Compounds for Medical Diagnostics and Treatment. *Molecules* **2022**, *27*, 1396. [CrossRef]
21. Mai, D.K.; Kim, C.; Lee, J.; Vales, T.P.; Badon, I.W.; De, K.; Cho, S.; Yang, J.; Kim, H.J. BODIPY nanoparticles functionalized with lactose for cancer-targeted and fluorescence imaging-guided photodynamic therapy. *Sci. Rep.* **2022**, *12*, 2541. [CrossRef]
22. Lv, F.; Li, H.; Wu, Q.; Guo, X.; Zhang, H.; Yu, C.; Jiao, L.; Hao, E. Silver-mediated, direct phosphorylation of BODIPY dyes at the 3- or 3,5-positions with H-phosphonates. *Chem. Commun.* **2022**, *58*, 3937–3940. [CrossRef] [PubMed]
23. Wang, L.; Ding, H.; Xiong, Z.; Ran, X.; Tang, H.; Cao, D. Design, synthesis and applications of NIR-emissive scaffolds of diketopyrrolopyrrole-aza-BODIPY hybrids. *Chem. Commun.* **2022**, *58*, 5996. [CrossRef] [PubMed]
24. Doria, S.; Taddei, M.; Cupellini, L.; Biagiotti, G.; Bartolini, P.; Bussotti, L.; Cicchi, S.; Foggi, P.; Mennucci, B.; Donato, M. Unravelling the ultrafast dynamics of a N-BODIPY compound. *Dyes Pigm.* **2022**, *200*, 110181. [CrossRef]
25. Maleckaite, K.; Narkevičius, D.; Žilenait, R.; Dodonova-Vaitkunien, J.; Toliautas, S.; Tumkevicius, S.; Vyšniauskas, A. Give or Take: Effects of Electron-Accepting/-Withdrawing Groups in Red-Fluorescent BODIPY Molecular Rotors. *Molecules* **2022**, *27*, 23. [CrossRef] [PubMed]
26. Piskorz, J.; Dlugaszewska, J.; Porolnik, W.; Teubert, A.; Mielcarek, J. Boron-dipyrromethene derivatives bearing N-alkyl phthalimide and amine substituents of potential application in the photoinactivation of bacteria. *Dyes Pigm.* **2020**, *178*, 108332. [CrossRef]
27. Júnior, J.; Holanda, V.N.; Gambôa, D.S.R.; Monte, T.V.S.; Araújo, H.D.A.; Araújo, J.A.A.N.J.V.F.S.; Callôu, M.A.M.; Assis, S.P.O.; Lima, V.L.M. Therapeutic Potential of Phthalimide Derivatives: A Review. *Am. J. Biomed. Sci. Res.* **2019**, *4*, 378–384. [CrossRef]
28. Koksoy, B.; Kaya, E.N.; Hacıvelioglu, F.; Yesilot, S.; Durmus, M. Effect of iodine substitution pattern on the singlet oxygen generation and solvent depended ketoenol tautomerization behavior of BODIPY photosensitizers. *Dyes Pigm.* **2017**, *140*, 384–391. [CrossRef]
29. Zhang, H.; Shin, B.G.; Lee, D.E.; Yoon, K.B. Preparation of PP/2D-Nanosheet Composites Using MoS2/$MgCl_{2-}$ and BN/$MgCl_{2-}$ Bisupported Ziegler–Natta Catalysts. *Catalysts* **2020**, *10*, 596. [CrossRef]
30. Singh, B.; Kaur, G.; Singh, P.; Singh, K.; Kumar, B.; Vij, A.; Kumar, M.; Bala, R.; Meena, R.; Singh, A.; et al. Nanostructured Boron Nitride with High Water Dispersibility for Boron Neutron Capture Therapy. *Sci. Rep.* **2016**, *6*, 35535.
31. Tang, C.; Bando, Y.; Huang, Y.; Zhi, C.; Golberg, D. Synthetic routes and formation mechanisms of spherical boron nitride nanoparticles. *Adv. Funct. Mater.* **2008**, *18*, 3653–3661. [CrossRef]
32. Advanced Organic Chemistry: Infrared Spectrum of 1-Iodobutane. Available online: https://docbrown.info/page06/spectra2/1-iodobutane-ir.htm (accessed on 9 November 2022).
33. Kumirska, J.; Czerwicka, M.; Kaczyński, Z.; Bychowska, A.; Brzozowski, K.; Thöming, J.; Stepnowski, P. Application of Spectroscopic Methods for Structural Analysis of Chitin and Chitosan. *Mar. Drugs* **2010**, *8*, 1567–1636. [CrossRef] [PubMed]
34. Fernandes, Q.M.; Melo, K.R.; Sabry, D.A.; Sassaki, G.L.; Rocha, H.A. Does the use of chitosan contribute to oxalate kidney stone formation? *Mar Drugs.* **2014**, *13*, 141–158. [CrossRef] [PubMed]
35. Staroszczyk, H.; Sztuka, K.; Wolska, J.; Wojtasz-Pająk, A.; Kołodziejska, I. Interactions of fish gelatin and chitosan in uncrosslinked and crosslinked with EDC films: FT-IR study. *Spectrochim. Acta A Mol. Biomol. Spectrosc.* **2014**, *3*, 707–712. [CrossRef] [PubMed]
36. Descalzo, A.B.; Ashokkumar, P.; Shen, Z.; Rurack, K. On the Aggregation Behaviour and Spectroscopic Properties of Alkylated and Annelated Boron-Dipyrromethene (BODIPY) Dyes in Aqueous Solution. *ChemPhotoChem.* **2020**, *4*, 120–131. [CrossRef]
37. Li, K.; Duan, X.; Jiang, Z.; Ding, D.; Chen, Y.; Zhang, G.-Q.; Liu, Z. J-aggregates of *meso*-[2.2]paracyclophanyl-BODIPY dye for NIR-II imaging. *Nat. Commun.* **2021**, *12*, 2376. [CrossRef]
38. Liu, Y.C.; Zhang, Y.; Liu, G.; Xing, G. J- and H-aggregates of heavy-atom-free Aza-BODIPY dyes with high 1O2 generation efficiency and photodynamic therapy potential. *Dyes Pigm.* **2022**, *208*, 110813. [CrossRef]
39. Kaufman, N.E.; Meng, Q.; Griffin, K.E.; Singh, S.S.; Dahal, A.; Zhou, Z.; Fronczek, F.R.; Mathis, J.M.; Jois, S.D.; Vicente, M.G.H. Characterization, and Evaluation of Near-IR Boron Dipyrromethene Bioconjugates for Labeling of Adenocarcinomas by Selectively Targeting the Epidermal Growth Factor Receptor. *J. Med. Chem.* **2019**, *62*, 3323–3335. [CrossRef]

40. Bobrov, A.V.; Kishalova, M.V.; Merkushev, D.A.; Marfin, Y.S. BODIPY in matrices: Brief review. *Phys. Chem. Chem. Phys.* **2021**, *23*, 12033–12044. [CrossRef]
41. dos Santos, A.F.; de Almeida, D.R.Q.; Terra, L.F.; Baptista, M.S.; Labriola, L. Photodynamic therapy in cancer treatment—An update review. *J. Cancer Metastasis Treat.* **2019**, *5*, 25. [CrossRef]
42. Marcus, Y. *The Properties of Solvents*; Marcel Dekker: New York, NY, USA, 2002; Volume 186.
43. Farris, S.; Introzzi, L.; Biagioni, P.; Holz, T.; Schiraldi, A.; Piergiovanni, L. Wetting of Biopolymer Coatings: Contact Angle Kinetics and Image Analysis Investigation. *Langmuir* **2011**, *27*, 7563–7574. [CrossRef]
44. Luo, Y.; Pan, X.; Ling, Y.; Wang, X.; Sun, R. Facile fabrication of chitosan active film with xylan via direct immersion. *Cellulose* **2014**, *21*, 1873–1883. [CrossRef]
45. Almeida, E.V.R.; Frollini, E.; Castellan, A.; Coma, V. Chitosan, sisal cellulose, and biocomposite chitosan/sisal, cellulose films prepared from thiourea/NaOH aqueoussolution. *Carbohydr. Polym.* **2010**, *80*, 655–664. [CrossRef]
46. Cunha, A.G.; Fernandes, S.; Freire, C.; Silvestre, A.; Neto, C.; Gandini, A. What Is the Real Value of Chitosan's Surface Energy? *Biomacromolecules* **2008**, *9*, 610–614. [CrossRef]
47. Liber-Kneć, A.; Łagan, S. Zastosowanie pomiarów kata zwilżania i swobodnej energii powierzchniowej do charakterystyki powierzchni polimerów wykorzystywanych w medycynie. *Polim. Med.* **2014**, *1*, 29–37.
48. Nosal, W.H.; Thompson, D.W.; Yan, L.; Sarkar, S.; Subramanian, A.; Woollam, J.A. UV–vis–infrared optical and AFM study of spin-cast chitosan films. *Colloids Surf. B Biointerfaces* **2005**, *43*, 131–137. [CrossRef]
49. Wu, S. *Polymer Interface and Adhesion*; Marcel Dekker: New York, NY, USA, 1982.
50. Wang, W.; Meng, Q.; Li, Q.; Liu, J.; Zhou, M.; Jin, Z.; Zhao, K. Chitosan Derivatives and Their Application in Biomedicine. *Int. J. Mol. Sci.* **2020**, *21*, 487. [CrossRef]
51. Rodrigues, S.; Dionísio, M.; Remuñán López, C.; Grenha, A. Biocompatibility of Chitosan Carriers with Application in Drug Delivery. *J. Funct. Biomater.* **2012**, *3*, 615–641. [CrossRef]
52. Jiang, W.; Cai, Y.; Hao-Ying, L. Chitosan-based spray-dried mucoadhesive microspheres for sustained oromucosal drug delivery. *Powder Technol.* **2017**, *312*, 124–132. [CrossRef]
53. Parhi, R. Drug delivery applications of chitin and chitosan: A review. *Environ. Chem. Lett.* **2020**, *18*, 577–594. [CrossRef]
54. Cui, W.; Lu, X.; Cui, K.; Wu, J.; Wei, Y.; Lu, Q. Fluorescent nanoparticles of chitosan complex for real-time monitoring drug release. *Langmuir* **2011**, *13*, 8384–8390. [CrossRef]
55. Chelminiak-Dudkiewicz, D.; Rybczynski, P.; Smolarkiewicz-Wyczachowski, A.; Mlynarczyk, D.T.; Wegrzynowska-Drzymalska, K.; Ilnicka, A.; Goslinski, T.; Marszałł, M.P.; Ziegler-Borowska, M. Photosensitizing potential of tailored magnetite hybrid nanoparticles functionalized with levan and zinc (II) phthalocyanine. *Appl. Surf. Sci.* **2020**, *524*, 146602. [CrossRef] [PubMed]
56. Gadgey, C.H.; Sharma, G.S. Investigation of mechanical properties of chitosan based films: A review. *Int. J. Adv. Res. Eng. Technol. IJARET* **2017**, *8*, 93–102.
57. Chen, P.; Xie, F.; Tang, F.; McNally, T. Influence of plasticiser type and nanoclay on the properties of chitosan-based materials. *Eur. Polym. J.* **2021**, *144*, 110225. [CrossRef]
58. Würth, C.; Grabolle, M.; Pauli, J.; Spieles, M.; Resch-Genger, U. Relative and absolute determination of fluorescence quantum yields of transparent samples. *Nat. Protoc.* **2013**, *8*, 1535–1550. [CrossRef] [PubMed]
59. Owens, D.; Wendt, R. Estimating the Surface Free Energy of Polymers. *J. Appl. Polym. Sci.* **1969**, *13*, 1741–1747. [CrossRef]

International Journal of
Molecular Sciences

MDPI

Article

Study of Elastin, Hydrolyzed Collagen and Collagen-like Products in a Tri-Layered Chitosan Membrane to Test Anti-Aging Skin Properties

Rocío Guerle-Cavero and Albert Balfagón-Costa *

Pharmaceutical Chemistry Research Group, Institut Químic de Sarrià, Universitat Ramon Llull, 08017 Barcelona, Spain; rocioguerlec@iqs.url.edu
* Correspondence: albert.balfagon@iqs.url.edu

Abstract: The use of animal testing in the cosmetic industry is already prohibited in more than 40 countries, including those of the EU. The pressure for it to be banned worldwide in the future is increasing, so the need for animal alternatives is of great interest today. In addition, using animals and humans in scientific research is ethically reprehensible. This study aimed to prove some of the anti-aging properties of elastin (EL), hydrolyzed collagen (HC), and two vegan collagen-like products (Veg Col) in a tri-layered chitosan membrane that was ionically crosslinked with sodium tripolyphosphate (TPP). In the first approach, as a way of representing different layers of a biological system, such as the epidermis and the two dermis sublayers, EL, HC, or Veg Col were independently introduced into the two inner layers ($2L_{(i+b)}$). Their effects were compared with those of their introduction into three layers (3L). Different experiments were performed on the membrane to test its elasticity, hydration, moisture retention, and pore reduction at different concentrations of EL, HC, and Veg Col, and the results were normalized vs. a blank membrane. This new alternative to animal or human testing can be suitable for proving certain efficacy claims for active ingredients or products in the pharmaceutical, nutritional, and cosmetic fields.

Keywords: chitosan; collagen; elastin; membrane; elasticity; moisture retention; pore

1. Introduction

Animals have been widely used for biomedical research as an alternative to human beings, although it is important to exercise caution when extrapolating findings to human outcomes. Apart from these aspects, ethical concerns regarding the use of animals, together with the high costs, time-consuming protocols, lack of effective extrapolations, and lack of reproducibility in results, are important drawbacks in all basic research.

An EU Directive is taking specific actions with the goal of fully replacing the use of animals for scientific purposes [1].

In March 2013, the European Union started the first full ban on animal testing for cosmetics. For this reason, there is a need for the replacement of animal skin models, since they are forbidden in more than 40 countries for cosmetic purposes and are ethically regulated for medical purposes [2,3].

Some synthetic or natural biomaterial-based scaffold skin models that mimic the complex and stratified structure of human skin have been developed to replace the use of animals. Some skin models represent the epidermis, including EpiSkin® (L'Oreal, Île-de-France, France), Epi-Derm® (MatTek Corporation, Ashland, MA, USA), SkinEthic® (SkinEthics, Lyon, France), epiCS® (CellSystems, Troisdorf, Germany), Holoderm®, and Kaloderm® (Seoul, Republic of Korea). Some other models represent the structure of the full thickness of human skin. Skin-on-a-chip models can also bring some other important aspects to light, such as the use of a vascular system, resulting in longer survival of the tissue [4]. However, these alternatives represent only some aspects of real human skin.

Some other aspects, such as the skin barrier function, are poorly represented. Some major drawbacks of these models are that they can be time-consuming, dependent on expertise, and costly [5]. Hence, there is still much research with the aim of finding an alternative system in which the complexity of human skin can be achieved [6].

In our previous work [7], a tri-layered chitosan membrane that was ionically crosslinked with sodium tripolyphosphate (TPP) was created as a physical model in an attempt to simulate different skin layers with the aim of providing similar results to those obtained with human skin, but with strong simplifications. The use of living cells was avoided, as there was no goal of reproducing the conditions and complexity of human skin. The materials employed were cost-effective, and they could mimic several skin characteristics and properties, such as the different layers, skin pore size, elasticity, hydration, and moisture retention. In this work, two kinds of membranes were tested: a base membrane and an activated membrane. In the second case, pores were mechanically created, as was already reported [7].

Ionic crosslinking was chosen, among other reasons, because this kind of crosslink is created between collagen fibers and proteoglycans in the extracellular matrix (ECM); thus, the components of our membranes had a similar structure [8]. These crosslinks confer structure to the skin. In contrast to ionic crosslinking, covalent crosslinking is formed in the skin between the amino acids that form proteins located in the major part of the ECM, which is mainly to give strength and elasticity to the skin.

Chitosan was chosen as the main scaffold biomaterial, and TPP was chosen as the crosslinker for many different reasons.

1. Chitosan is derived from chitin, and it is probably the second most abundant polymer in nature, behind only cellulose.
2. Chitosan has a similar structure to that of glycosaminoglycans (GAGs), an essential component of the ECM [9].
3. Chitosan is characterized by the presence of primary amino groups along its polymer backbone, which causes its structure to interact with different proteins, cells, and living organisms.
4. Chitosan can have different degrees of deacetylation (DDAs), from 50 to 100%, which determines its final properties [7].
5. Chitosan is obtained from sustainable chitin, as chitin is a byproduct of crustaceans in the food industry.
6. Chitosan is affordable compared to other biomaterials that can be naturally found in skin, such as collagen (Col), EL, and GAGs, such as hyaluronic acid.
7. TPP is non-toxic, simple to use, widely available, affordable, and an excellent ionic crosslinker of chitosan.
8. Chitosan is a well-known and well-studied biomaterial for artificial skin scaffolds [10].
9. This kind of membrane based on chitosan could have potential for use in biomedical applications. Thanks to its hemostatic, antimicrobial, and antifungal properties, chitosan could be a suitable option for burn and wound treatments [11,12].

In this research, EL and HC were selected with the aim of testing their efficacy in a Ch-based membrane [13,14]. EL and Col, together with GAGs, are some of the key components of the ECM. The ECM is a non-cellular, three-dimensional macromolecular network in which cells reside, and it is found in all tissues and organs [15,16].

EL and HC were introduced into the 3L of the membrane to test their efficacy in the entire membrane. The three layers of the membrane were intended to represent the epidermis and the two dermis sublayers, namely, the papillary dermis and the reticular dermis. For this reason, EL and HC were also independently introduced into the two inner layers, where they are naturally found in human skin [6].

EL, with a molecular weight of around $\approx$70 KDa, is the main component of elastic fibers. Its role is closely linked with that of Col; it provides stretching, recoil, and elasticity to the skin. It is also located in the dermal layer of the skin and makes up approximately

2–4% of the dry weight of the dermis in the skin of adults [17]. It is composed of alternating hydrophilic and hydrophobic segments [18–20].

In contrast to HC, EL, whether or not it is in its hydrolyzed form, has not been extensively employed in cosmetics or as a nutraceutical, despite some of its beneficial effects that have been found in skin, such as its improvement of skin elasticity in combination with Col and with some other biomaterials [21]. EL is rarely used in bioscaffolds, and it is mainly used for blending with other polymers, such as Col or GAGs [22–24]. It is mainly used as an additive for other scaffold materials, such as Col, due to its poor mechanical strength and availability. The MatriDerm® matrix (Billerberck, Germany) is an example of this, where EL is used to modify Col to mimic elastic fibers in the native dermis [12].

The addition of EL to a Ch membrane can increase its elasticity and mechanical strength, improve hydrophilicity, and increase degradation rates [25]. Hydration of EL is necessary for obtaining elasticity. EL monomers are disordered and flexible in water. They can act as plasticizers in water by interacting with the water molecules that are bound to the main chain, favoring its movement. A hydrated elastin chain can form transient hydrogen-bonded turns. These turns are not random but, rather, dynamic conformation structures that confer flexibility and contribute to high disorder. Furthermore, the hydrophobic segments prevent the formation of large secondary structures that could block chain motion [26].

Col, with a molecular weight around ≈300 KDa, is produced by fibroblasts. It provides tensile strength to the skin [15]. It is located in the dermis, provides physical support to tissues, plays an important role in the structural and biological integrity of the ECM, and represents 75% of the dry weight of the skin. It is also an amphoteric macromolecule [27].

It was already reported that Ch was used in combination with Col due to its antibacterial properties and because it reduces the biodegradation of the scaffold [12]. Col can improve the tensile strength and elasticity of chitosan scaffolds depending on the dosage used [10,16,28].

Col, especially in its hydrolyzed form, is a well-known anti-aging ingredient that is used in cosmetics, pharmaceuticals, healthcare, and the beverage and food industries [29]. It is also a widely used biomaterial scaffold, and chemical crosslinking is the most suitable method for type I Col. However, most crosslinkers are expensive, difficult to manage, and cytotoxic, as with glutaraldehyde; in most cases, they cannot be applied alone, as in the case of N-hydroxysuccinimide (NHS) and 1-ethyl-3-(3-dimethylamino-propyl) (EDC) [30]. Col is used as a skin substitute in different commercial products, such as Integra® (Princetown, NJ, USA), OrCel® (Forticell Bioscience, New York, NY, USA), Promogran Prisma® (3M, Two Harbors, MN, USA), and PuraPly® (Organogenesis, Inc., Canton, MA, USA), to name a few [12].

By combining Col with Ch, two kinds of interactions take place—hydrogen bonding and electrostatic interactions—and these reinforce the mechanical strength [31]. According to Taravel et al. [20,32], there is a weak interaction between Ch and Col that forms polyanion-polycation complexes (–NH_3^+ from Ch with the –COO^- group from Col) in slightly acidic solutions, although this is obstructed by Col gel formation.

In Col, different hydrogen bonds can be formed, such as those between chains by hydroxyproline –OH groups, those between other side groups, those involved in the formation of fibrils, and those with the –OH and NH_2 groups from Ch. Additionally, hydrogen bonds can be formed between the end groups of –COOH and NH_2 of Col and the –OH and –NH_2 groups of Ch, as Ch contains large numbers of –OH groups [33].

With an excess of Ch, a hydrogen-bonding-type complex can be formed in addition to an electrostatic complex, whereas a self-crosslinking phenomenon is induced with an excess of Col [16,34]. However, the ionic interaction of TPP with Col chains is small since the number of exposed amino groups in Col chains is very small.

In this research, different tests were performed to test the anti-aging efficacy of EL, HC, and two Veg Col products.

- Elasticity is the ability of the skin to return to its resting state after displacement [35].
- Hydration is a complex process composed of two mechanisms—the barrier effect or moisture retention—and is the ability of a system to avoid water loss. In the skin, this is known as transepidermal water loss (TEWL). On the other hand, humidity absorption is the capacity to absorb water from the environment [36–38].
- The pore reduction effect is the capacity of a material to reduce its pore size compared to that in its initial state.

From the results of this study, it can be stated that these membranes are sensitive to specific properties.

1. These properties are sensitive to the concentrations of the active ingredients and Veg Col products.
2. It was demonstrated that the effects of some active ingredients, namely EL, HC, and two Veg Col products, were evidenced in terms of pore reduction, water permeation, elasticity tests, swelling, and moisture retention. Hence, anti-aging claims could be proven.
3. The effects of the active ingredients and Veg Col products on the previous properties varied depending on the layer in which they were introduced.

2. Results and Discussion

2.1. Pore Quantification

Pores have different sizes depending on age, sex, ethnicity, and body area. Facial skin pores are a great concern for the beauty market, and their size may vary for different reasons, such as high sebum excretion, decreased elasticity around pores, an increase in hair follicle volume, and dehydration [39–41].

Table 1 shows the pore areas that were measured for activated membranes with extreme concentrations in order to see their differences with respect to blank activated membranes. It can be seen the different mean pore areas for some EL and HC studied concentrations and the percentage of pore reduction versus blank containing the active ingredient in 3L or $M2L_{(i+b)}$ according to Equation (4). All concentrations appearing in this work are expressed as *w*/*w*. Other expressions of concentrations will be specifically detailed.

Table 1. Mean pore areas (μm^2) with the standard deviation (SD) and mean percentage of pore reduction (%) in blank activated membranes and in activated membranes containing different concentrations of EL and HC and in different layers vs. blank.

ID M	N° Pores	Area	SD	Pore Reduction
M3L—Ch (blank)	35	39,045	2975	—
M3L—0.085EL	35	37,120	3910	−4.9
M3L—0.28EL	35	32,204	4452	−18
$M2L_{(i+b)}$—0.085EL [1]	35	38,856	3743	−0.48
$M2L_{(i+b)}$—0.28EL [1]	35	31,978	2678	−18
M3L—2.6HC	35	33,091	4467	−15
M3L—10HC	35	30,508	4867	−22
$M2L_{(i+b)}$—2.6HC [1]	35	38,216	4122	−2.1
$M2L_{(i+b)}$—10HC [1]	35	35,722	4032	−8.5

[1] The top layer of membranes containing the active ingredient in the two inner layers ($M2L_{(i+b)}$) was composed of only Ch 2.6% (stock solution).

As previously reported by Flament et al. [39], facial pore size varies depending on ethnicity, from 15,000 μm^2 for the Chinese ethnicity to 92,500 μm^2 for the Brazilian ethnicity. In the case of Asiatic facial pores, the pores are the smallest. For Caucasians, the skin values are in the middle, ranging between 40,000 and 55,000 μm^2 (calculated as the area of a circle) [42]. All of these pore sizes can be studied with these kinds of membranes; the mean pore size was 39,045 μm^2 for the blank activated membrane, which had even lower values than those of Caucasian skin at some concentrations of EL and HC.

When EL was included in the activated membrane, the effect on pore size reduction was most evident at the highest concentration, 0.28%, with the highest effect in both 3L and $2L_{(i+b)}$; there was a pore reduction of −18%. A possible explanation for this having the same effect in both kinds of membranes could be that EL was able to diffuse from the inner to the top layer in $2L_{(i+b)}$, thus producing the same effect as that in 3L. When the EL content was low (0.085%), the effect was a bit more pronounced when EL was included in 3L (−4.9%), but its effect at this low concentration was hardly perceived when it was included in 2L (−0.48%). In any case, the pore size reduction was directly proportional for both 3L and $2L_{(i+b)}$.

In the case of HC, the effect on pore size reduction was also evidenced the most with the highest concentration (10%); again, the greatest effect on pore reduction was in 3L (−22%), which was in contrast to when it was included in $2L_{(i+b)}$ (−8.5%). The pore size reduction was also directly proportional to the concentration for both 3L and $2L_{(i+b)}$.

Nowadays, noticeable pore size is a great concern for consumers [43]. Hence, these results can be helpful for claiming the beneficial effects of these active ingredients or products on reducing pore size. The effect of HC on pore size reduction was also evidenced in previous scientific reports [44].

In Figure 1, there is visual evidence of the most extreme cases of pore reduction for both the EL- and HC-activated membranes in comparison with the activated blank.

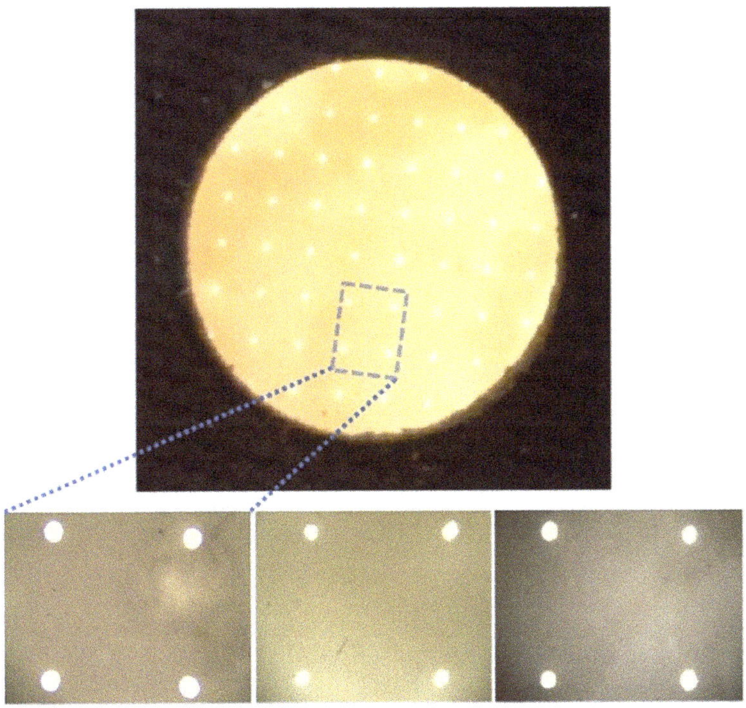

(A) (B) (C)

Figure 1. Pore visualization with an optical microscope in the activated membranes with a semi-plan 4 × 0.10 oil-immersion objective with a wide field of 0.45: (**A**) blank; (**B**) with EL included at 0.28% in 3L; (**C**) with HC included at 10% in 3L.

When looking at the images obtained with the microscope, the pore size reduction was visually evident, and it can also be stated that the shapes of the pores varied arbitrarily.

A possible reason for these two phenomena could be the kinds of interactions between Ch, HC, or EL and TPP. When layers were composed of only Ch, well-shaped and firm channels were formed. Nevertheless, when HC or EL were included, some interactions could occur, as they could be located between adjacent molecules of Ch and TPP to produce arbitrary links. As a consequence, different effects could be promoted, such as increasing plasticity or disrupting the Ch structure. In addition, an increase in swelling could be achieved because of the nature of Col and EL, as the membranes were in contact with water during their formation. Consequently, water molecules could diffuse into the membrane and into the pores' surroundings. This fact could promote the disruption of the intermolecular forces that tightly held the polymer chains. Therefore, as a possible consequence of these two phenomena, the formation of irregularities in the channels and reductions in pore size could take place.

2.2. Permeation Tests

As a means of studying the permeability, 30 g of water was permeated through an activated membrane in a Franz diffusion cell. During the permeation tests, the blank activated membranes reached the highest permeation of water in the first minute, and the 30 g of water was completely permeated in 2.9 min on average. For this reason, the quantity of water permeated within the first minute was used to compare the pore sizes. The reference value for the experimental results was the mean water permeation through the blank activated membranes for one minute (15 g). The mean water permeation (%) of the different experiments was calculated according to Equation (5).

2.2.1. Permeability of the EL-Activated Membranes

Three different concentrations of EL were studied for the water permeation tests. The lowest one, 0.085%, fell within the margins of the EL content in dermal skin.

In Figures 2 and 3, the differences that were observed when EL was included in 3L and $2L_{(i+b)}$, respectively, can be seen.

Figure 2. Mean reduction (%) in water permeation at 1 min for the activated membranes containing EL at 0.085, 0.23, and 0.28% in 3L vs. the blank activated membrane.

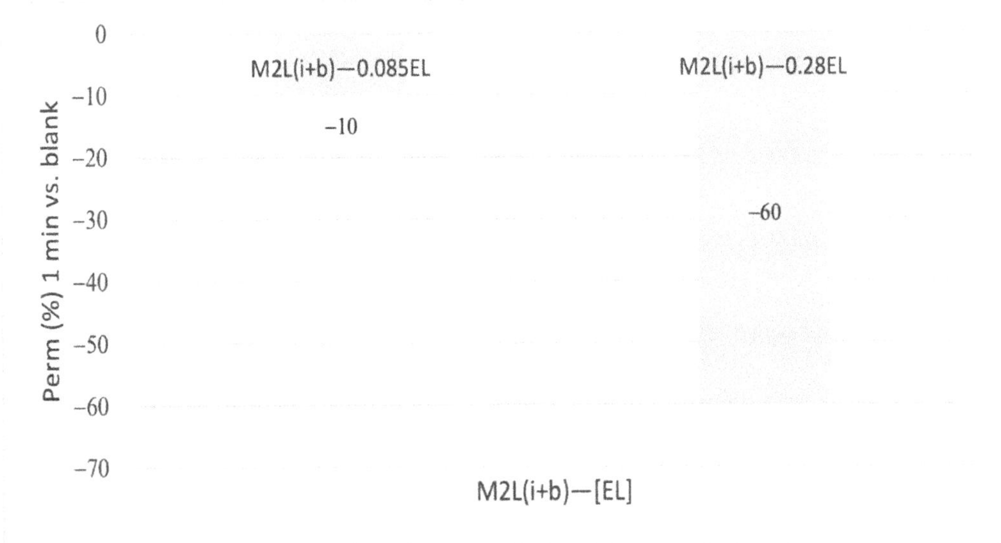

Figure 3. Mean reduction (%) in water permeation at 1 min for the activated membranes containing EL at 0.085 and 0.28% in $2L_{(i+b)}$ vs. the blank activated membrane.

As shown in Figure 2, a 65% reduction in permeation was observed with 0.283% EL, which was a reduction in permeation by more than half with respect to the blank.

As shown in Figure 3, when EL was included in $2L_{(i+b)}$, a 60% water permeation reduction was observed at the highest concentration of EL (0.283%), showing the high power of EL in water permeation reduction, despite only being included in $2L_{(i+b)}$.

These results show that EL had a very strong effect on the reduction in water permeation. It worked in a concentration-dependent manner for both kinds of membranes (3L and $2L_{(i+b)}$). The effect of water permeation reduction was slightly clearer when EL was introduced in 3L than in $2L_{(i+b)}$, as there was no EL in the first layer. Thus, the water permeation reduction was less evident.

2.2.2. Mean Permeability of the HC- and Veg-Col-Activated Membranes

In the case of HC, eight different concentrations were studied to see their effects on water permeation. In addition, two different Veg Col products were also studied at the concentration recommended by the supplier (2%).

In Figures 4 and 5, the differences observed when HC was just studied in 3L or $2L_{(i+b)}$ at some concentrations can be seen.

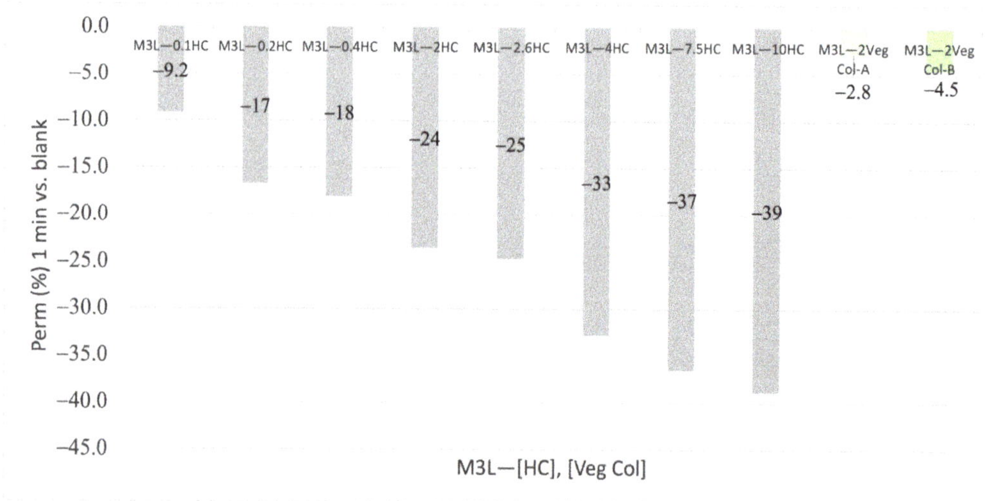

Figure 4. Mean reduction (%) in water permeation at 1 min for the activated membranes containing HC at 0.1, 0.2, 0.4, 2, 2.6, 4, 7.5, and 10% represented in dark grey or, Veg Col type A and B represented in different green intensities at 2.0%, in 3L vs. the blank activated membrane.

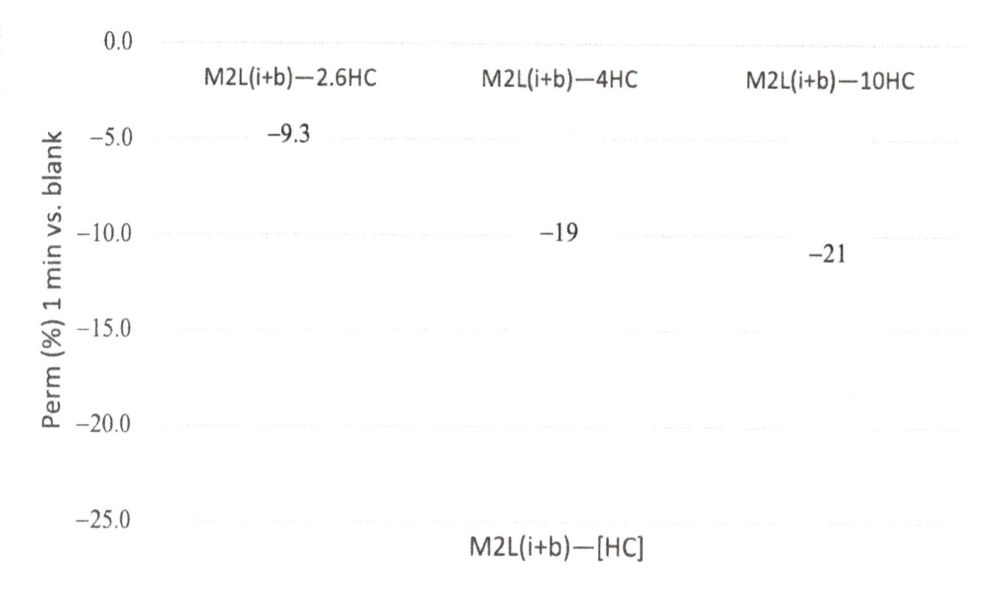

Figure 5. Mean reduction (%) in water permeation at 1 min for the activated membranes containing HC at 2.6, 4, and 10% in $2L_{(i+b)}$ vs. the blank activated membrane.

As shown in Figure 5, the reduction in water permeation was, again, not as strong as when HC was introduced in 3L, but the effect was also concentration dependent.

Well-arranged values were observed for the eight different concentrations that were studied, as the water permeation decreased from −9.2% to −39%. It should be noted that HC is a small molecule compared to EL. Hence, although the concentration of HC was higher than that of EL, it was foreseeable that the water permeation reduction would not have as much of an impact as that of EL (from −23% to −65%), as the latter has longer hydrophobic segments.

In the case of Veg Col, the effect of water permeation reduction was barely noticeable in comparison with the HC of animal origin.

2.3. Relationship between the Mean Water Permeability and the Mean Pore Reduction

In Table 2, the relationship between the mean percentage of water permeation reduction and the mean percentage of pore reduction is shown for both active ingredients (EL and HC) that were studied.

Table 2. Mean permeated grams of water at 1 min with SD and relationship between the mean percentage of water permeation reduction (%) and the mean percentage of pore size reduction (%) for EL and HC vs. blank at 1 min.

ID M	Water Permeation ± SD	Water Permeation Reduction	Pore Reduction
M3L—Ch (blank)	15 ± 2.6	—	—
M3L—0.085EL	12 ± 2.8	−23	−4.9
M3L—0.28EL	5.3 ± 0.07	−65	−18
$M2L_{(i+b)}$—0.085EL [1]	14 ± 1.7	−10	−0.48
$M2L_{(i+b)}$—0.28EL [1]	6.1 ± 1.1	−60	−18
M3L—2.6HC	11 ± 3.1	−25	−15
M3L—10HC	9.3 ± 1.7	−39	−22
$M2L_{(i+b)}$—2.6HC [1]	14 ± 5.5	−9.3	−2.1
$M2L_{(i+b)}$—10HC [1]	12 ± 1.4	−21	−8.5

[1] The top layer of membranes containing the active ingredient in the two inner layers ($M2L_{(i+b)}$) was composed of only Ch 2.6% (stock solution).

When comparing both active ingredients, it can be seen that in the case of EL, the water permeation reduction was greater than that with HC for similar values of pore reduction. The difference increased at high concentrations of both active ingredients. An explanation for this divergence could be because of the nature of EL. EL contains more hydrophobic segments as it has a longer chain compared to that of Col, which is hydrolyzed. This issue may impact the water permeability by slowing down the permeation process, despite the fact that the effective pore size obtained with EL was not much smaller than that obtained with HC.

From the results obtained, it can be seen that the greatest pore reduction was achieved when 10% HC was added to 3L, and the greatest water permeation reduction was obtained when 0.28% EL was added to 3L.

2.4. Rheological Tests

All elastic modulus results are mean values of G′ (Pa) that were obtained within 2 min. For blank rheological tests, it was evident that the elasticity results were very sensitive to the room temperature during membrane formation. All membranes studied for the rheological tests were base membranes.

As can be observed in Figure 6, the elasticity of the blank membranes followed a good model of linear regression (R^2 = 0.9844) with respect to temperature in the range of room temperatures studied (from 17.5 to 25 °C).

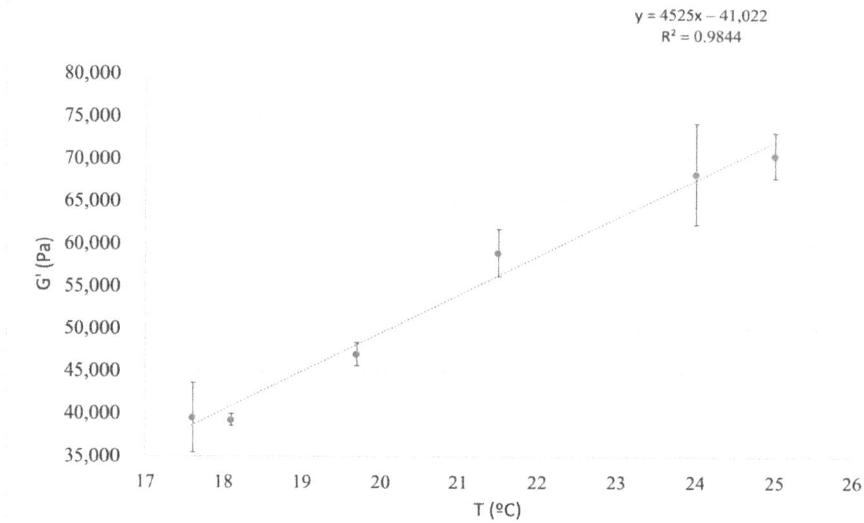

Figure 6. Mean elastic moduli of the blank membranes at different temperatures.

All elastic modulus results, which are shown in %, of the membranes containing the different active ingredients or products were normalized (Equation (7)) with respect to their blank membrane interpolation in the linear regression model (Equation (6)), as shown in Figures 7–10.

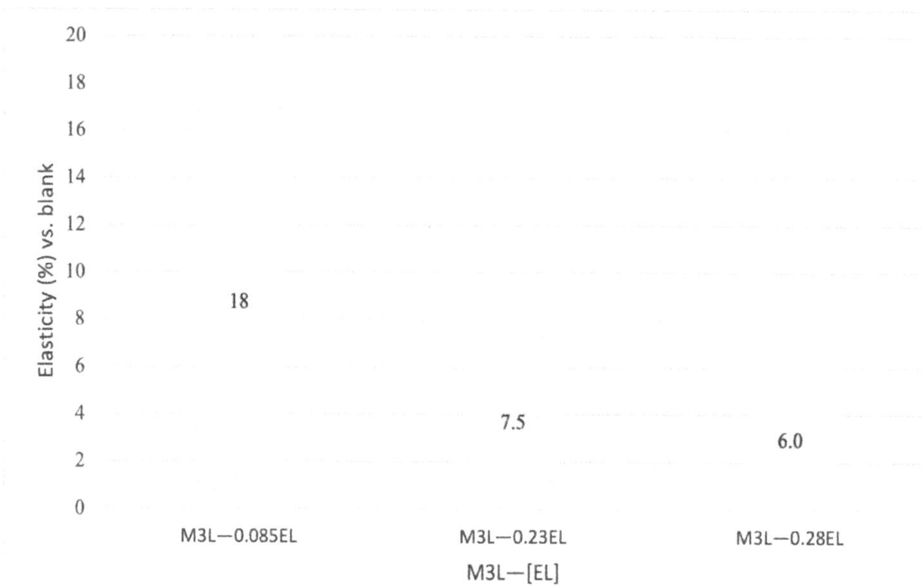

Figure 7. Mean elasticity (%) for EL membranes normalized vs. blank membranes with EL included at 0.085, 0.23, and 0.28% in 3L.

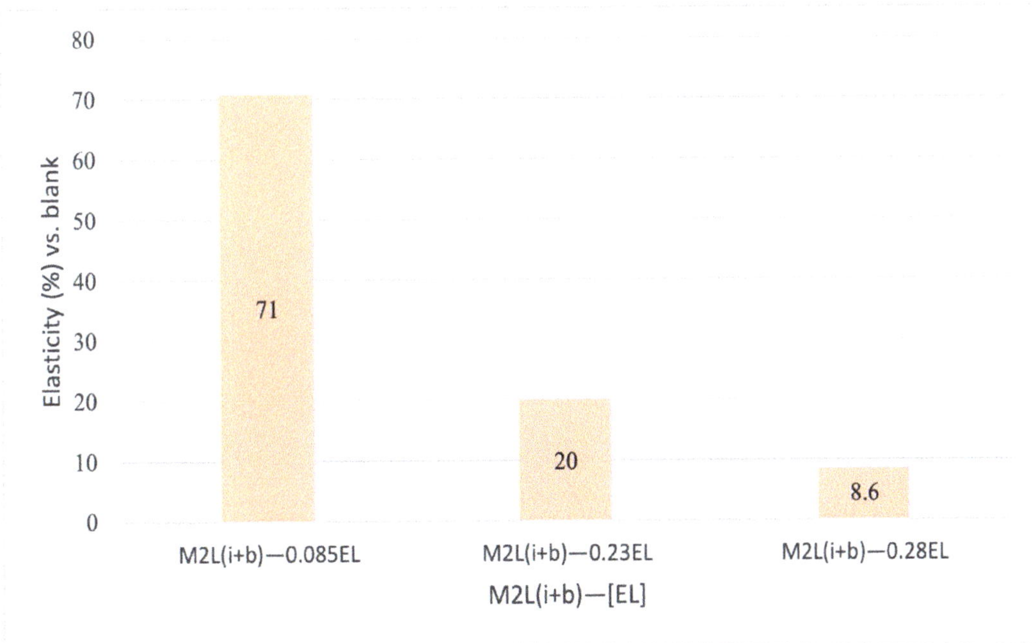

Figure 8. Mean elasticity (%) for EL membranes normalized vs. blank membranes with EL at 0.085, 0.23, and 0.28% included in $2L_{(i+b)}$.

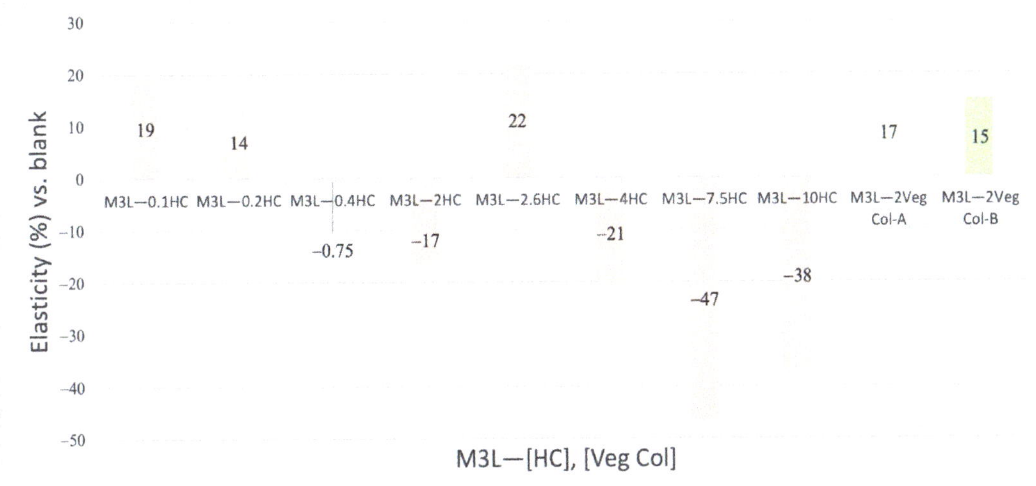

Figure 9. Mean elasticity results (%) for HC and Veg Col type A and B membranes normalized vs. blank membranes with HC at 0.1, 0.2, 0.4, 2, 2.6, 4, 7.5, and 10%, represented in orange, and with Veg Col type A and B at 2% represented in different green intensities, in 3L.

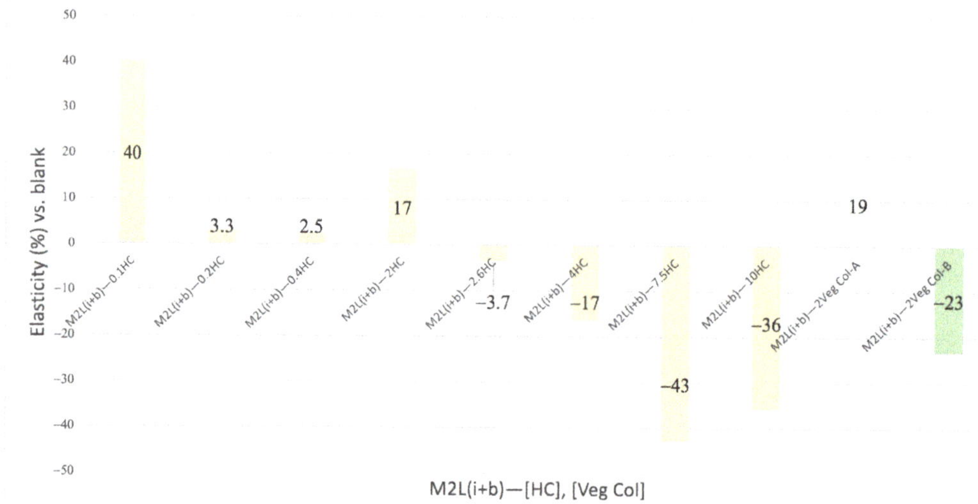

Figure 10. Mean elasticity results (%) for HC and Veg Col type A and B membranes normalized vs. blank membranes with HC at 0.1, 0.2, 0.4, 2, 2.6, 4, 7.5, and 10%, represented in yellow, or with Veg Col type A and B at 2%, represented in two green intensities, in $2L_{(i+b)}$.

In Figure 7, it can be observed that, when EL was added to 3L, the elasticity decreased as the EL concentration increased.

In Figure 8, it is shown that, when studying these three concentrations of EL in $2L_{(i+b)}$, the tendency was the same, causing the percentage of elasticity to decrease as the EL concentration increased.

However, when comparing the same concentration of EL in 3L and $2L_{(i+b)}$, the effect on elasticity was increased in $2L_{(i+b)}$; the greatest difference was at the lowest concentration in the solution (0.085%), and this was from 18% (3L) to 71% ($2L_{(i+b)}$). This concentration (0.085% EL) was equal to 2.6% EL in dry weight. This concentration fell within the margins of the content of EL in human dermal skin (2–4% in dry weight). This fact suggests that the EL content in human skin has the optimal concentration for achieving the highest elasticity capacity. For the last concentration that was studied (0.28%), there was almost no difference between the two kinds of membranes (3L or $2L_{(i+b)}$).

From the results obtained for EL, it can be seen that for the three concentrations studied, EL always produced an increase in elasticity, although it was inversely proportional to the concentration. An elastic booster effect was obtained when EL was included in $2L_{(i+b)}$ at all concentrations studied.

A possible explanation for the increase in elasticity could be that when EL is in contact with Ch, a unique hydrogen binding site between EL and Ch is achieved. Thus, it could break down the inter- and intramolecular hydrogen bonding networks of Ch. So, hydrophobic segments of EL that can form disordered assemblies' structures may hinder the formation of intermolecular hydrogen bonding networks. Thus, this would favor the motion of Ch chains. This effect reached its maximum at the lowest concentration of EL, 0.085%. When the EL content was increased, what could have happened is that fewer hydrogen bonding sites could be achieved between Ch and EL, as there could be saturation of EL, so a decrease in elasticity was observed.

As shown in Figure 9, at 0.1% and 0.2% HC, according to literature, when there is an excess of Ch, a hydrogen-bonding-type complex and an electrostatic complex could form between Ch and HC, thus producing some elasticity in the membrane. Apart from these two kinds of complexes, a small interaction between TPP and HC could be produced, which may have also helped to increase the elasticity.

For higher concentrations, we have considered the following hypothesis. When the concentration of HC began to increase from 0.4% to 2%, a more disorganized network was produced, resulting in a reduction in elasticity. When the concentration of Ch was almost equal to that of HC, it seemed that a more ordered network was formed, and this was able to create more bonds between Ch, TPP, and HC. Hence, an increase in elasticity was observed. At higher concentrations, from 4 to 10% HC, a more disorganized network was achieved, as the excess of HC disturbed the links between Ch and TPP. Hence, a disruption in molecules' alignment was caused, leading to a decrease in elasticity. At 7.5% HC, the greatest effect on elasticity reduction (−47%) was obtained.

An inversely proportional relationship between Ch and HC concentrations was also observed when the peak of elasticity found when both concentrations were almost equal was excluded.

For both kinds of Veg Col that were studied, a noticeable effect on elasticity was found at the concentration of 2% recommended by the supplier. In the case of Veg Col-A, the "collagen-like" active ingredient was combined with glycerin. Glycerin is known to have good plastic properties in the skin [45]. When it is combined with Ch, its single hydrogen bonding site can break down the inter- and intramolecular hydrogen bonding networks of, thus leaving hydrophobic C-H ending groups to limit intermolecular hydrogen bond formation and allowing free motion of the chitosan chains [46]. According to the supplier of Veg Col-A, glycerin was found in the product at 1.4% when using 2% Veg Col-A. Some elasticity tests were performed by adding 1.4% glycerin to the blank solution in order to check its influence on elasticity. There was a 16% increase in elasticity with respect to the blank. Hence, it seems that almost all increases in elasticity with Veg Col-A (17%) could be attributed to the plastifying effect of glycerin. In the case of Veg Col-B, the elasticity value of 15% could only be attributed to the active ingredient "vegan Col".

As shown in Figure 10, an even greater increase in elasticity was observed at 0.1% HC (40%) in $2L_{(i+b)}$ than with the same concentration in 3L (19%). Both hydrogen bonding and electrostatic complexes could also be formed here. Our hypothesis in this case was that, apparently, when HC was included in $2L_{(i+b)}$ at the lowest concentration (0.1%), it did not disturb the interaction of Ch and TPP. At the same time, HC could also form links with Ch from the same layer and from the top layer. A possible explanation could be that, in this case, the molecules could be organized in a way that favored linkages, thus acting as a dominant force for some other kinds of links in opposite directions, which could disturb the interaction. It seems that the HC located in the intermediate layer may still have had carboxylic groups that were able to react with free amino groups and –OH from the Ch stock solution in the last layer by forming more hydrogen bonds. It seems that in 3L at 0.1% HC, the molecules were a bit more disorganized than those in $2L_{(i+b)}$, which had lower elasticity values (19%).

However, when the concentration of HC was 0.2 or 0.4%, the effect on elasticity was reduced to almost negligible. At 2% HC, a peak in elasticity was observed. A possible explanation of this effect could be that the diffusion of HC to the top layer may have had a similar effect on elasticity to that seen for the concentrations of 0.1 or 0.2% HC in 3L, suggesting that a similar concentration could diffuse to the top layer. Hence, the elasticity had a similar value to that obtained with those two concentrations (17%). For cases in which there was an excess of HC, the same effect on 3L was also visualized in $2L_{(i+b)}$, but with slightly lower values of elasticity, as the concentration of HC in the whole membrane was not as high as in 3L. In the case of Veg Col-A, the elasticity was found to be similar for both kinds of membranes, and the positive effect of elasticity due to glycerin was observed in both cases (17 and 19%, respectively). For Veg Col-B, based only on the effect of the "collagen-like" active ingredient, the elasticity decreased when it was included in the two inner layers, and it was even lower than in the blank membrane (−23% in $2L_{(i+b)}$ in contrast to 15% in 3L).

In the $2L_{(i+b)}$ membranes, as shown in Figure 10, an inversely proportional relationship between the Ch and HC concentrations was found, except for the peak of elasticity that was found when Ch was at 2.5% and HC was at 2%.

These elasticity results clearly demonstrate that EL, HC, and the two Veg Col products had an influence on the elasticity of the membrane, as they increased or decreased its value depending on the concentration used. It was also evidenced that when the active ingredient or Veg Col product was included in $2L_{(i+b)}$, the elasticity values for the lowest concentration were boosted in comparison with the membrane in which the active ingredient or Veg Col product was included in 3L.

Some similarities with our results were found in the elasticity results obtained by Hydalgo-Vicelis et al. [28], where membranes were prepared in the form of films with the solvent evaporation technique and were composed of Ch and Col, which were crosslinked by a covalent crosslinker (EDC). In their case, different proportions of both Ch and Col were combined, unlike in our study, where the concentration of Ch was almost constant, but only the HC concentration varied. When a low concentration of Col was used, the elasticity obtained had the highest value, but, as the Col concentration increased, the elasticity values decreased. In this case, although they did not keep the Ch concentration constant, a tendency similar to that in our study was found despite the use of different types of crosslinkers.

According to Martínez et al. [16], they also tested EDC and TPP as crosslinkers for Ch membranes. In their case, the membranes were prepared by freeze-drying and combining three different proportions of Ch and Col. For TPP-crosslinked membranes with Col in the minority, the highest elasticity value was obtained. When the concentration of Col was equal to that of Ch, they found a reduction in the elasticity value, which was not in concordance with the values that we obtained. When the quantity of Col was in the majority, the elasticity increased, but not as much as when it was in the minority. For EDC-crosslinked membranes and when using a combination of both TTP and EDC, they showed a directly proportional concentration-dependent behavior as the Col concentration increased with the three concentrations that were studied.

These results show that, despite the use of the same crosslinker, depending on the technique of preparation of the membrane and the combination of concentrations of both Ch and Col, the tendencies of the elasticity results can substantially vary.

2.5. Swelling Index

The swelling index (SI) of the different kinds of tri-layered crosslinked base membranes was calculated according to Equation (8) for different lengths of PBS baths, and the dry state of each membrane was taken as a reference. The SI obtained for the blank membranes was 114%. All SIs of the membranes containing different active ingredients or Veg Col products studied were normalized (Equation (9)) with respect to the blank membrane, as shown in Figures 11–14.

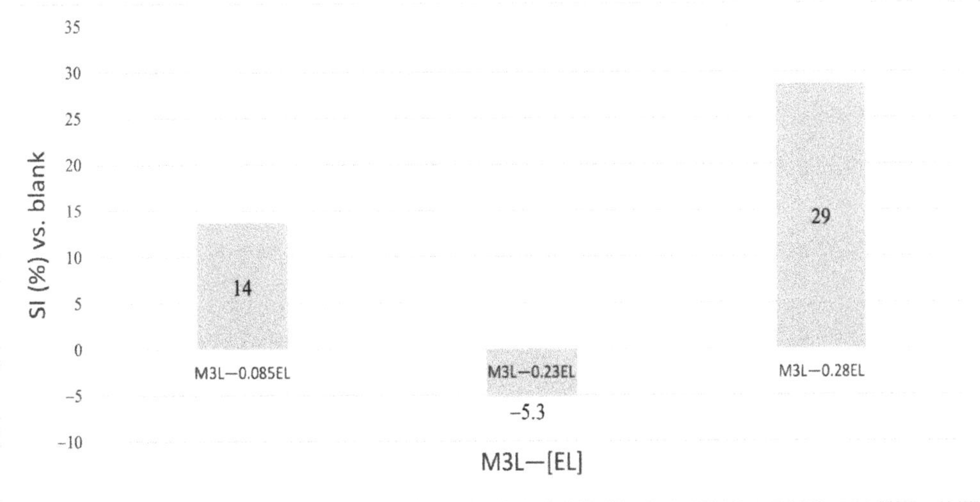

Figure 11. Mean swelling indexes (%) for EL membranes normalized vs. blank membranes with EL at 0.085, 0.23, and 0.28% in 3L.

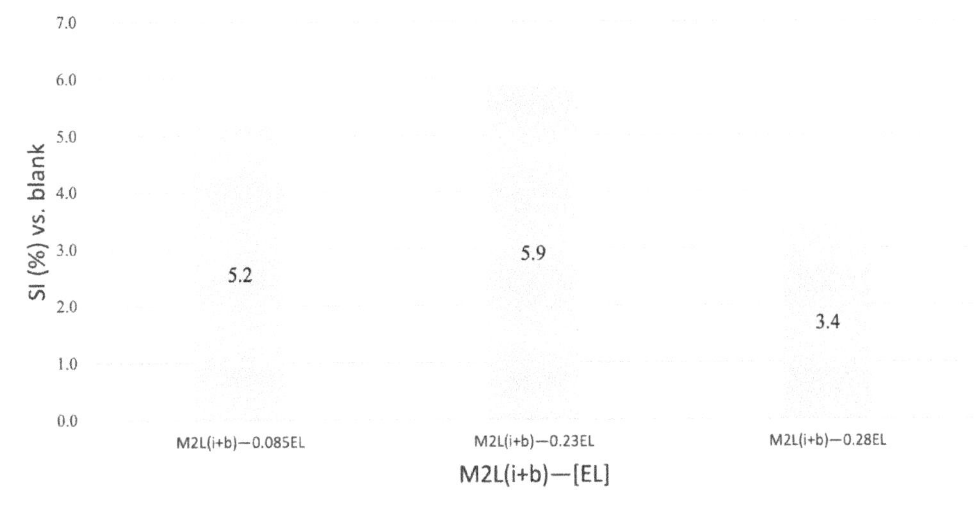

Figure 12. Mean swelling indexes (%) for EL membranes normalized vs. blank membranes with EL at 0.085, 0.23, and 0.28% in $2L_{(i+b)}$.

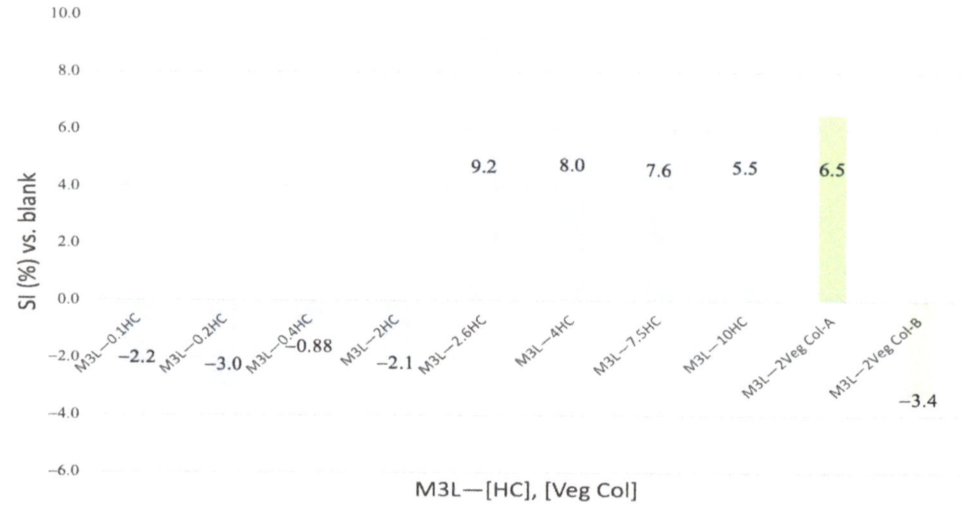

Figure 13. Mean swelling indexes (%) of HC and Veg Col type A and B membranes normalized vs. blank membranes with HC at 0.1, 0.2, 0.4, 2, 2.6, 4, 7.5, and 10% represented in blue, and Veg Col type A and B represented in different green intensities at 2% in 3L.

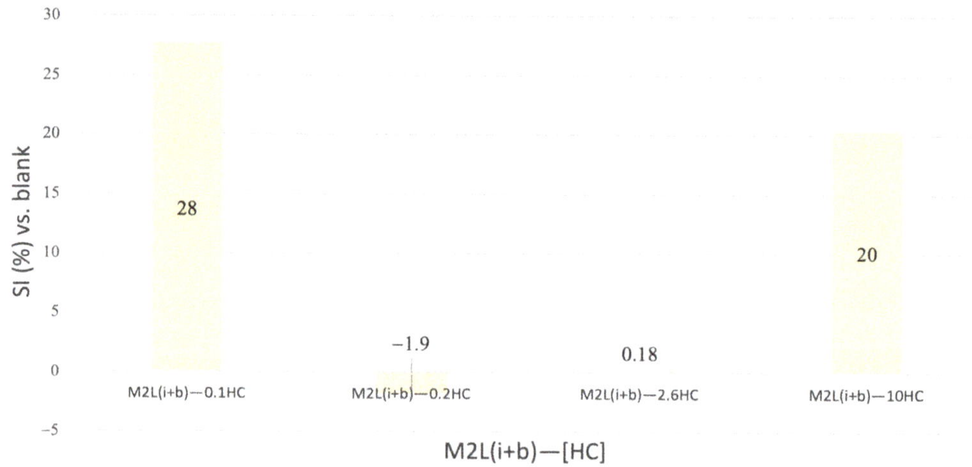

Figure 14. Mean swelling indexes (%) of HC normalized vs. blank membrane with HC at 0.1, 0.2, 2.6, and 10% in $2L_{(i+b)}$.

In Figure 11, it can be seen that the swelling effect of EL was evidenced at 0.085%, and at 0.283% EL, the highest swelling effect was obtained (29%) in 3L. The intermediate concentration of 0.23% EL in 3L did not have any positive effect on swelling (−5.3%).

As shown in Figure 12, when EL was only included in $2L_{(i+b)}$, the swelling effect at extreme concentrations was reduced compared to that in 3L, but it was higher at the intermediate concentration compared to that in 3L (5.9%). For the three concentrations of EL included in $2L_{(i+b)}$, similar SIs were observed.

As shown in Figure 13, swelling effects were not observed at low concentrations—from 0.1 to 2% HC—when this ingredient was introduced into 3L. However, a certain SI was obtained at concentrations at which there was an almost equal or greater concentration of HC with respect to Ch—from 2.6 to 10% HC. It should also be clearly stated that very similar values were obtained in that range, and they were slightly higher when the concentrations of Ch and HC were almost equal (SI of 9.2%). There was also a tiny tendency toward swelling reduction from 2.6 to 10% HC.

A positive SI was obtained for Veg Col-A but not for Veg Col-B. It should be considered that 1.4% glycerin was contained in the 2% Veg Col-A used. Glycerin is an ingredient that is commonly used in topical products due to its good properties as a humectant [47]. A swelling test—not shown in Figure 14—was performed for a membrane containing 1.4% glycerin in 3L to determine the effect of glycerin. An SI of 7.0% with respect to the blank was obtained. Thus, it seems that the SI of Veg Col-A was only achieved because of glycerin's effect.

As shown in Figure 14, a noticeable effect on swelling was achieved for extreme concentrations (0.1 and 10% HC: 28 and 20%, respectively). No effects were observed for the two intermediate concentrations studied (0.2 and 2.6% HC). It can be highlighted that the effect of 10% HC was much higher in $2L_{(i+b)}$ (20%) than in 3L (5.5%).

As has already been stated in the scientific literature, the swelling behavior of a Ch membrane crosslinked with TPP and with the inclusion of EL or Col could be influenced by the following:

1. The degree of crosslinking; increasing the crosslinking density, which produces an increase in elasticity, could lead to a decrease in swelling due to the formation of a more compact structure [48,49].
2. Increasing the concentrations of both EL and Col could also lead to an increase in swelling, as they contain some hydrophilic regions that can interact with water molecules, thus promoting swelling [16,50,51]. Another aspect could be that both active ingredients could produce a more porous structure in the membrane and a more organized and structured network that could favor water absorption [52].

2.6. Moisture Retention

Moisture retention (MR) was measured in the base membranes containing the active ingredient or Veg Col products in 3L due to the conditions of this test. The water loss (in %) for the blank membranes was 60% at 15 min and 80% at 30 min. The moisture retention percentage for each membrane under study, normalized versus blank membrane, was calculated according to Equation (10).

As shown in Figure 15, strong moisture retention abilities were observed for both concentrations of EL at 15 min; the value was higher (21%) at the highest concentration (0.27%). A lower moisture retention was observed at 30 min than at 15 min. These results suggest that EL has good short-term water retention abilities but a poor ability to retain moisture in the long term.

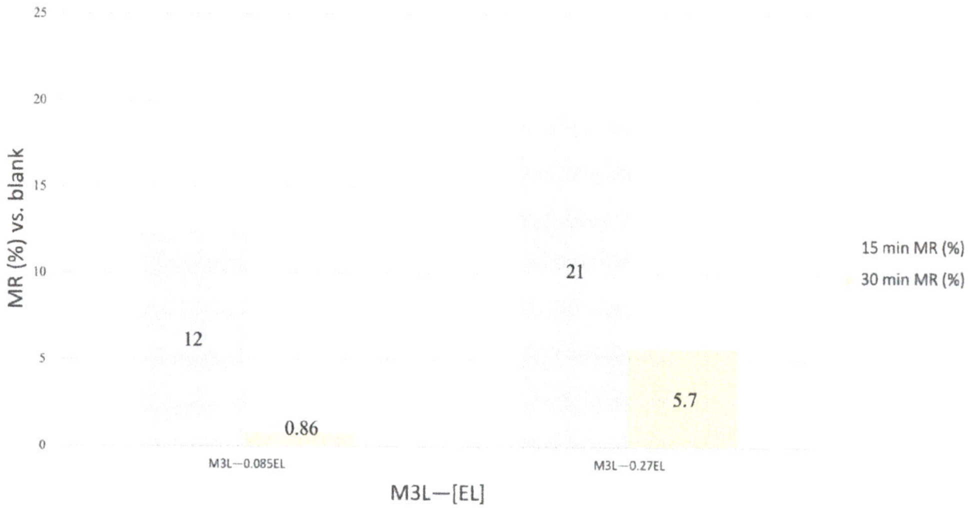

Figure 15. Mean moisture retention (%) at 15 and 30 min of the EL vs. blank membrane with EL at 0.085 and 0.27% in 3L.

As shown in Figure 16, there was a negligible or negative effect on moisture retention for the studied concentrations of 2, 2.6, and 4% HC. This could mean that HC was able to disrupt the Ch structure, thus favoring water release.

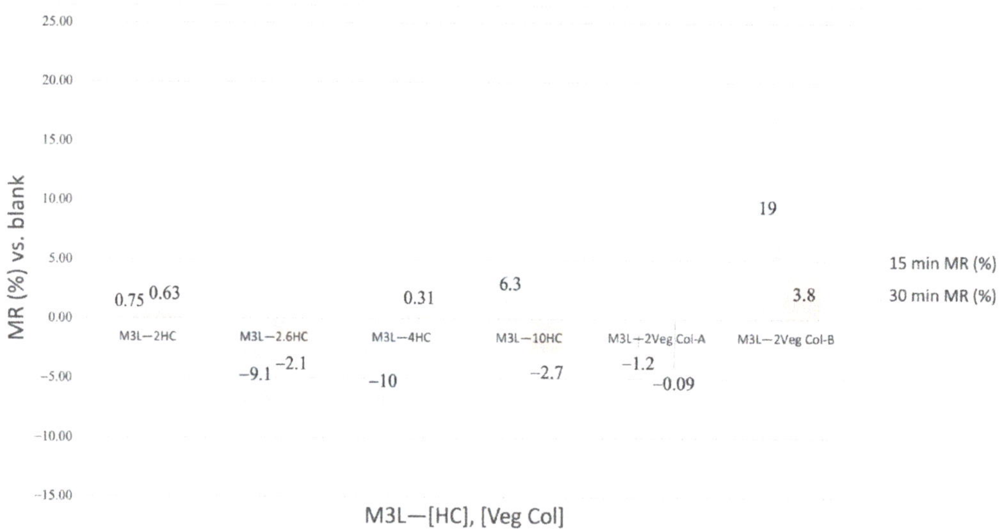

Figure 16. Mean moisture retention (%) of HC at 2, 2.6, 4, and 10% and both kinds of Veg Col type A and B at 2% vs. the blank membrane at 15 and 30 min, in 3L.

However, some moisture retention was observed for 10% HC at 15 min (6.3%). This could mean that this kind of membrane was more compact and closed than that in which

Ch was in the majority. Water could diffuse easily because a more porous structure could be formed at 10% HC, and it was kept inside, so water retention was favored.

For the two types of vegan Col studied, Veg Col-B presented a strong moisture retention power in the short term (19%), unlike Veg Col-A, which had no positive effects (−1.2%). The moisture retention power of Veg Col-B (19%) was similar to that of the highest concentration of EL studied (21%).

As was shown, HC had a short-term capacity for water retention but almost no capacity for retaining water in the long term.

2.7. Comparison of the Studied Properties with Results in Human Skin Reported in the Scientific Literature

Some scientific papers have already stated claims based on efficacy tests that were performed with volunteers for HC and EL ingestion or topical application of cosmetic products containing these two active ingredients. Different clinical trials demonstrated that properties such as hydration, elasticity, transepidermal water loss (TEWL), reductions in wrinkles, and some other aspects of skin aging can be addressed by using these active ingredients. HC was used as both a nutraceutical and a cosmetic ingredient. The oral intake of a hydrolysate of EL showed beneficial effects on human skin, such as improvements in skin elasticity and wrinkles [21,53].

In most of the reported studies of oral ingestion of HC at different dosages, it was shown that improvements in skin elasticity and hydration were more evident than reductions in transepidermal water loss (TEWL) [44,54–61]. However, no effects on increased skin elasticity with the use of topical products with HC were reported. According to Berardesca et al. [62], topical application of HC produced significant improvements in skin hydration, surface smoothness, and luminosity. Ohara et al. [63] reported a dose-dependent effect of oral ingestion of two and a half or ten grams of HC on stratum corneum hydration, but it must be noted that only two dosages were studied. They also highlighted that stratum corneum hydration is the unique beneficial aspect of skin, but they found no significant differences in elasticity or TEWL improvement. In our study, hydration could be related to two properties: pore reduction and moisture retention. A directly proportional relationship was found between pore reduction and the concentrations of HC and EL. Although EL had positive effects on moisture retention in the short term, not enough concentrations were studied to see a clear tendency. In the case of HC, no tendencies were found, but a positive result was obtained for only the highest concentration: 10% HC.

As reported by Maia Campos et al. [64], the combination of topical products and oral supplementation of Col peptides can bring the best benefits for keeping skin in good condition, as their effects can be complementary. Topical products produce a short-term effect, and nutraceuticals have a long-term effect. They also demonstrated the pore miniaturization effect of HC.

3. Materials and Methods

3.1. Materials

Medium-molecular-weight chitosan extracted from seashell skeletons with a DDA of 85% was acquired from Aldrich, Merck Life Science S.L. (Madrid, Spain) (product number 448877) with a viscosity of 340 cps (200–800 cps) (C = 1%, 1% acetic acid). Acetic acid (glacial) and sodium hydroxide pellets were acquired from PanReac AppliChem, Panreac Química S.L.U. (Barcelona, Spain). Sodium tripolyphosphate with a technical grade of 95.2% was supplied by Alfa Aesar, Thermofisher GmbH (Kandel, Germany). Agarose BioReagent for molecular biology, low electroendosmosis (EEOO), was supplied by Sigma-Merck Life Science S.L. (Madrid, Spain). Phosphate-buffered saline tablets were supplied by Sigma-Merck Life Science S.L. (Madrid, Spain). One tablet of phosphate-buffered saline dissolved in 200 mL of deionized water yielded 0.01 M phosphate buffer, 0.0027 M potassium chloride, and 0.137M sodium chloride with a pH of 7.4 at 25 °C. Calcium chloride from Quimivita, S.A. (Barcelona, Spain) was used in granular form. Teflon molds with a diameter

of
6.0 cm were fabricated at the university and were specially designed for the creation of these membranes. The microneedle device was a 140 DRS DermaStamp® system acquired from IBeauyMachine.com GBS International Holding Ltd. (Guangzhou, China). In this work, a Dermastamp® with needles with a depth of 3.0 mm was employed [7]. This depth allowed us to achieve complete perforation of the three layers of the membrane. Customized Franz cells of a special size were designed and fabricated by Fisher Scientific (S.L., part of Thermofisher Scientific (Madrid, Spain)). Bovine elastin from the neck ligament—also in powdered form—was bought from Sigma-Aldrich®. Refrigerated storage (2–8 °C) was important for preventing EL degradation. HC is a group of peptides with a low molecular weight (3–6 KDa) [27]. Type I Col is the main Col found in skin, representing 80–90% of skin Col. Type I HC at 5 KDa in powder from fish skin was used. It was bought from Rousselot as the Peptan F5000 HD. According to Rousselot, it presents a random coiled conformation due to its denaturation process. Two commercial vegan "collagen-like" products for cosmetic use with the following compositions (expressed in inci names) were also studied: Veg Col-A: glycerin: 70.0099%, aqua (water): 27.99%, pentylene glycol: 2%, and nicotiana benthamiana hexapeptide-40 sh-polypeptide-47: 0.0001% [65]. Veg Col-B: aqua (water): 93.5%, collagen amino acids: 4.5%, and leuconostoc/radish root ferment filtrate: 2% [66]. Veg Col-A is a type I Col fragment produced in Wild plants as biofactories. According to the supplier's information on Veg Col-A, a molecule identical to a fragment of the human Col type I sequence could be obtained with the proper post-translational hydroxylations required to be fully functional. Veg Col-B was made from the enzymatic hydrolysis of corn, wheat, and soy proteins.

3.2. Methods

3.2.1. Support Preparation

The support for membrane manipulation was based on agarose. The preparation of this support was performed according to a procedure that has been described elsewhere [7]. A 2% solution of agarose was prepared under magnetic stirring and heated up to 80 °C until turning transparent. Then, 9 g of this solution was placed in the Teflon molds for 5 min. Finally, supports are extracted from molds and soaked in a 10% TPP bath for at least 15 min to passivate agarose reactivity, followed by a water bath for another 15 min.

3.2.2. Stock Solution Preparation

Three independent solutions were prepared: acetic acid, NaOH, and TPP were separately dissolved in water, and Ch was added posteriorly to the acetic solution. The following quantities are described for obtaining 116 g of stock solution. The first solution was acetic acid 2% (v/v) sol. The second solution was 1 g of NaOH dissolved in 25 mL of water. The third solution was a 10% sodium tripolyphosphate (TPP) solution, which needed to be mixed with magnetic stirring until it was solubilized. A total of 3 g of chitosan powder was added to 97 mL of acetic acid solution, followed by the addition of a known quantity of the second solution. Firstly, the solution was stirred manually for about one minute until jellification was obtained. Hereafter, it was stirred automatically with an anaxial flow agitator for 3 h at 542 rpm until homogenization. The final solution pH should be between 5.5 and 5.8 to assure crosslinking. In order to always have 116 g of stock solution weight, a weight correction was made at the end by adding water. It was again stirred with an anaxial flow agitator for 15 min at 542 rpm until homogenization. Finally, a transparent solution was obtained [7].

3.2.3. EL, HC and Veg Col Solution Preparation

The previous stock solution was used to obtain the EL, HC, and Veg Col solutions. Different solutions with different concentrations were used.

EL Solution Preparation

Different solutions of several concentrations were used by adding different concentrations of Bovine elastin from the neck ligament in powder to a known quantity of stock solution. Once the powder was added, automatic stirring was performed for 30 min at 542 rpm for all EL concentrations studied. The resultant solution was homogeneous yet translucid. No increase in the viscosity of the solutions was observed.

HC and Veg Col Solution Preparation

Different solutions were prepared by adding several concentrations of Peptan F5000 HD (identified as HC) in powder to a known quantity of stock solution. Consecutively, automatic stirring was performed for 30 min at 542 rpm for all HC concentrations studied. The obtained solution was transparent for all concentrations of HC. However, an increase in solution viscosity was observed in a concentration-dependent manner for the different solutions. Still, it was observed that the viscosity decreased over time. As an increase in viscosity could create some difficulties during membrane creation, the following procedure was performed: For low concentrations—from 0.1 to 2% HC—the solutions were ready to use from the moment that they were created. From 2.6 to 10% HC, the solutions were left to evolve for three days to two weeks, depending on the concentration, so the viscosity decreased enough to make them easy to pour.

Two vegan "collagen-like" solutions were prepared by adding 2% of each to a known quantity of stock solution. Automatic stirring was performed for 30 min at 542 rpm. A transparent solution was obtained.

EL, HC, and Veg Col solutions at different concentrations were independently tested in different layers of a membrane in order to see their effects on 3L and $2L_{(i+b)}$.

3.2.4. Membrane Preparation

The membrane preparation was conducted at room temperature (21 ± 2 °C).

The preparation of a tri-layered membrane consisted of creating three layers, as previously described [7]. The difference in this work was the addition of the solution containing the active ingredient, or Veg Col product, to 3L or $2L_{(i+b)}$. Finally, a different curation process was performed to determine whether the base or the activated membrane. The activated membrane differed from the base membrane because some physical pores were created with a microneedle device during the curation process [7].

In Scheme 1, it can be observed the detailed procedure of preparing the different tri-layered membranes with the inclusion of the active ingredients or Veg Col products independently in the three or two inner layers. The first layer, bottom, was prepared by weighting 4 g of Ch or I solution, depending on whether a blank membrane (Ch solution) or a membrane containing EL, HC, or Veg Col (I solution) was performed. The second layer, the inner, was composed of 3 g of Ch (for blank) or I solution. The third layer, the top, was also composed of 3 g of Ch (for blank or $M2L_{(i+b)}$) or I solution. All layers were crosslinked between them with a 10% TPP solution, as detailed elsewhere [7]. Finally, the resulting membrane was left to rest in the mold for 4 h to get consistency, with a Teflon lid on top to obtain flatness. Afterwards, the membrane was weighted and kept in a Petri dish.

It should be stated that the addition of EL, HC, or Veg Col products was made in solution form and always combined with Ch. Hence, these solutions were used as the scaffold for the layer development process when creating the tri-layered membrane and were not added after the tri-layered membrane formation.

In Scheme 2, it can be observed that the curing process was performed for all membranes obtained after finishing the process described in Scheme 1. For base membranes, they were soaked in a 10% TPP bath for 1 h 5 min followed by a 15 min water bath, in order to stop the TPP coagulation reaction. For activated membrane, after 5 min of TPP bath, the membrane gained some consistency. Following this, the membrane was punched with the microneedles, and it was left in the TPP bath for 1 h. Then, a 10 min water bath was performed with the Dermastamp®. Finally, the Dermastamp® was removed, and a 5 min water bath was performed to neutralize the whole activated membrane.

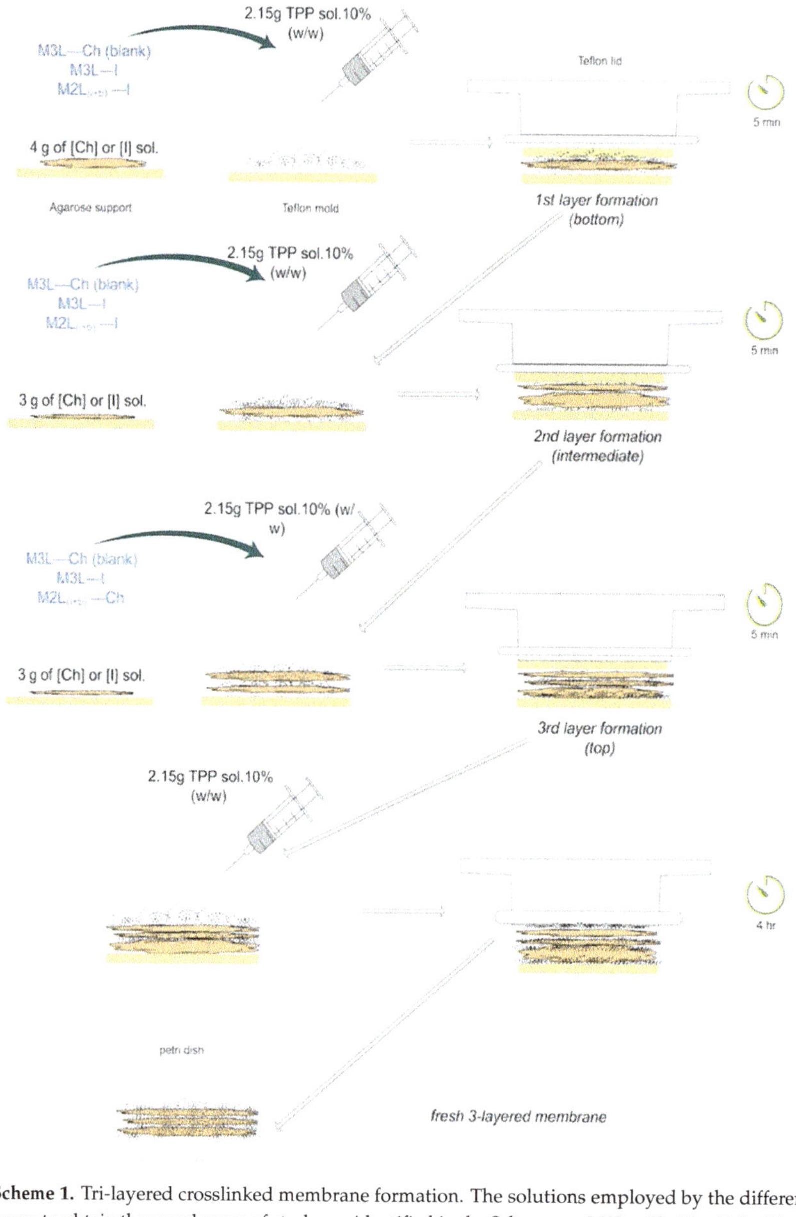

Scheme 1. Tri-layered crosslinked membrane formation. The solutions employed by the different layers to obtain the membranes of study are identified in the Scheme as: M3L—Ch (blank) for blank membranes containing the stock solution in the 3L; M3L—I for membranes containing EL, HC, or Veg Col solutions in the 3L; and $M2L_{(i+b)}$—I for membranes containing EL, HC, or Veg Col solutions in the two inner layers. The top layer of $M2L_{(i+b)}$—I was identified by $M2L_{(i+b)}$—Ch, which always contained 2.6% of Ch solution (stock solution).

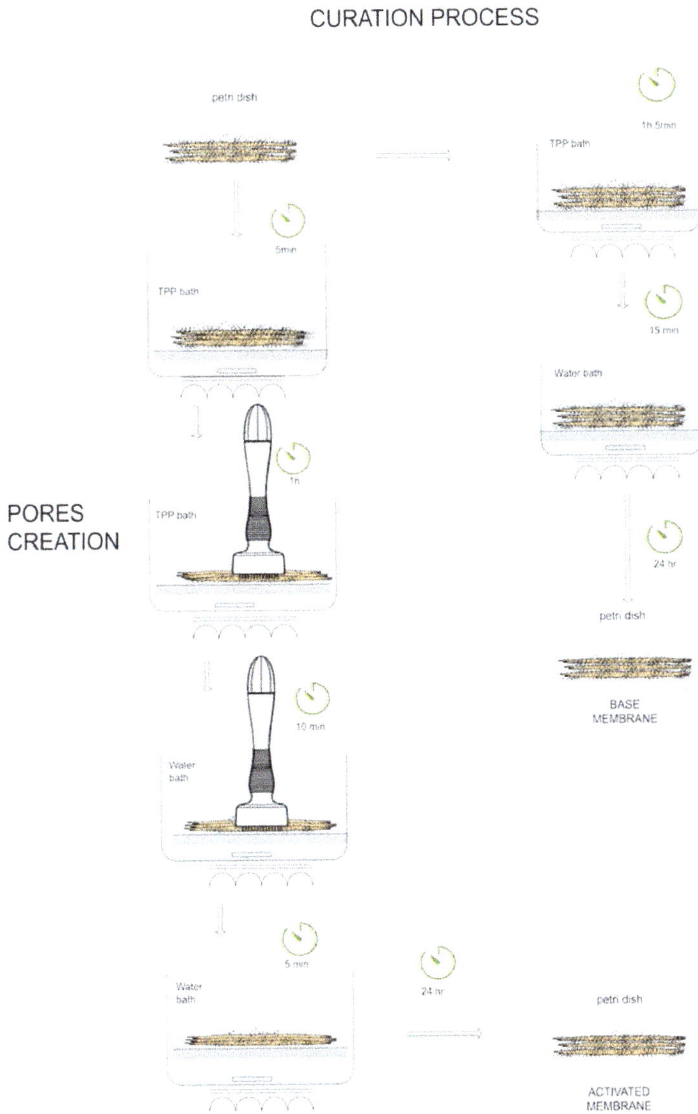

Scheme 2. The curing process of activated and base tri-layered membranes is performed for all kinds of membranes [7].

After 24 h, the base, or activated membrane, was ready to use.

Finally, the membranes were cut in order to obtain a round membrane with a diameter of 2 cm.

3.2.5. Characterization of the Different Membranes

The different kinds of membranes were identified as follows: $M2L_{(i+b)}$-I%. First position: M = membrane. Second position: 3L, 2L = number of layers containing the active ingredient or product of study (3 layers or 2 layers); subindex: i = intermediate layer; b = bottom layer if the active ingredient or product was introduced into the two inner layers. Third position I%: active ingredient or product (in this case, HC, EL, or Veg Col product) with the concentration % (w/w) added to the Ch stock solution. All of the membranes were crosslinked with TPP.

All characterization experiments presented below were carried out with tri-layered crosslinked base membranes, except for the permeation tests and pore quantification tests, where only activated membranes were used.

3.2.6. Concentrations of the Active Ingredients and Veg Col Products

The stock solution contained 2.6% Ch and 0.56% sodium hydroxide. Subsequently, different concentrations of HC or EL were added to this stock solution.

The Ch concentration (%) in the solution when different concentrations of the active ingredient or Veg Col product (I) had been posteriorly added was calculated with Equation (1).

$$[\mathrm{Ch}]\%\ \mathrm{sol} = \frac{2.6\ \mathrm{g\ Ch}}{(100\ \mathrm{g\ sol} + \mathrm{I\ g})} \times 100 \quad (1)$$

where I is the quantity of the active ingredient or product added.

The dry weight of the concentration in the solutions was calculated by considering the following residues in the percentage of the HC or EL solutions: concentration of Ch in %, 0.56% sodium hydroxide, and concentration of the active ingredient in %. The acetic acid and TPP used between the layers had almost negligible amounts of residue. The TPP residue from the bath used during the curation process of the membrane was not calculated, as it could not be determined how much TPP was trapped during the curation process of the membrane. Hence, the TPP absorbed from the bath was not considered.

The concentration of Ch (%) in dry weight, [Ch]% dry weight, when different concentrations of active ingredients were added to the solution was calculated with Equation (2).

$$[\mathrm{Ch}]\%\ \mathrm{dry\ weight} = \frac{[\mathrm{Ch}]\%\ \mathrm{sol}}{([\mathrm{Ch}]\%\ \mathrm{sol} + 0.56\%\ \mathrm{NaOH\ sol} + [\mathrm{I}]\%)\ \mathrm{sol}} \times 100 \quad (2)$$

The concentration of the active ingredient [I] (%) in dry weight when different concentrations of active ingredients were added to the solution was calculated with Equation (3).

$$[\mathrm{I}]\%\ \mathrm{dry\ weight} = \frac{[\mathrm{I}]\%\ \mathrm{sol}}{([\mathrm{Ch}]\%\ \mathrm{sol} + 0.56\%\ \mathrm{NaOH\ sol} + [\mathrm{I}]\%)\mathrm{sol}} \times 100 \quad (3)$$

Tables 3 and 4 show the identification of membranes with the inclusion of I in 3L or $2L_{(i+b)}$ at the different concentrations studied, with the equivalent of their concentrations in dry weight. When I was included in the $2L_{(i+b)}$, it was always introduced into the two inner layers, as it is naturally found in skin. The top layer was always a solution of 2.6% Ch (stock solution).

Table 3. Identification (ID) of different membranes with the percentage of EL inclusion in 3L or $2L_{(i+b)}$ and the concentrations (%) (w/w) of Ch and EL in solution and in dry weight.

ID M3L	ID M2L	Ch %	EL %	Ch % (Dry w)	EL % (Dry w)
M3L—Ch (blank)	—	2.6	—	82	—
M3L—0.085EL	$M2L_{(i+b)}$—0.085EL [1]	2.6	0.085	80	2.6
M3L—0.23EL	$M2L_{(i+b)}$—0.23EL [1]	2.6	0.23	77	6.7
M3L—0.27EL	$M2L_{(i+b)}$—0.27EL [1]	2.6	0.27	76	7.8
M3L—0.28EL	$M2L_{(i+b)}$—0.28EL [1]	2.6	0.28	75	8.3

[1] The top layer of membranes containing the active ingredient in the two inner layers ($M2L_{(i+b)}$) was composed of only Ch 2.6% (stock solution).

Table 4. Identification (ID) of different membranes with the percentage of HC inclusion in 3L or $2L_{(i+b)}$ and the concentrations (%) (w/w) of Ch and HC studied in solution and in dry weight.

ID M3L	ID M2L	Ch %	HC %	Ch % (Dry w)	HC % (Dry w)
M3L—Ch (blank)	—	2.6	—	82	—
M3L—0.1HC	$M2L_{(i+b)}$—0.1HC [1]	2.6	0.1	80	3.1
M3L—0.2HC	$M2L_{(i+b)}$—0.2HC [1]	2.6	0.2	77	6.0
M3L—0.4HC	$M2L_{(i+b)}$—0.4HC [1]	2.6	0.4	73	11
M3L—2HC	$M2L_{(i+b)}$—2HC [1]	2.5	2	50	39
M3L—2.6HC	$M2L_{(i+b)}$—2.6HC [1]	2.5	2.6	44	46
M3L—4HC	$M2L_{(i+b)}$—4HC [1]	2.5	4	35	57
M3L—7.5HC	$M2L_{(i+b)}$—7.5HC [1]	2.4	7.5	23	72
M3L—10HC	$M2L_{(i+b)}$—10HC [1]	2.3	10	18	78

[1] The top layer of membranes containing the active ingredient in the two inner layers ($M2L_{(i+b)}$) was composed of only Ch 2.6% (stock solution).

As it was already stated, the concentration of EL in the dry weight of the skin is around 2–4%. This concentration was also studied in this work, with a concentration of 0.085% EL in the solution.

As was previously described, the concentration of Col in the dry weight of the skin is around 75%. This concentration was covered with concentrations of 7.5 and 10% HC.

In order to see if the reduction in Ch concentration when the concentration of HC increased would affect the results, a lowest concentration of Ch sol., 2.1%, was used as a blank to see the effects on elasticity, and no differences in the results were found. For the swelling, permeation, and moisture retention tests, 2.3% Ch blank membranes were obtained to prove that there were no variations in the results in comparison with a 2.6% Ch membrane.

Two commercial products of vegan origin with a "collagen-like" effect were also studied at the concentrations used in the efficacy studies performed by the two different suppliers of the products, as shown in Table 5.

Table 5. Identification (ID) of different membranes with the percentage of inclusion of Veg Col types A and B in 3L or $2L_{(i+b)}$ and the concentrations (%) (w/w) of Ch, Veg Col-A, and Veg Col-B in solution.

ID M3L	ID M2L	Ch %	ID I	I %
M3L—Ch (blank)	-	2.6	-	-
M3L—2Veg Col-A	$M2L_{(i+b)}$—2Veg Col-A [1]	2.5	Veg Col-A	2
M3L—2Veg Col-B	$M2L_{(i+b)}$—2Veg Col-B [1]	2.5	Veg Col-B	2

[1] The top layer of membranes containing the Veg Col product in the two inner layers ($M2L_{(i+b)}$) was composed of only Ch 2.6% (stock solution).

3.2.7. Pore Quantification

The quantification of pores was conducted on activated membranes, as mechanical pore channels were created in these kinds of membranes.

The pores of the tri-layered crosslinked membranes were observed with a Euromex Bioblue Optical Microscope. A semi-plan 4 × 0.10 oil-immersion objective with a wide field of 0.45 was used. The ImageFocus 4 software v. 3.1.2 for Windows, developed by Euromex, from Papenkamp, 6836 BD Arnhem, The Netherlands, was employed to measure the area of the pores. Calibration correction was applied on a computer with the objective that was used. The resolution that was chosen was 2040 × 1528. The pores that were observed were located in the diffusion area of 1.06 cm^2 of the Franz cell rings, and the number of pores was 52. Periodical pores were optically checked throughout the activated tri-layered crosslinked membrane, as visualized in Figure 1.

Activated membranes employed for permeation tests were first checked with the Optical Microscope.

The mean percentage of pore reduction in comparison with the blank membrane was calculated according to Equation (4).

$$\text{Av pore \%} = \frac{\text{Pore area [I]} - \text{Pore area blank}}{\text{Pore area blank}} \times 100 \quad (4)$$

where Pore area [I] represents the area (μm^2) of pore size of the activated membrane containing the active ingredient or Veg Col product, and Pore area blank represents the area (μm^2) of pore size of the blank activated membrane.

3.2.8. Permeation Tests

The permeation tests were conducted at room temperature (21 ± 2 °C).

Customized Franz cells from Fisher Scientific, S.L. (Madrid, Spain) were especially designed for the membranes in order to perform the water permeation tests.

The diffusion area was 1.06 cm^2. The membrane was fixed between two permeation rings that were 6.0 cm in diameter, and they were collocated between the donor and receptor compartments of the Franz cells [7]. Afterwards, the system was sealed and clamped until tight. Then, 30 g of deionized water was introduced into the donor compartment to determine the permeation over time by collecting the permeated water in the receptor compartment.

The mean percentage of permeation at 1 min in comparison with the blank membrane was calculated according to Equation (5).

$$\text{Av \% perm} = \frac{\text{G perm [I]} - \text{G perm blank}}{\text{G perm blank}} \times 100 \quad (5)$$

where G perm [I] represents the grams that permeated through the membrane containing the specific concentration of the active ingredient or Veg Col product, and G perm blank represents the grams of water that permeated through the blank membrane.

3.2.9. Rheological Tests

Rheological tests were only carried out for base membranes with the different layers combined with an AR 2000 ex Rheometer and Rheology Advantage Instrument Control AR, which were both acquired from TA Instruments (New Castle, DE, USA). The conditions were a frequency of 1 Hz, a strain between 0.01 and 100%, a 20 mm steel cross-hatched (CH) plate, and a temperature of 25 °C. In the test, the membrane was placed on top of the lower geometry, and the rheometer's motor decreased the upper geometry until the normal force applied was around 1.5 N. The elastic modulus of the material was obtained with an oscillatory procedure.

All elasticity results for the membranes with different concentrations of active ingredients or Veg Col products obtained at different temperatures were normalized with respect to an elasticity result obtained from an interpolated blank membrane in Equation (6), which was obtained from Figure 9, according to the same room temperature of the membrane under study:

$$G' \text{blank(Pa)} = 4525 \times T(°C) - 41{,}022 \quad (6)$$

The percentage difference in the elasticity of the active ingredient or Veg Col membrane at the different concentrations [I] with respect to the blank interpolated in Equation (6) was obtained with Equation (7):

$$\text{Elasticity[I](\%)} = \frac{\text{Av } G'[I] \text{ (Pa)} - \text{Av } G' \text{ blank (Pa)}}{\text{Av } G' \text{ blank (Pa)}} \quad (7)$$

3.2.10. Swelling Index

After the creation of the membranes, the samples were dried with granular calcium chloride for 24 h in a container. In our previous research, we used 24 h after the creation of the membranes until they were dried with calcium chloride [7]. In this work, it was seen that this time could be avoided, as it did not affect the results. After 24 h in the dryer, the membrane weight was constant, and then it was assured that the samples were completely dried.

The swelling tests were conducted at room temperature (21 ± 2 °C).

The dried membranes containing the different concentrations of active ingredient or Veg Col product were weighed and then immersed in phosphate-buffered solution (PBS, pH 7.4) for different lengths of time (0, 30 min, 1 h, 2 h, 12 h, and 24 h) until they reached the maximum weight [7]. The phosphate-buffered solution was prepared as previously reported. The phosphate-buffered solution was under constant agitation during the entire experiment.

An electronic balance was used to weigh the swollen membranes after the surface water was removed with filter paper. Then, the swelling index (SI) of the membranes was calculated according to Equation (8):

$$\mathrm{SI}(\%) = \frac{\mathrm{Ws} - \mathrm{Wd}}{\mathrm{Wd}} \times 100 \tag{8}$$

where Ws and Wd were the weights, in grams, of the swollen and dried membranes, respectively.

The swollen weight value was obtained as the maximum weight obtained during the different periods of measurement—up to 24 h (Ws).

The percentage difference in swelling of the active ingredient or Veg Col membrane [I] with respect to the blank membrane was obtained with Equation (9):

$$\mathrm{SI}\ [\mathrm{I}](\%) = \frac{\mathrm{SI}\ [\mathrm{I}] - \mathrm{SI}\ (\mathrm{blank})}{\mathrm{SI}\ (\mathrm{blank})} \times 100 \tag{9}$$

3.2.11. Moisture Retention

A DBS moisture analyzer from Kern was used to determine the retention of humidity in each membrane. The following program was used: A ramp in the temperature was applied for the first 5 min until 100 °C was reached. The defined temperature was the lamp temperature. The weight loss of the membrane was evaluated every minute for 30 min. The result could be directly related to the capacity of the membrane to retain water. In human skin, this is known as the barrier effect.

In this test, the moisture retention of the whole membrane was evaluated, not just the water retention from the surface (barrier effect), as the membrane was wrinkled when it dried. Therefore, the bottom, which simulated the inner layer of skin and was not exposed to air, was also exposed to the same environmental dryness. Hence, due to the conditions of this test, only the membranes containing the ingredient in 3L were tested.

The moisture retention (MR) (%) was normalized versus the blank membrane according to Equation (10).

$$\mathrm{MR}(\%) = \frac{(\mathrm{Av\ blank\ WL\%_t} - \mathrm{Av\ [I]\ WL\%_t})}{\mathrm{Av\ blank\ WL\%_t}} \times 100 \tag{10}$$

where Av blank WL and Av [I] WL were the water-loss values from the blank membrane and from the membrane containing the specific concentration of the active ingredient, or Veg Col product, respectively. The subindex t was the time of measurement—15 or 30 min—of the water-loss percentage of the membrane.

4. Conclusions

In this work, some properties of tri-layered membranes containing different active ingredients—specifically, elastin, hydrolyzed collagen, and two different commercial vegan alternatives to collagen—were studied. Based on the results of this study, it can be stated that these membranes were sensitive to the studied properties: pore and water permeation reduction, elasticity, swelling, and moisture retention.

1. These properties were sensitive to the concentration of the active ingredient and the vegan collagen-like product. For the different properties that were studied, different results were obtained. Hence, the effects of the active ingredients and the vegan collagen-like product could be quantified.
2. It was demonstrated that the effects of some active ingredients, namely EL, HC, and two Veg Col products, were shown in the form of pore reduction, water permeation reduction, elasticity, swelling, and moisture retention. Hence, anti-aging claims could be proven.
3. The effects of the active ingredients and vegan collagen-like products on the previous properties varied depending on the layers into which they were introduced. Depending on whether EL, HC, or Veg Col was introduced into the three layers or into the target layers, in our case, the two inner layers, boosted or reduced efficacy was observed for the properties that were studied.

The main results obtained for the different measured properties were:
For permeation tests and its relationship with Pore Reduction:

- The permeation reduction was found to be directly proportional to the concentrations of EL and HC.
- Pore reductions were measured for some concentrations of EL and HC, and a correlation with permeation reduction was found. For similar values of reductions in pores between the two active ingredients, a more significant permeation reduction was observed for EL because of its hydrophobic segments.

For Elasticity tests:

- In general, an inversely proportional relationship between concentration and elasticity was found for both EL and HC.
- The greatest booster effect in the elasticity was found when including a 0.085% of EL in $2L_{(i+b)}$ (71%). This 0.085% concentration of EL fell within the margins of EL weight in human dermal skin.
- A boost in elasticity was found for the lowest concentration of HC studied (0.1%), and this was greater when HC was included in $2L_{(i+b)}$ (40%).

For Swelling tests:

- In the case of EL, at extreme concentrations, a boost in swelling was obtained in 3L with respect to $2L_{(i+b)}$.
- In the case of HC, at extreme concentrations, a boost in swelling was obtained in $2L_{(i+b)}$ with respect to 3L.

For Moisture retention tests:

- For the two concentrations of EL that were studied, some noticeable moisture retention was observed, and the highest value was observed at the higher concentration (0.27%) at 15 min (short term) (21%).
- For the four concentrations of HC that were studied, an increase in moisture retention was obtained only for the highest concentration studied (10% HC) at 15 min (short term) (19%).

Author Contributions: Conceptualization, R.G.-C. and A.B.-C.; methodology, R.G.-C. and A.B.-C.; validation, R.G.-C. and A.B.-C.; investigation, R.G.-C. and A.B.-C.; writing—original draft preparation, R.G.-C.; writing—review and editing, R.G.-C. and A.B.-C.; supervision, A.B.-C. All authors have read and agreed to the published version of the manuscript.

Funding: This research received no external funding.

Institutional Review Board Statement: Not applicable.

Informed Consent Statement: Not applicable.

Data Availability Statement: No new data were created or analyzed in this study. Data sharing is not applicable to this article.

Conflicts of Interest: The authors declare no conflict of interest.

References

1. European Commission. Animals in Scientific Research-Environment. Available online: https://ec.europa.eu/environment/chemicals/lab_animals/index_en.htm (accessed on 20 April 2023).
2. Sreedhar, D.; Manjula, N.; Pise, A.; Pise, S.; Ligade, V.S. Ban of Cosmetic Testing on Animals: A Brief Overview. *Int. J. Curr. Res. Rev.* **2020**, *12*, 113–116. [CrossRef]
3. Silva, R.J.; Tamburic, S. A State-of-the-Art Review on the Alternatives to Animal Testing for the Safety Assessment of Cosmetics. *Cosmetics* **2022**, *9*, 90. [CrossRef]
4. Risueño, I.; Valencia, L.; Jorcano, J.L.; Velasco, D. Skin-on-a-Chip Models: General Overview and Future Perspectives. *APL Bioeng.* **2021**, *5*, 030901. [CrossRef]
5. Suhail, S.; Sardashti, N.; Jaiswal, D.; Rudraiah, S.; Misra, M.; Kumbar, S.G. Engineered Skin Tissue Equivalents for Product Evaluation and Therapeutic Applications. *Biotechnol. J.* **2019**, *14*, 1900022. [CrossRef]
6. Yun, Y.E.; Jung, Y.J.; Choi, Y.J.; Choi, J.S.; Cho, Y.W. Artificial Skin Models for Animal-Free Testing. *J. Pharm. Investig.* **2018**, *48*, 215–223. [CrossRef]
7. Guerle-Cavero, R.; Lleal-Fontàs, B.; Balfagón-Costa, A. Creation of Ionically Crosslinked Tri-Layered Chitosan Membranes to Simulate Different Human Skin Properties. *Materials* **2021**, *14*, 1807. [CrossRef]
8. Scott, J.E. Proteoglycan-Fibrillar Collagen Interactions. *Biochem. J.* **1988**, *252*, 313–323. [CrossRef]
9. Domalik-Pyzik, P.; Chłopek, J.; Pielichowska, K. *Chitosan-Based Hydrogels: Preparation, Properties, and Applications BT—Cellulose-Based Superabsorbent Hydrogels*; Mondal, M.I.H., Ed.; Springer International Publishing: Cham, Switzerland, 2019; pp. 1665–1693; ISBN 978-3-319-77830-3.
10. Indrani, D.J.; Lukitowati, F.; Yulizar, Y. Preparation of Chitosan/Collagen Blend Membranes for Wound Dressing: A Study on FTIR Spectroscopy and Mechanical Properties. In Proceedings of the 4th International Conference on Advanced Materials Science and Technology, Malang, Indonesia, 27–28 September 2016; Volume 202, p. 012020. [CrossRef]
11. Da, L.C.; Huang, Y.Z.; Xie, H.Q.; Zheng, B.H.; Huang, Y.C.; Du, S.R. Membranous Extracellular Matrix-Based Scaffolds for Skin Wound Healing. *Pharmaceutics* **2021**, *13*, 1796. [CrossRef]
12. Sheikholeslam, M.; Wright, M.E.E.; Jeschke, M.G.; Amini-Nik, S. Biomaterials for Skin Substitutes. *Adv. Healthc. Mater.* **2018**, *7*, 1700897. [CrossRef]
13. Cosmile. Elastin. Available online: https://cosmileeurope.eu/inci/detail/5210/elastin/ (accessed on 20 April 2023).
14. Cosmile. Hydrolyzed Collagen. Available online: https://cosmileeurope.eu/inci/detail/6824/hydrolyzed-collagen/ (accessed on 20 April 2023).
15. Reilly, D.M.; Lozano, J. Skin Collagen through the Lifestages: Importance for Skin Health and Beauty. *Plast. Aesthetic Res.* **2021**, *8*, 2. [CrossRef]
16. Martínez, A.; Blanco, M.D.; Davidenko, N.; Cameron, R.E. Tailoring Chitosan/Collagen Scaffolds for Tissue Engineering: Effect of Composition and Different Crosslinking Agents on Scaffold Properties. *Carbohydr. Polym.* **2015**, *132*, 606–619. [CrossRef]
17. Hult, A.M.; Goltz, R.W. The Measurement of Elastin in Human Skin and Its Quantity in Relation To Age. *J. Investig. Dermatol.* **1965**, *44*, 408–412. [CrossRef]
18. Wilharm, N.; Fischer, T.; Hayn, A.; Mayr, S.G. Structural Breakdown of Collagen Type I Elastin Blend Polymerization. *Polymers* **2022**, *14*, 4434. [CrossRef]
19. dos Santos, L.A.L. Natural Polymeric Biomaterials: Processing and Properties. In *Reference Module in Materials Science and Materials Engineering*; Elsevier: Amsterdam, The Netherlands, 2017; pp. 1–6. [CrossRef]
20. Taravel, M.N.; Domard, A. Collagen and Its Interaction with Chitosan: II. Influence of the Physicochemical Characteristics of Collagen. *Biomaterials* **1995**, *16*, 865–871. [CrossRef]
21. Shiratsuchi, E.; Nakaba, M.; Shigemura, Y.; Yamada, M.; Sato, K. Fish-Elastin Hydrolysate: Development and Impact on the Skin and Blood Vessels. *Mar. Proteins Pept. Biol. Act. Appl.* **2013**, *2*, 467–486. [CrossRef]

22. Daamen, W.F.; Van Moerkerk, H.T.B.; Hafmans, T.; Buttafoco, L.; Poot, A.A.; Veerkamp, J.H.; Van Kuppevelt, T.H. Preparation and Evaluation of Molecularly-Defined Collagen-Elastin-Glycosaminoglycan Scaffolds for Tissue Engineering. *Biomaterials* **2003**, *24*, 4001–4009. [CrossRef] [PubMed]
23. Aprahamian, M.; Lambert, A.; Balboni, G.; Lefebvre, F.; Schmittaeusler, R.; Damge, C.; Rabaud, M. A New Reconstitued Connective Tissue Matrix: Preparation, Biochemical, Structural and Mechanical Studies. *J. Biomed. Mater. Res.* **1987**, *21*, 965–977. [CrossRef] [PubMed]
24. Del Prado-Audelo, M.L.; Mendoza-Muñoz, N.; Escutia-Guadarrama, L.; Giraldo-Gomez, D.M.; González-Torres, M.; Florán, B.; Cortes, H.; Leyva-Gómez, G. Recent Advances in Elastin-Based Biomaterials. *J. Pharm. Pharm. Sci.* **2020**, *23*, 314–332. [CrossRef] [PubMed]
25. Su, H.; Fujiwara, T.; Bumgardner, J.D. A Study of Combining Elastin in the Chitosan Electrospinning to Increase the Mechanical Strength and Bioactivity. *Mar. Drugs* **2021**, *19*, 169. [CrossRef]
26. Trębacz, H.; Barzycka, A. Mechanical Properties and Functions of Elastin: An Overview. *Biomolecules* **2023**, *13*, 574. [CrossRef]
27. Le, A.; Morales-Peñaloza, A.; Mart, M. Hydrolyzed Collagen—Sources and Applications. *Molecules* **2019**, *24*, 4031.
28. Hidalgo-Vicelis, J.L.; Alvarez-Perez, M.A.; Miranda-Castro, S.P.; Piña-Barba, M.C. Type I Collagen-Chitosan Membranes Crosslinked Chemically with N-(3-Dimethylaminopropyl)-N′-Ethylcarbodiimide Hydrochloride for Guided Bone Regeneration: A Comparative Study. *Fibers Polym.* **2020**, *21*, 262–272. [CrossRef]
29. Vollmer, D.L.; West, V.A.; Lephart, E.D. Enhancing Skin Health: By Oral Administration of Natural Compounds and Minerals with Implications to the Dermal Microbiome. *Int. J. Mol. Sci.* **2018**, *19*, 3059. [CrossRef] [PubMed]
30. Jiang, Y.-H.; Lou, Y.-Y.; Li, T.-H.; Liu, B.-Z.; Chen, K.; Zhang, D.; Li, T. Cross-Linking Methods of Type I Collagen-Based Scaffolds for Cartilage Tissue Engineering. *Am. J. Transl. Res.* **2022**, *14*, 1146–1159.
31. Parenteau-Bareil, R.; Gauvin, R.; Berthod, F. Collagen-Based Biomaterials for Tissue Engineering Applications. *Materials* **2010**, *3*, 1863–1887. [CrossRef]
32. Taravel, M.N.; Domard, A. Relation between the Physicochemical Characteristics of Collagen and Its Interactions with Chitosan: I. *Biomaterials* **1993**, *14*, 930–938. [CrossRef] [PubMed]
33. Sionkowska, A.; Wisniewski, M.; Skopinska, J.; Kennedy, C.J.; Wess, T.J. Molecular Interactions in Collagen and Chitosan Blends. *Biomaterials* **2004**, *25*, 795–801. [CrossRef]
34. Taravel, M.N.; Domard, A. Collagen and Its Interactions with Chitosan: III. Some Biological and Mechanical Properties. *Biomaterials* **1996**, *17*, 451–455. [CrossRef]
35. Hussain, S.H.; Limthongkul, B.; Humphreys, T.R. The Biomechanical Properties of the Skin. *Dermatol. Surg.* **2013**, *39*, 193–203. [CrossRef]
36. Draelos, Z.D. The Science behind Skin Care: Moisturizers. *J. Cosmet. Dermatol.* **2018**, *17*, 138–144. [CrossRef]
37. Verdier-Sévrain, S.; Bonté, F. Skin Hydration: A Review on Its Molecular Mechanisms. *J. Cosmet. Dermatol.* **2007**, *6*, 75–82. [CrossRef]
38. Balfagón-Costa, A.; Guerle-Cavero, R.; Monteagudo-Quintás, A.; Marimón-Maragarit, N. Cuantificación de La Capacidad Hidratante de Las Cremas. Clasificación de Las Cremas Hidratantes. Available online: https://www.e-seqc.org/revista/22/revista-ncp-n-326/ (accessed on 20 May 2023).
39. Flament, F.; Francois, G.; Qiu, H.; Ye, C.; Hanaya, T.; Batisse, D.; Cointereau-Chardon, S.; Seixas, M.D.G.; Dal Belo, S.E.; Bazin, R. Facial Skin Pores: A Multiethnic Study. *Clin. Cosmet. Investig. Dermatol.* **2015**, *8*, 85–93. [CrossRef]
40. Lee, S.J.; Seok, J.; Jeong, S.Y.; Park, K.Y.; Li, K.; Seo, S.J. Facial Pores: Definition, Causes, and Treatment Options. *Dermatol. Surg.* **2016**, *42*, 277–285. [CrossRef]
41. Miyamoto, K.; Dissanayake, B.; Omotezako, T.; Takemura, M.; Tsuji, G.; Furue, M. Daily Fluctuation of Facial Pore Area, Roughness and Redness among Young Japanese Women; Beneficial Effects of Galactomyces Ferment Filtrate Containing Antioxidative Skin Care Formula. *J. Clin. Med.* **2021**, *10*, 2502. [CrossRef] [PubMed]
42. Sugiyama-Nakagiri, Y.; Sugata, K.; Hachiya, A.; Osanai, O.; Ohuchi, A.; Kitahara, T. Ethnic Differences in the Structural Properties of Facial Skin. *J. Dermatol. Sci.* **2009**, *53*, 135–139. [CrossRef] [PubMed]
43. L'Oreal Paris Faces the Facts about Pores. Available online: https://www.prnewswire.com/news-releases/loreal-paris-faces-the-facts-about-pores-262699361.html (accessed on 29 April 2023).
44. Campos MBG, P.M. An Oral Supplementation Based on Hydrolyzed Collagen and Vitamins Improves Skin Elasticity and Dermis Echogenicity: A Clinical Placebo-Controlled Study. *Clin. Pharmacol. Biopharm* **2015**, *4*, 2. [CrossRef]
45. Fluhr, J.W.; Darlenski, R.; Surber, C. Glycerol and the Skin: Holistic Approach to Its Origin and Functions. *Br. J. Dermatol.* **2008**, *159*, 23–34. [CrossRef]
46. Chen, M.; Runge, T.; Wang, L.; Li, R.; Feng, J.; Shu, X.; Shi, Q. Hydrogen Bonding Impact on Chitosan Plasticization. *Carbohydr. Polym.* **2018**, *200*, 115–121. [CrossRef]
47. Hara, M.; Verkman, A.S. Glycerol Replacement Corrects Defective Skin Hydration, Elasticity, and Barrier Function in Aquaporin-3-Deficient Mice. *Proc. Natl. Acad. Sci. USA* **2003**, *100*, 7360–7365. [CrossRef]
48. Pieróg, M.; Gierszewska-Drużyńska, M.; Ostrowska-Czubenko, J. Effect of Ionic Crosslinking Agents on Swelling Behavior of Chitosan Hydrogel Membranes. *Carbohydr. Polym.* **2009**, *77*, 590–598. [CrossRef]
49. Ma, L.; Gao, C.; Mao, Z.; Zhou, J.; Shen, J.; Hu, X.; Han, C. Collagen/Chitosan Porous Scaffolds with Improved Biostability for Skin Tissue Engineering. *Biomaterials* **2003**, *24*, 4833–4841. [CrossRef] [PubMed]

50. Wise, S.G.; Mithieux, S.M.; Weiss, A.S. *Engineered Tropoelastin and Elastin-Based Biomaterials*, 1st ed.; Elsevier: Amsterdam, The Netherlands, 2009; Volume 78.
51. Trabbic-Carlson, K.; Setton, L.A.; Chilkoti, A. Swelling and Mechanical Behaviors of Chemically Cross-Linked Hydrogels of Elastin-like Polypeptides. *Biomacromolecules* **2003**, *4*, 572–580. [CrossRef] [PubMed]
52. Fernandes, L.L.; Resende, C.X.; Tavares, D.S.; Soares, G.A.; Castro, L.O.; Granjeiro, J.M. Cytocompatibility of Chitosan and Collagen-Chitosan Scaffolds for Tissue Engineering. *Polimeros* **2011**, *21*, 1–6. [CrossRef]
53. Shiratsuchi, E.; Nakaba, M.; Yamada, M. Elastin Hydrolysate Derived from Fish Enhances Proliferation of Human Skin Fibroblasts and Elastin Synthesis in Human Skin Fibroblasts and Improves the Skin Conditions. *J. Sci. Food Agric.* **2016**, *96*, 1672–1677. [CrossRef]
54. Sibilla, S.; Godfrey, M.; Brewer, S.; Budh-Raja, A.; Genovese, L. An Overview of the Beneficial Effects of Hydrolysed Collagen as a Nutraceutical on Skin Properties: Scientific Background and Clinical Studies. *Open Nutraceuticals J.* **2015**, *8*, 29–42. [CrossRef]
55. Kim, D.U.; Chung, H.C.; Choi, J.; Sakai, Y.; Lee, B.Y. Oral Intake of Low-Molecular-Weight Collagen Peptide Improves Hydration, Elasticity, and Wrinkling in Human Skin: A Randomized, Double-Blind, Placebo-Controlled Study. *Nutrients* **2018**, *10*, 826. [CrossRef]
56. Jhawar, N.; Wang, J.V.; Saedi, N. Oral Collagen Supplementation for Skin Aging: A Fad or the Future? *J. Cosmet. Dermatol.* **2020**, *19*, 910–912. [CrossRef]
57. Proksch, E.; Segger, D.; Degwert, J.; Schunck, M.; Zague, V.; Oesser, S. Oral Supplementation of Specific Collagen Peptides Has Beneficial Effects on Human Skin Physiology: A Double-Blind, Placebo-Controlled Study. *Ski. Pharmacol. Physiol.* **2013**, *27*, 47–55. [CrossRef]
58. Miyanaga, M.; Uchiyama, T.; Motoyama, A.; Ochiai, N.; Ueda, O.; Ogo, M. Oral Supplementation of Collagen Peptides Improves Skin Hydration by Increasing the Natural Moisturizing Factor Content in the Stratum Corneum: A Randomized, Double-Blind, Placebo-Controlled Clinical Trial. *Ski. Pharmacol. Physiol.* **2021**, *34*, 115–127. [CrossRef]
59. Kamel, H.K. Oral Collagen Supplementation: A Systematic Review of Dermatological Applications. *Nutr. Rev.* **2003**, *61*, 157–167. [CrossRef]
60. Inoue, N.; Sugihara, F.; Wang, X. Ingestion of Bioactive Collagen Hydrolysates Enhance Facial Skin Moisture and Elasticity and Reduce Facial Ageing Signs in a Randomised Double-Blind Placebo-Controlled Clinical Study. *J. Sci. Food Agric.* **2016**, *96*, 4077–4081. [CrossRef] [PubMed]
61. Aguirre-Cruz, G.; León-López, A.; Cruz-Gómez, V.; Jiménez-Alvarado, R.; Aguirre-Álvarez, G. Collagen Hydrolysates for Skin Protection: Oral Administration and Topical Formulation. *Antioxidants* **2020**, *9*, 181. [CrossRef] [PubMed]
62. Berardesca, E.; Abril, E.; Serio, M.; Cameli, N. Effects of Topical Gluco-Oligosaccharide and Collagen Tripeptide F in the Treatment of Sensitive Atopic Skin. *Int. J. Cosmet. Sci.* **2009**, *31*, 271–277. [CrossRef] [PubMed]
63. Ohara, H.; Ito, K.; Iida, H.; Matsumoto, H. Improvement in the Moisture Content of the Stratum Corneum. *J. Jpn. Soc. Food Sci. Technol.* **2009**, *56*, 137–145. [CrossRef]
64. Campos, P.M.B.G.M.; Melo, M.O.; César, F.C.S. Topical Application and Oral Supplementation of Peptides in the Improvement of Skin Viscoelasticity and Density. *J. Cosmet. Dermatol.* **2019**, *18*, 1693–1699. [CrossRef]
65. Cosmile. Nicotiana Benthamiana Hexappetide-40 Sh-Polypeptide-47. Available online: https://cosmileeurope.eu/inci/detail/30617/nicotiana-benthamiana-hexapeptide-40-sh-polypeptide-47/ (accessed on 20 April 2023).
66. Cosmile. Collagen Amino Acids. Available online: https://cosmileeurope.eu/inci/detail/3744/collagen-amino-acids/ (accessed on 20 April 2023).

International Journal of *Molecular Sciences*

MDPI

Article

Chitosan Composites with Bacterial Cellulose Nanofibers Doped with Nanosized Cerium Oxide: Characterization and Cytocompatibility Evaluation

Valentina A. Petrova [1], Iosif V. Gofman [1], Natallia V. Dubashynskaya [1], Alexey S. Golovkin [2], Alexander I. Mishanin [2], Elena M. Ivan'kova [1], Dmitry P. Romanov [3], Albert K. Khripunov [1], Elena N. Vlasova [1], Alexandra V. Migunova [4], Alexander E. Baranchikov [5], Vladimir K. Ivanov [5], Alexander V. Yakimansky [1] and Yury A. Skorik [1,*]

1 Institute of Macromolecular Compounds of the Russian Academy of Sciences, Bolshoi VO 31, St. Petersburg 199004, Russia
2 Almazov National Medical Research Centre, Akkuratova 2, St. Petersburg 197341, Russia
3 Institute of Silicate Chemistry of the Russian Academy of Sciences, Adm. Makarova emb. 2, St. Petersburg 199034, Russia
4 Department of Microbiology, Faculty of Biology, St. Petersburg State University, Universitetskaya emb. 7-9, St. Petersburg 199034, Russia
5 Kurnakov Institute of General and Inorganic Chemistry of the Russian Academy of Sciences, Leninskii 31, Moscow 119071, Russia
* Correspondence: yury_skorik@mail.ru

Abstract: In this work, new composite films were prepared by incorporating the disintegrated bacterial cellulose (BCd) nanofibers and cerium oxide nanoparticles into chitosan (CS) matrices. The influence of the amount of nanofillers on the structure and properties of the polymer composites and the specific features of the intermolecular interactions in the materials were determined. An increase in film stiffness was observed as a result of reinforcing the CS matrix with BCd nanofibers: the Young's modulus increased from 4.55 to 6.3 GPa with the introduction of 5% BCd. A further increase in Young's modulus of 6.7 GPa and a significant increase in film strength (22% increase in yield stress compared to the CS film) were observed when the BCd concentration was increased to 20%. The amount of nanosized ceria affected the structure of the composite, followed by a change in the hydrophilic properties and texture of the composite films. Increasing the amount of nanoceria to 8% significantly improved the biocompatibility of the films and their adhesion to the culture of mesenchymal stem cells. The obtained nanocomposite films combine a number of favorable properties (good mechanical strength in dry and swollen states, improved biocompatibility in relation to the culture of mesenchymal stem cells), which allows us to recommend them for use as a matrix material for the culture of mesenchymal stem cells and wound dressings.

Keywords: nanocomposites; biopolymers; bacterial cellulose; chitosan; ceria nanoparticles; stem cells' proliferation

1. Introduction

Natural polysaccharides are promising materials for biomedical applications due to their biocompatibility, biodegradability, non-toxicity, low immunogenicity, and biological activity. The various polysaccharide-based polymer composites can be obtained in the form of films, sponges, electrospun mats, hydrogels, and nano- and microparticles. The design of polymer composites involves a combination of polymers with different chemical structures and physical forms, as well as a combination of polymers with nanoscale modifiers [1]. Polymer composites of chitosan (CS) and bacterial cellulose (BC) that combine the structure-forming ability, mechanical properties, and hydrophilicity of BC with the antimicrobial properties of CS are considered promising [2].

CS is a polyaminosaccharide derived from the deacetylation of chitin; CS is biodegraded in the human body under the action of lysozyme to N-acetylglucosamine and glucosamine, natural metabolic products [3]. CS has been widely used to design various tissue engineering structures [4,5]. A disadvantage of CS is the low strength of the formed films in the wet state. In order to produce materials with the enhanced complex of properties (the optimal combination of mechanical properties, surface charge, and biological activity), the strategy of introducing different nanofillers, including nanofibers of chitin or cellulose, into the CS matrix is widely used [6–9].

One of the most promising natural polymers for biomedical applications is BC, which is a linear, unbranched polysaccharide consisting of 1,4-glucopyranose units that is biosynthesized by microorganisms (e.g., *Komagataeibacter rhaeticus*) [10,11]. BC is characterized by a complex hierarchical structure and a high degree of crystallinity [12]. In BC, nanofibrils are highly oriented nanocrystals embedded in an amorphous matrix. These individual nanofibrils are further assembled to form highly oriented microfibrils or even fiber bundles, conferring exceptional mechanical properties to biological materials. Because of its structural similarity to the components of the extracellular matrix (e.g., collagen), BC is capable of interacting with biological tissues [13,14]. This complex of beneficial properties makes BC a promising biomaterial for the development of biocompatible matrices for tissue engineering [15–20]. However, the use of BC has a number of drawbacks, including the lack of antimicrobial properties, difficulty in regulating the pore size, and slow degradation [21].

The introduction of BC nanofibers into CS matrices can lead to the formation of composite materials with a number of valuable properties, such as improved mechanical properties, an optimized surface charge, and useful morphological features. Both the starting material and the methods of BC nanofiber isolation can influence the efficiency of nanofibrillation and the properties of the products, including morphology and crystallinity, which further affect the physical properties of the material. BC nanofibers obtained by the mechanical treatment of BC are typically characterized by both long length and high aspect ratio (length up to 500 μm, and thickness up to 50 nm). Since BC nanofibers have a large number of hydroxyl groups on their surface, and nanofibers isolated from BC have a high specific surface area, their surface can be easily functionalized and their interaction with the polymer matrix components can lead to a change in the polymer structure to form a nanocomposite with precisely tuned properties [9,12,22,23]. For example, Fernandes et al. [24] developed composites based on CS and BC nanofibers (5–40% of the CS weight) as a reinforcing agent. The surface morphology of the composite films showed the presence of a three-dimensional fibrillar network of BC. The observed crystallization of CS in the obtained nanocomposites is explained by the deposition of CS on the surface of the crystalline domains of BC nanofibers. Phisalaphong and Jatupaiboon [25] obtained composite films of BC with the addition of CS, which have excellent mechanical properties in wet and dry states, high water absorption and water retention capacity, and a high surface area. These materials have bacteriostatic and bactericidal activities.

Another important factor is the relationship between mechanical properties, geometry, and biological properties of scaffolds for tissue engineering. In this regard, the development of biocompatible composites containing nanoparticles is being considered [10,12,24–26]. Ceria nanoparticles are attractive inorganic fillers that strongly influence the physicochemical and biological properties of polymer systems. The introduction of ceria into the polymer matrix imparts specific biological activity to the composite and also changes its physical and mechanical properties; on the other hand, the polymer environment can regulate the properties of the nanoceria (e.g., reduce their toxicity). Thus, CeO2-containing polymeric nanocomposites combine the advantages and reduce the disadvantages of both polymer matrices and nanoparticles [27]. In addition, ceria has beneficial effects and provides cell and tissue protection through in vitro and in vivo inhibition of reactive oxygen species, suppression of cytokine levels, and reduction in inflammation [28–31].

In our previous studies, we obtained BC-based composites by impregnating the squeezed BC gel film with polysaccharide solutions containing citrate-stabilized ceria nanoparticles (CeONPs) [12] and CS-based composites containing CeONPs [32]. These studies have demonstrated the beneficial effects of CeONPs on the culture of human mesenchymal stem cells (MSCs), as evidenced by changes in stem cell behavior such as their migration, proliferation, and differentiation [12,32].

The next phase of our research was to develop a multi-component CS-BC-CeONP composite. This composite would combine the advantages of both previously developed scaffolds. Thus, the aim of this study was to obtain CS-based nanocomposites containing ensembles of nanoparticles with different architectures (disintegrated BC nanofibers and ceria nanoparticles), to study their interactions and the influence of the amount of nanoparticles present in the composite to modulate the physical and mechanical properties and network architecture of the nanocomposite, and to improve its biomedical potential for MSC culture.

2. Results and Discussion

2.1. Infrared Spectroscopy

The comparative Fourier-transform infrared (FTIR) spectra of the original BC and disintegrated bacterial cellulose (BCd) (Figure 1a) show that in the spectrum of BCd, the 1570 cm^{-1} band appears and the 1430 cm^{-1} band changes (the band change is most obvious in the difference spectrum, Figure 1b). These bands correspond to the antisymmetric and symmetric vibrations of the –COO^- group. At the same time, a change in the shape and mutual intensities of the bands in the region of 1150–1000 cm^{-1} corresponding to the vibrations of the COC, C-C, C-OH, CH, and OH groups is observed. It can be assumed that during the preparation of disintegrated BCd, the BC molecules are degraded with the formation of terminal carboxylate groups.

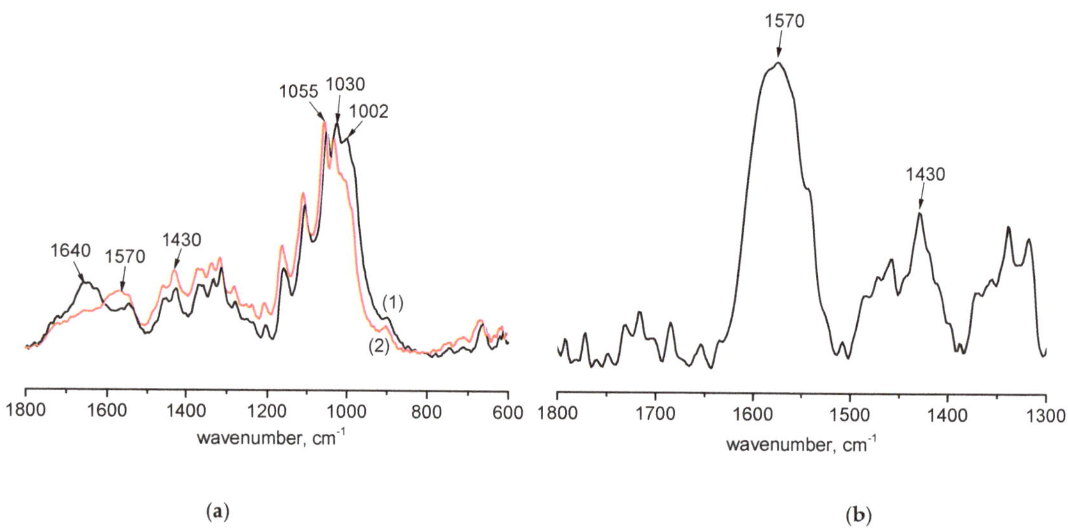

Figure 1. FTIR spectra: (**a**) the original bacterial cellulose (BC) (1) and disintegrated bacterial cellulose (BCd) (2); (**b**) difference spectrum of BCd and BC.

The presence of carboxylate groups on the BCd surface is confirmed by the negative ζ-potential of the BCd dispersion (−15 mV). To study the interaction of BCd with CS, we titrated the aqueous dispersion of BCd with a solution of CS in 2% acetic acid and measured the ζ-potential of the resulting particles (Figure 2). The change in the sign of the ζ-potential indicates the interaction between the negatively charged BCd nanofibers and the positively charged CS molecules.

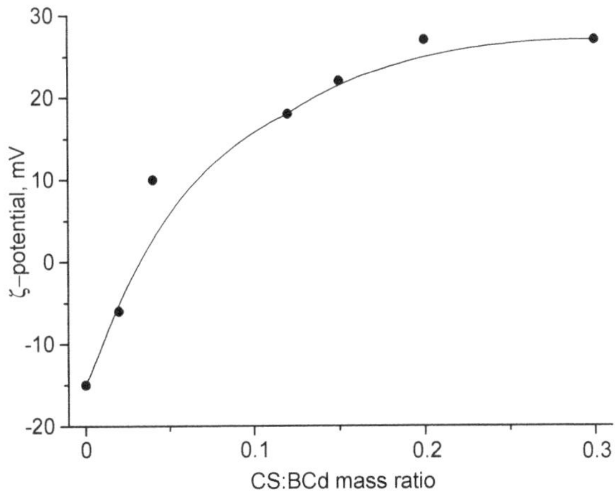

Figure 2. Changes in the ζ-potential of aqueous BCd dispersion (0.48 mg/mL) when adding a solution of chitosan (CS) in 2% acetic acid (0.48 mg/mL).

The presence of these charges promotes the mutual repulsion of BCd nanofibers and prevents their aggregation during dispersion in water, which further ensures uniform distribution of BCd in the CS matrix.

2.2. Wide-Angle X-ray Scattering

Cellulose is semicrystalline polymer with two different crystal structures in the native state, namely Iα and Iβ. Cellulose Iα crystallizes in a triclinic unit cell, space group P1 with a = 6.717 Å, b = 5.962 Å, fiber repeat c = 10.400 Å, α = 118.08°, β = 114.80°, γ = 80.37°, and Iβ in a monoclinic unit cell, space group P21 with a = 7.784 Å, b = 8.201 Å, fiber repeat c = 10.38 Å, γ = 96.5° [33].

Figure 3a shows the wide-angle X-ray scattering (WAXS) pattern obtained for the BC film, which is typical of cellulose I. Three diffraction peaks at 2θ = 14.5°, 16.6°, and 22.7° are attributed to the (100), (010), and (110) planes of cellulose Iα (triclinic), or the $(1\bar{1}0)$, (110), and (200) planes of cellulose Iβ (monoclinic) [34]. Because the positions of these peaks corresponding to two allomorphic phases are too close together, it is difficult to distinguish between them. Usually BC consists mainly of the Iα modification.

It should be noted that the two WAXS patterns are similar to each other, but the X-ray diffraction peaks revealed for the BCd film have a lower intensity compared to those for BC. In other words, the intensity of the amorphous halo in the BCd pattern is much higher than that in the BC pattern. The reason is believed to be a partial amorphization (i.e. a decrease in a degree of crystallinity) of BCd.

It should also be noted that in the BCd film, an increase in the intensity of the (010) peak and a simultaneous disappearance of the $(\bar{1}\bar{1}4)$ peak can be observed [35]. This means that some changes in the texture (i.e., preferential orientation of the crystallites) take place in the BCd film compared to the BC one. It is also clearly seen that the X-ray diffraction peaks detected in the BCd sample are narrower than those in the BC film. This could be explained by the fact that the crystallites of the BCd film are less defective.

The diffractogram of CS (Figure 3b(1)) shows reflexes at around 15° and 22°, corresponding to the anhydrous polymorphic modification of CS [36]. The CS(80)-BCd(20) film (Figure 3b(2)) shows a weak reflex in the region of 2θ = 10.2° and reflexes at 14.8° and 17.0° as well as a blurred reflex in the angle range of 20–23°. Thus, the diffractogram of the control CS(80)-BCd(20) sample contains both the reflexes typical for the hydrated polymorphic modification of CS (10°, 15°, and 20°) [36] and the reflexes typical for BCd.

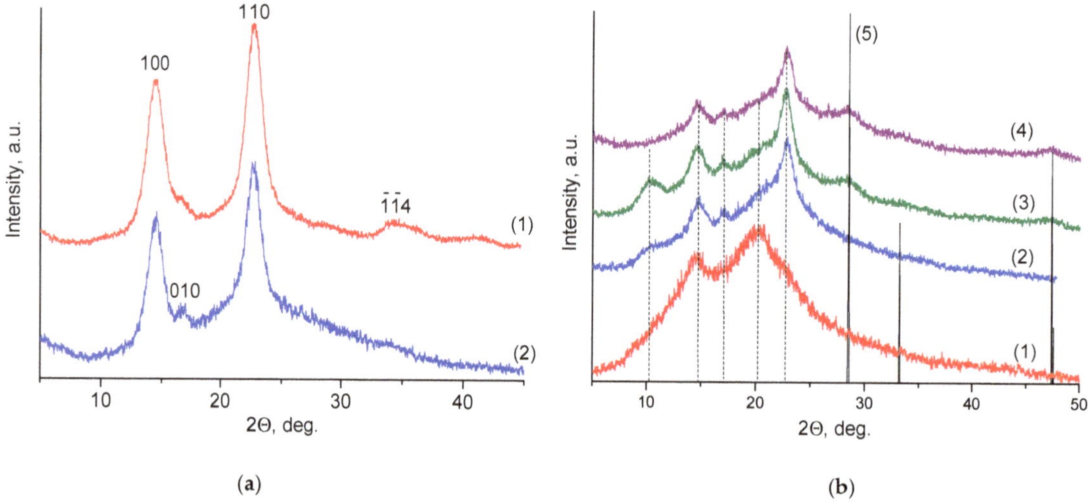

Figure 3. Wide-angle X-ray scattering patterns of: (**a**) BC (1), BCd (2); (**b**) CS (1), CS(80)-BCd(20) (2), CS(80)-BCd(20)-CeONP(4) (3), Cs(80)-BCd(20)-CeONP(8) (4), CeO_2 ICDD PDF card #34-394 (5).

The diffractogram of CS(80)-BCd(20)-CeONP(4) (Figure 3b(3)) shows a more intense reflex in the 10° region compared to the control, corresponding to the hydrated polymorphic modification of CS. Increasing the CeONP content in CS(80)-BCd(20)-CeONP(8) (Figure 3b(4)) leads to the disappearance of this reflex (anhydrous CS polymorphic modification).

For CS(80)-BCd(20)-CeONP(4) and Cs(80)-BCd(20)-CeONP(8) (Figure 3b(3,4)) both reflexes are observed typical of the control film and reflexes in the regions $2\theta = 28.7°$, 33°, and 47.5° corresponding to reflexes of the CeO_2 crystal lattice planes (111), (200), (220), and (311), respectively (Cubic crystal structure of fluorite: ICDD PDF card #34-394, data from National Institute of Standards and Technology, Gaithersburg, MD, USA) [37].

2.3. Swelling Properties

A comparative study of swelling properties of the composite films (Table 1) indicates that the introduction of BCd into the CS film promotes an increase in the swelling degree in water, with the swelling degree increasing with the increasing BCd content (Table 1, the CS(95)-BCd(5) and CS(80)-BCd(20) samples). The incorporation of 4% CeONPs into the composites also leads to an increase in the swelling degree in water compared to the control films. However, increasing the CeONP content up to 8% leads to a decrease in the swelling degree of the nanocomposite film both in water and in saline solution. The changes in the hydrophilicity of the nanocomposite films are due to the modification of the nanocomposite structure and the change in the nature of the intermolecular interactions.

Table 1. The swelling degree of the CS and composite films.

Sample	Swelling Degree in Water (g/g)	Swelling Degree in 0.9% NaCl Solution, (g/g)
CS	2.6	-
CS(80)-BCd(20)	10.8	6.2
CS(80)-BCd(20)-CeONP(4)	13.0	6.8
CS(80)-BCd(20)-CeONP(8)	4.7	2.9
CS(95)-BCd(5)	6.6	5.1
CS(95)-BCd(5)-CeONP(4)	10.2	8.9

2.4. Mechanical Properties

To compare the contribution of both polysaccharides to the nanocomposite film properties, the mechanical properties of individual CS and BCd films were investigated (Table 2). The BC-based film is 2.5 times superior to the CS film in Young's modulus value, but it is almost 10 times inferior in deformation value. Therefore, the introduction of BCd into the CS films and the increasing concentration of BCd in the CS-BCd composition leads to a sequential increase in the stiffness of the material and a decrease in the elongation at break (Table 2). At the same time, due to the increase in Young's modulus, the decrease in ε_b does not lead to a decrease in the strength of the nanocomposite film with increasing BCd content.

Table 2. Mechanical properties of the polysaccharides and nanocomposite films.

Sample	E (GPa)	σ_b (MPa)	ε_b (%)
CS	4.55 ± 0.12	114 ± 5	30 ± 2
BCd	11.7 ± 0.3	460 ± 17	3.6 ± 0.3
CS(95)-BCd(5)	6.3 ± 0.4	110 ± 6	11 ± 2
CS(80)-BCd(20)	6.7 ± 0.3	140 ± 6	4.5 ± 0.6
CS(95)-BCd(5)-CeONP(4)	6.2 ± 0.5	118 ± 5	7 ± 1
CS(80)-BCd(20)-CeONP(4)	7.0 ± 0.4	111 ± 6	3.7 ± 0.4
CS(80)- BCd (20)-CeONP(8)	6.8 ± 0.2	110 ± 6	6 ± 1
CS(80)-BCd(20)-CeONP(8) in swollen state	$(73 \pm 3) \times 10^{-3}$	22 ± 2	15 ± 2

The incorporation of CeONPs into the polysaccharide matrices does not significantly change the Young's modulus of the films formed (Table 2). This fact shows that the formation of such nanocomposites does not lead to a significant strengthening of the intermolecular bonds in the material. At the same time, the X-ray diffraction data show certain interactions of CeONPs with the polysaccharide composite film. These apparently contradictory results indicate that the introduction of nanoparticles into the polymer composition leads to a rearrangement of the intermolecular binding, but that this rearrangement has a complex nature. On the one hand, some interactions of CS macrochains with the surface of nanoparticles are formed, which may have the function of interchain cross-links that increase the stiffness of the material. However, the introduction of nanoparticles ~3 nm in size [38] into the polymer matrix leads to a local increase in interchain distances in the polymer matrix and, consequently, to a certain decrease in the density of intrinsic intermolecular interactions in the polymer base of the composite. Apparently, these two opposite processes balance each other in the formation of the nanocomposite material.

We can also note that the incorporation of nanoparticles into the polysaccharide matrix, in both concentrations used in this work (4% and 8%), leads to a decrease in the ultimate deformation typical of polymer-inorganic nanocomposites.

It is of special interest to characterize the mechanical properties of the developed nanocomposite material in the swollen equilibrium state. Indeed, for biomedical applications, film materials are used not in the dry but in the wet form. The mechanical tests of the swollen film CS(80)-BCd(20)-CeONP(8) with ~470% water relative to the weight of the dry material (Table 2 and Figure 4) showed that even in this state the nanocomposite material is superior in strength to, for example, polyethylene films and is suitable for practical use.

2.5. Thermal Analysis

The thermal degradation of CS(80)-BCd(20) and CS(80)-BCd(20)-CeONP(8) in air proceeds in two stages (Figure 5a), which is typical of many polysaccharides [39,40]. In the first stage (200–350 °C), both the control sample and the CS(80)-BCd(20)-CeONP(8) composite lose ~40% of their weight. At the second high temperature stage (400–550 °C), the complete decomposition of the CS-BCd film occurs with the transition of thermal-oxidative degradation products into the gas phase, and the weight of the nanocomposite decreases to a residual value of 7.3%, after which it remains constant during further heating. This residual mass corresponds to the concentration of CeO_2 in the film. Taking into account the concentration of water in the material (weight loss at the initial stage of heating to 100–150 °C), the content of oxide nanoparticles is 8.1%.

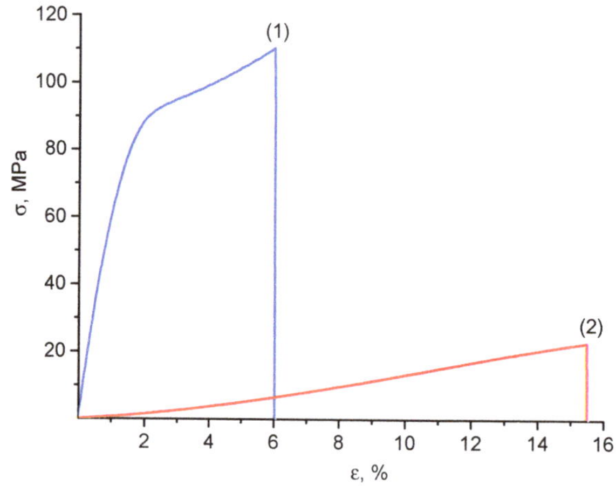

Figure 4. Stress–strain curves of the CS(80)-BC(20)-CeONP(8) film in dry (1) and swollen (2) states.

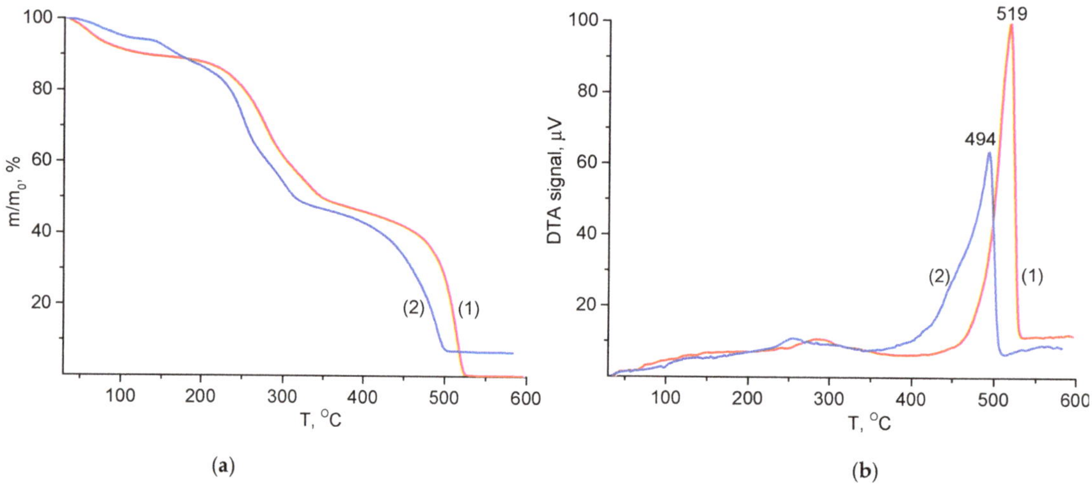

Figure 5. TGA (**a**) and and differential thermal analysis (DTA) (**b**) curves of (1), CS(80)-BCd(20), (2), CS(80)-BCd(20)-CeONP(8) films in air atmosphere.

The thermo-oxidative degradation processes in the first, low-temperature step correspond to the initial stage of pyrolysis when dehydration, depolymerization, and decomposition of acetylated and deacetylated units occur simultaneously. At this stage, pyrolysis causes the random cleavage of glycosidic bonds, followed by further decomposition, resulting in the formation of C2, C3, and C6 fatty acids, including acetic acid and butyric acid. The further process of deep destruction of materials takes place at the high temperature stage, up to 500–550 °C. At this stage, the oligomeric and monomeric products formed at the first stage decompose in an oxygen-containing atmosphere resulting in the formation of gaseous substances [41]. A similar pattern of thermo- oxidation of BCd was reported in our previous studies [37].

The thermogravimetric analysis (TGA) results show that the incorporation of CeONPs into the polysaccharide matrix did not lead to a significant change in the temperature characteristics of the thermo-oxidative degradation of the material over a wide temperature range (up to 450–550 °C). In particular, the temperature resistance indices of CS-BCd and CS-BCd-CeONP films (Table 3) were almost identical. It was only in the last stage of degradation (in the region above 450 °C) that an increase in degradation of the CeONPs containing nanocomposite was observed compared to the original CS-BCd film (Figure 5a).

Table 3. The thermogravimetric analysis (TGA) data of the polysaccharides and nanocomposite films.

Sample	τ5, °C	τ10, °C
CS(80)-BCd(20)	243	256
CS(80)-BCd(20)-CeONP(8)	239	255

This trend is well illustrated by the differential thermal analysis (DTA) curves of the degradation processes of CS-BCd and CS-BCd-CeONP films (Figure 5b). The heat release peak for the CS-BCd-CeONP film (maximum at 521 °C) is shifted by 10 °C to lower temperatures with respect to the heat release maximum for the CS-BCd film. It should also be noted that the DTA curves show a significant difference in the energy characteristics of the material degradation occurring in two stages of thermal-oxidative degradation (low and high temperature). The first stage proceeds without significant heat release: the DTA curves show a low exothermic effect with the maximum intensity of heat release at 282 °C, and in the second stage, an intense exothermic peak is registered (Figure 5b).

2.6. Scanning Electron Microscopy

The CS film shows a uniform surface, while BCd shows a random network of dried nanofibers with a thickness of 20–30 nm (Figure 6). The surface morphology of the control CS(80)-BCd(20) film is characterized by a random distribution of BCd nanofibers in the CS matrix. In the CS(80)-BCd(20)-CeONP(8) film, the BCd nanofibers are less visualized than in the CS(80)-BCd(20) film, which may indicate a stronger interaction between the BCd nanofibers and the CS-CeONP.

The scanning electron microscopy (SEM) image of the CS film cryocleavage (Figure 7) shows no ordered fragments. At the same time, the BCd film contains layers of ordered nanofibers. The ordered BCd nanofibers were observed in the cleavage of CS(80)-BCd(20) films, and they were covered with CS solution and formed a bonded structure, which may further indicate the interaction between CS and BCd nanofibers. The incorporation of CeONPs (the CS(80)-BCd(20)-CeONP(8) sample, Figure 7e,f) changes this pattern: the BCd nanofibers are not visible and the SEM shows a denser packing. The white dots in the SEM images of the cryocleavages of the CS-BCd-CeONP composites (Figure 7c–f) are the broken ends of the BCd nanofibers and the fracture surfaces of these nanofibers formed during the cryocleavage of the materials.

In addition, the surface of the samples was examined using energy-dispersive X-ray spectroscopy (EDX), which allows the presence and distribution of chemical elements in a sample to be determined (in this case, the distribution of cerium). EDX confirmed the incorporation of cerium-containing species into CS(80)-BCd(20)-CeONP(8). Cerium distribution maps showed the uniform distribution of this element in the samples (Figure 8).

(a)

(b)

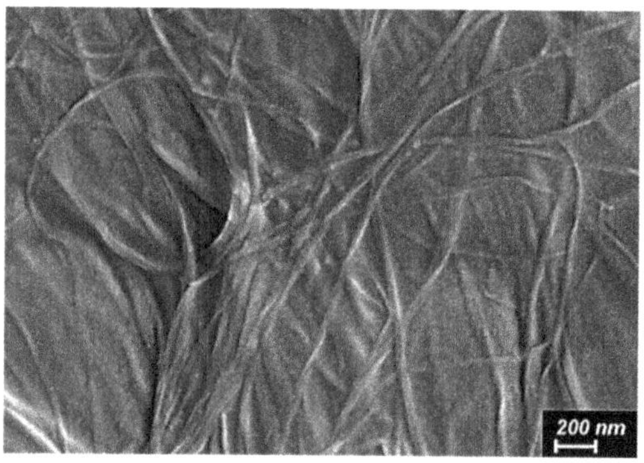

(c)

(d)

Figure 6. Scanning electron microscopy (SEM) images of the surfaces of the CS film (**a**), dried dispersion of BCd (**b**), the CS(80)-BC(20) (**c**) and BCd-CS-CeONP(8) (**d**) films.

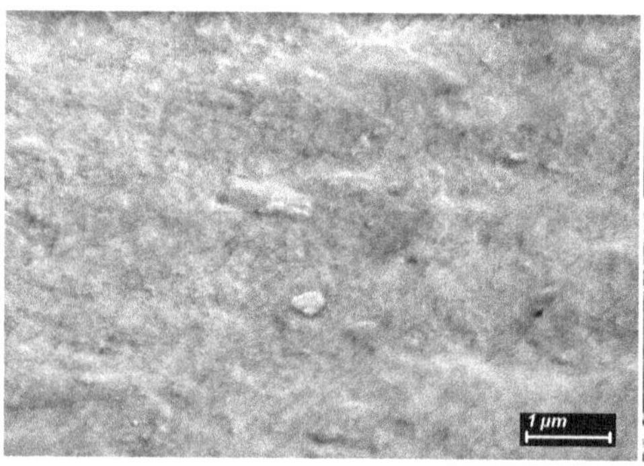

(a)

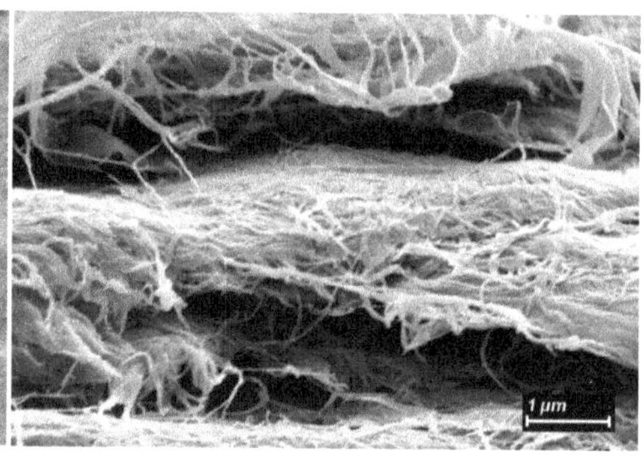

(b)

Figure 7. *Cont.*

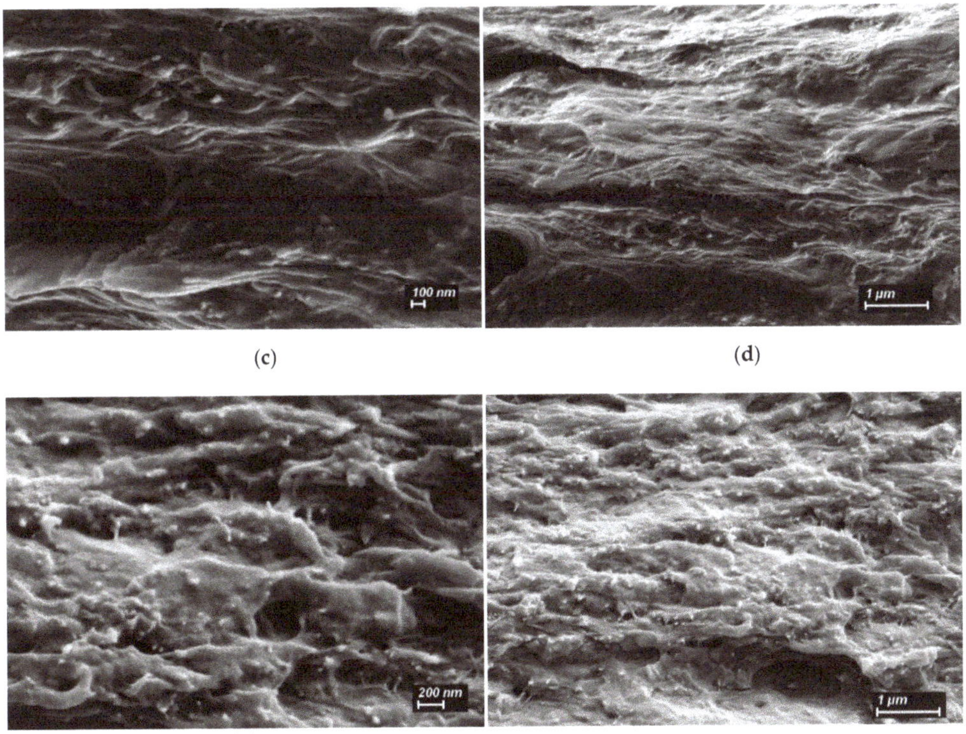

(c) (d)

(e) (f)

Figure 7. SEM images of cryocleaved surfaces of the CS film (**a**), dried dispersion of BCd (**b**), the CS(80)-BC(20) (**c**,**d**) and CS(80)-BCd(20)-CeONP(8) (**e**,**f**) films.

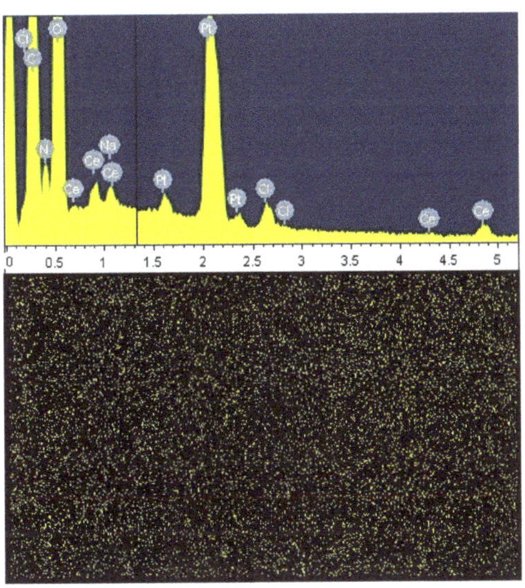

Figure 8. Energy dispersive X-ray spectrum and map of Ce distribution on the surface of a CS(80)-BCd(20)-CeONP(8) film.

2.7. Culture of Multipotent Mesenchymal Stem Cells

2.7.1. Quantitative Analysis

The number of isolated MSCs and cells in the flat colonies was significantly higher in CS(80)-BCd(20)-CeONP(8) compared to the control ($p < 0.05$). In the other experimental groups, no significant differences in the number of nuclei were found compared to the control ($p > 0.05$). The number of nuclei on the surface of CS(80)-BCd(20)-CeONP(8) was significantly higher than in CS(95)-BC(5)-CeONP(4), while there were no significant differences with CS(80)-BCd(20)-CeONP(4) ($p>0.05$) (Table 4).

Table 4. Adhesion of the human mesenchymal stem cells (MSCs) on the samples surface.

Sample	Adhered Cells, cells/mm^2	Number of Spheroids	Spheroids' Size, μm
CS(80)-BCd(20)—control	216 ± 16	6.6 ± 0.9	64 ± 7
CS(80)-BCd(20)-CeONP(4)	218 ± 22	8.2 ± 1.5	66 ± 6
CS(95)-BCd(5)-CeONP(4)	199 ± 10	6.7 ± 0.6	141 ± 17 *
CS(80)-BCd(20)-CeONP(8)	300 ± 33 *	6.8 ± 0.7	136 ± 12 *

Significance of differences compared to the control sample (Mann–Whitney), * $p < 0.05$

There were no significant differences in the number of spheroid colonies in the experimental groups compared to the control and each other ($p > 0.05$). The spheroids in CS(95)-BCd(5)-CeONP(4) and CS(80)-BCd(20)-CeONP(8) were larger in size compared to the control and the other experimental groups ($p < 0.05$). In addition, there were no significant differences in the size of spheroidal colonies in control and CS(80)-BCd(20)-CeONP(4) ($p > 0.05$) (Table 4).

2.7.2. Qualitative Analysis

On coverslips, the cells were evenly arranged and spread on the glass surface, forming a confluent/sub-confluent monolayer similar to syncytium, and had a typical elongated shape with multiple processes; in addition, some cells were in the process of division. Longitudinal linear structures stained red, actin microfilaments, were clearly seen in the cells.

In all experimental groups, the cells were irregularly arranged on the surface of the samples in the form of single cells, flat, and spherical colonies (Figure 9). Some of the solitary cells had a typical elongated shape with processes; another part had a round/near round shape with or without cytoplasmic protrusions at the periphery (probably in the state of blebbing). Flat colonies were visualized as large or smaller groups of cells connected by syncytial-like processes or as star-shaped colonies. Spheroid colonies were more often rounded/oval in shape, less often oblong/non-rectangular in shape. Some spheroids showed evidence of cell migration along the periphery of the colonies in the form of individual cells protruding along the periphery without connection to surrounding cells/colonies (+), cell bridges with other flat/spheroidal colonies (++), or broad fusion with flat colonies (spheroid-to-monolayer transition). No evidence of cell migration was observed in some spheroid colonies (Figure 10, Table 5).

In CS(80)-BCd(20)-CeONP(4) and CS(80)-BCd(20) (as a control), multiple separately arranged cells of both typical elongated shape and rounded/nearly rounded shape were detected (Figure 9). Flat stellate and syncytial colonies consisted of a small number of cells. Single dividing MSCs were detected among the separated cells as well as within the flat colonies. Most of the spheroid colonies showed signs of cell migration: in CS(80)-BCd(20) and CS(80)-BCd(20)-CeONP(4), migration was observed as single cells at the periphery of the spheroids and single bridges with the surrounding cells/colonies. Single spheroidal colonies without cell migration were visualized on the surface of the above sample groups (Figure 10, Table 5).

In CS(95)-BCd(5)-CeONP(4), individually arranged cells of typical elongated shape and rounded/nearly rounded cells were observed in small numbers (Figure 9). Most of the adherent cells formed flat syncytium-like colonies, which occupied a larger area compared to the colonies of CS(80)-BCd(20). Single, dividing MSCs were detected among the cells in the flat colonies. Most of the spheroidal colonies showed signs of cell migration in the form of cell bridging and fusion with the flat colonies. Single spheroidal colonies without signs of cell migration were visualized on the surface of the above samples (Figure 10, Table 5).

In CS(80)-BCd(20)-CeONP(8), the separately arranged cells of typical elongated shape and rounded/nearly rounded cells were detected in small numbers (Figure 9). Most of the adherent cells formed flat syncytium-type colonies, which occupied a larger area compared to the colonies in the other samples. Among the cells in the flat colonies, we detected individual MSCs in the process of division. Most of the spheroid colonies showed signs of cell migration in the form of multiple cell bridges and fusion with the flat colonies. Single spheroidal colonies without signs of cell migration were visualized on the surface of the above samples (Figure 10, Table 5).

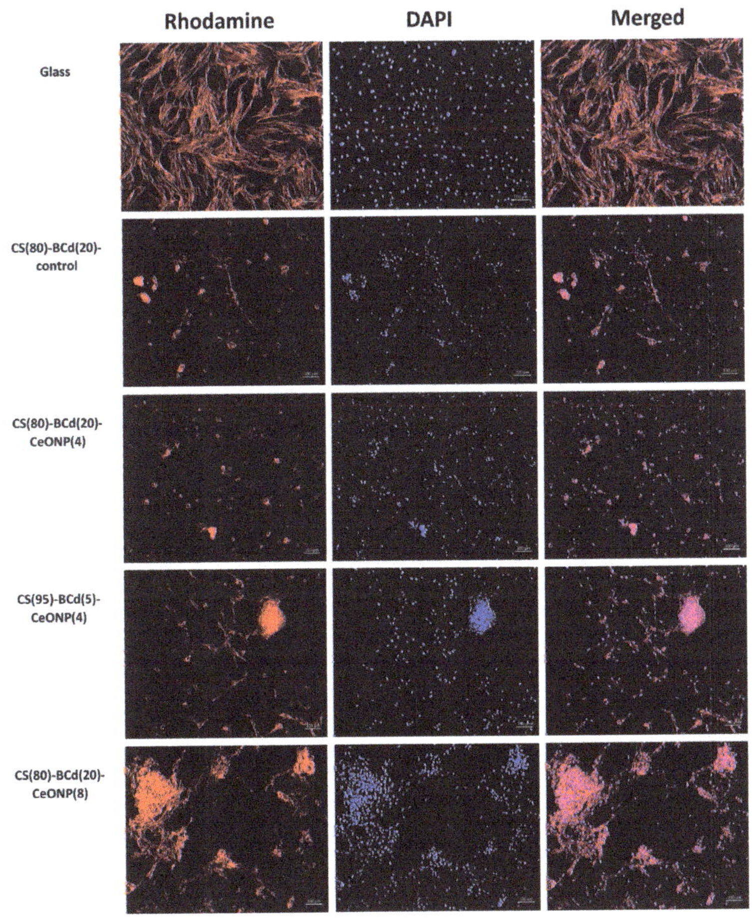

Figure 9. MSCs adhered to the surface of glasses and CS-BCd composite films. Staining of cell cytoskeleton fibrillar actin with Rhodamine-channel fluorochrome visualization, staining of cell nuclei with DAPI (4,6-diamidino-2-phenylindole), combined two-channel image. Magnification ×100.

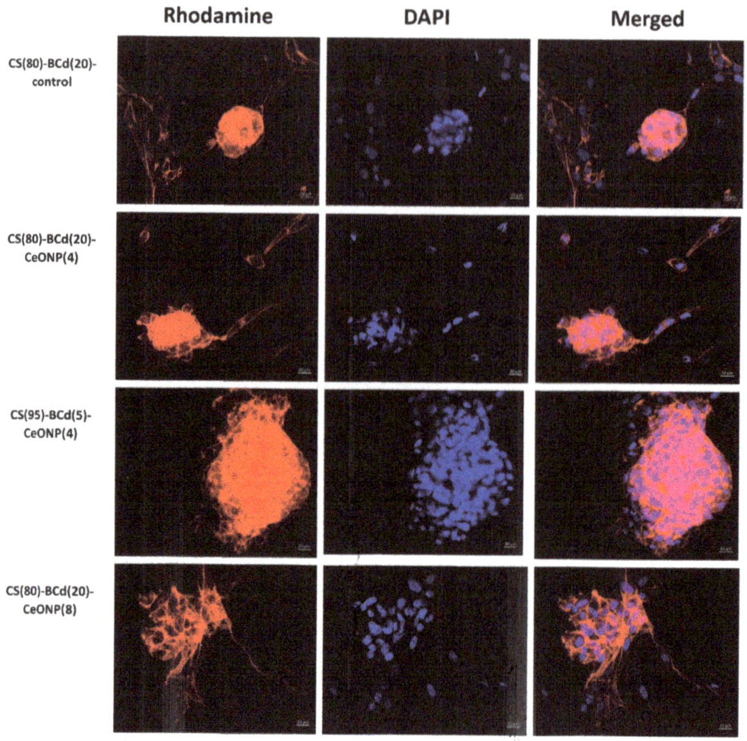

Figure 10. Spheroid colonies on the surface of the CS-BCd composite films. Staining of cell cytoskeleton fibrillar actin with Rhodamine-channel fluorochrome visualization, staining of cell nuclei with DAPI, combined two-channel image. Magnification ×400.

Table 5. Characterization of separately located cells and cell colonies on the surface of samples.

Sample	Separate cells		Type of Colonies	Cell Migration from Spheroids
	Elongated Cells	Rounded Cells		
Glass	multiple	-	flat colonies/monolayer	-
CS(80)-BCd(20)—control	multiple	multiple	flat colonies + spheroids	++
CS(80)-BCd(20)-CeONP(4)	multiple	multiple	flat colonies + spheroids	++
CS(95)-BCd(5)-CeONP(4)	single	single	flat colonies + spheroids	+++
CS(80)-BCd(20)-CeONP(8)	single	single	flat colonies + spheroids	+++

Therefore, the greatest number of adherent cells located separately and as part of planar colonies (outside of spheroids) was detected on the surface of CS(80)-BCd(20)-CeONP(8). In addition, the number of isolated cells (outside the planar and spheroidal colonies), including those in a state of apoptosis, was greater on the surface of both the control CS(80)-BCd(20) and CS(80)-BCd(20)-CeONP(4) samples.

On the surface of CS(95)-BCd(5)-CeONP(4) and CS(80)-BCd(20)-CeONP(8), most adherent cells were found in flat and spheroidal colonies. Spheroidal colonies were observed in all samples. No significant differences were found in the number of spheroidal colonies on the surface of the samples, with the colonies in CS(95)-BCd(5)-CeONP(4) and CS(80)-BCd(20)-CeONP(8) being larger. Maximum cell migration was observed in CS(95)-BCd(5)-CeONP(4) and CS(80)-BCd(20)-CeONP(8). Finally, the proliferation of adherent cells with primary formation of squamous colonies was more pronounced in

CS(80)-BCd(20)-CeONP(4). CS(80)-BCd(20)-CeONP(8) showed the best adhesion properties and biocompatibility.

In summary, cell adhesion, spreading, proliferation, differentiation, and function are highly dependent on scaffold properties such as mechanical properties, chemical structure, nano- and microtopography, surface charge, porosity, wettability, etc. [42,43]. CS-based scaffolds have many advantages, including biocompatibility, biodegradability, and non-toxicity. However, CS films have high deformability and insufficient strength in the swollen state. The incorporation of BCd into the CS film not only improves the mechanical properties, but also may contribute to the regulation of the surface charge due to the ionic bonding of negatively charged BCd nanofibers with positively charged protonated amino groups of CS. In addition, CeONPs are known to influence intracellular signaling pathways and overall cellular metabolism [44,45].

The developed composite containing CS, BCd, and CeONPs showed an altered topography with a stronger interaction between CS, BCd nanofibers, and CeONPs. Apparently, the altered topography improved the adhesive properties of the composite scaffolds. The combination of complex physicochemical interactions, including hydrophobic, coulombic, and van der Waals forces between the cell membrane and the material surface, occurs at the very beginning of the cell adhesion process [43,46]. Thus, the presence of negatively charged BCd nanofibers and positively charged CS molecules not only promotes the mutual repulsion of BCd nanofibers and prevents their aggregation during dispersion in water, but also influences the charge of the material and the initiation of the cell adhesion process.

3. Methods and Materials

3.1. Polysaccharides

The CS sample (Ennagram, Pantin, France) from crab shells with a molecular weight of 160,000 (capillary viscometry) and a degree of deacetylation of 0.80 (conductometric back titration) was used [47].

BC was produced by *Kommagateibacter xylinus* (acetic acid bacteria, a VKM-880 strain) in aqueous solutions containing 2 wt% of glucose, 0.3 wt% of yeast extract, and 2 wt% of ethanol at 30 °C for 14–21 days [48,49]. BCs were isolated by boiling in 6% NaOH followed by multiple rinses with water to neutral pH. The as-prepared BC samples were gel-like pellicles up to 25 mm thick. The pellicles were partially dehydrated using a hand press. The pressed BC was disintegrated in a high-speed blender (15,000 rpm, 15 min) in an aqueous medium (300 mL of water per 1 g of dry BC); the resulting sample was a dispersion of disintegrated BC (BCd).

3.2. Preparation of Composite Films

The CS film (control) was prepared by extruding a 3% solution of CS in 2% acetic acid through a spinneret onto a glass substrate, followed by drying at room temperature and further heating at 80 °C for 4 h.

The detailed protocol of BCd film formation has been described elsewhere [50].

CeONPs used in the experiments were synthesized according to the protocol described elsewhere [38]. CeONPs had a ζ-potential of –25.0 mV and a hydrodynamic diameter (D_h) of 9 nm.

CS-stabilized CeONPs were prepared according to the procedure described elsewhere [32]. CS-coated CeONPs had a ζ-potential of +20.3 mV and a D_h of 244 nm.

The film-casting mixtures were prepared by mixing a 2% solution of CS in 2% acetic acid, a 0.3% dispersion of BCd, and a dispersion of CeONPs. The resulting mixtures were homogenized by mechanical stirring for 1 h. Films were prepared from the composite mixtures by dry casting; the mixtures were extruded through a spinneret onto a glass substrate and dried at room temperature (film thickness was 30–40 μm). The composite films were then heated at 80 °C for 2 h, which caused them to lose their solubility in water [51]. The composite films of different compositions were prepared (the CS/BCd ratio was 80%/20% or 95%/5%); the amount of CeONPs (4% or 8%) was calculated from the total polymer mass. The resulting samples were as follows: CS(80)-BCd(20)-CeONP(4); CS(80)-BCd(20)-CeONP(8); CS(95)-BCd(5)-CeONP(4); and control films: CS(80)-BCd(20); CS(95)-BCd(5).

3.3. Characterization of Composites

FTIR spectra of the films were recorded on a Vertex 70 IR Fourier spectrometer (Bruker Optik, Ettingen, Germany). A Pike MIRacle attenuated total reflection sampling accessory (Pike Technologies, Madison, WI, USA) with a ZnSe working element was used to preserve the structure of the films. In the registration of the spectra, a correction was introduced that takes into account the penetration depth depending on the wavelength.

The ζ-potential of BCd nanofibers was determined using a Photocor Compact-Z instrument (Photocor Ltd., Moscow, Russia) at a laser wavelength of 659 nm and a detection angle of 90°.

To study the interaction between BCd and CS, a solution of CS in 2% acetic acid (C = 0.48 mg/mL) was added dropwise to a dispersion of BCd (2.5 mL, C = 0.48 mg/mL) and the ζ-potential of the resulting systems was measured.

The equilibrium swelling degree of the films (the swelling degree after 24 h exposure to water) was determined by the gravimetric method.

The resulting films were characterized by SEM using a SUPRA-55VP scanning electron microscope (Zeiss, Oberkochen, Germany) and by WAXS using a D8 DISCOVER X-ray diffractometer (Bruker, Karlsruhe, Germany) with CuKα radiation.

SEM images were obtained using both a secondary electron detector and a backscattered electron detector. To visualize the distribution of CeONPs in the samples, the films were frozen and split in liquid nitrogen, then glued to a conductive tape, sputtered through a thin layer of platinum, and analyzed EDX to obtain the maps of Ce distribution. The elemental maps were obtained using an EDX-Max 80 mm^2 detector (Oxford Instruments, Oxford, UK). The analysis was performed over the whole visible range of the samples.

An AG-100kNX Plus setup (Shimadzu, Kyoto, Japan) operating in uniaxial extension mode was used to investigate the mechanical properties of the films. Strip-like samples (2 × 20 mm) were stretched at room temperature at a rate of 2 mm/min according to the requirements of ASTM D638. The stress–strain curves of the samples were recorded during the tests. Young's modulus (E), yield stress (σ_b), and ultimate strength (ε_b) were determined.

TGA and DTA were carried out to determine the residual water concentration in the films, the content of cerium oxide in the nanocomposite materials, and to characterize the effect of the nanofiller on the thermal properties of the composite films. TGA curves were used to determine the thermal stability indices of the samples, τ5 and τ10 (the temperatures at which a polymer or composite loses 5% and 10% of its initial weight, respectively, as a result of thermal destruction processes). A DTG-60 thermal analyzer (Shimadzu, Kyoto, Japan) was used and samples (~5 mg) were heated in air to 600 °C at a rate of 5 °C/min.

3.4. Biocompatibility Testing

Adhesive properties of the materials were investigated using MSCs derived from the adipose tissue of healthy donors. The study was conducted in accordance with the Declaration of Helsinki, and approval was obtained from the local Ethical Committee of the Almazov National Medical Research Centre (№12.26/2014; 1 December 2014). Written informed consent was obtained from all subjects prior to adipose tissue biopsy. The study was conducted as previously described [10,42,52].

The biocompatibility of cell culture with nanocomposite samples was investigated using CS(80)-BC(20) and 12 mm diameter coverslips as controls. Cell culturing was performed in α-MEM medium supplemented with 10% fetal bovine serum, 1% L-glutamine, and 1% penicillin/streptomycin solution in a CO_2 incubator at 37 °C and 5% CO_2.

Samples of 12 × 8 mm materials (depending on the size of the well) were kept in phosphate-buffered saline (PBS) with the addition of 2% penicillin/streptomycin solution for 30 min, followed by three washes with PBS. The coverslips were sterilized in 70%

ethanol for 10 min, followed by three washes with PBS. The samples and coverslips were then placed in the wells of a 24-well plate. Then, 1 mL of medium was added to the wells and incubated for 24 h in a CO_2 incubator (to ensure uniform distribution of the components of the medium in the structure of the samples). After 24 h, the medium was removed and a suspension of MSCs at a concentration of 50,000 cells/mL was added to the wells and co-cultured for 72 h in a CO_2 incubator. The experiment was performed in triplicate.

After 3 days, the samples and coverslips were transferred to the wells of a new plate, washed with PBS, and fixed with 4% paraformaldehyde solution for 10 min.

After fixation, samples and coverslips were washed from paraformaldehyde with PBS and stained with rhodamine-labeled phalloidin according to a previously developed protocol. According to the protocol, samples and coverslips containing cells were first treated with 0.05% Triton X-100 solution for 3 min and then washed three times with PBS. Next, a solution of rhodamine-conjugated phalloidin at a dilution of 1:500 in 1% fetal bovine serum solution in PBS was added to the wells, incubated for 20 min at room temperature, and then washed five times with PBS. Finally, cell nuclei were stained with DAPI (4,6-diamidino-2-phenylindole) at a dilution of 1:40,000, incubated for 40 s, and then the samples were thoroughly washed from DAPI with PBS.

After staining, the samples were stored in PBS in the dark at +4 °C. The coverslips containing cells from the control wells were mounted on glass slides with mounting medium and stored in the dark at room temperature.

Fluorescence microscopy with qualitative and quantitative analysis of adherent cells was performed on the stained MSCs on film samples and glasses. An Axiovert inverted fluorescence microscope (Zeiss, Oberkochen, Germany) and a compatible Canon camera were used to visualize the cells. Pieces of material containing cells were placed between two coverslips. DAPI fluorescence was recorded using the DAPI channel, and rhodamine-phalloidin fluorescence was recorded using the rhodamine channel. Ten different fields of view were photographed at ×100 and ×400 magnifications for each technical replicate. Quantitative analysis consisted of counting the number of nuclei on the surface of the film samples (separately located cells and cells in flat colonies in fields of view without spheroids), counting the number of spheroid colonies per unit area (1 mm^2), and estimating spheroid size (maximum longitudinal size). The number of nuclei in spheroids was not assessed because it is technically impossible to accurately count DAPI-positive signals in colonies with a 3D structure. The morphology of cells and their colonies was assessed qualitatively by staining the cytoskeleton, the type of colonies (flat colonies, spheroids), the presence and extent of cell migration along the periphery of spheroid colonies (+—single protruding cells along the periphery of spheroid not in contact with surrounding cells, ++—single cell bridges with surrounding cells, +++—fusion with surrounding flat colonies and transition of spheroid into monolayer). Statistical processing of the obtained data was performed with the GraphPad Prism 6.01 software (GraphPad Software, San Diego, CA, USA) using the non-parametric Mann–Whitney U criterion. The results were presented as mean ± standard error.

4. Conclusions

In this work, biocompatible CS-based composites containing a combination of two types of nanofillers, namely BCd nanofibers and CeONPs, were developed. The mechanical disintegration of the pressed gel film of BC leads to a change in the structural organization of the polymer and is accompanied by the formation of $-COO^-$ groups on the surface of the resulting BCd nanofibers. The results of titration of the aqueous dispersion of BCd with a solution of CS in acetic acid suggest the possibility of interaction of CS with BCd nanofibers. CeONPs precoated with CS to impart a positive charge to the particles showed high compatibility with the polymer matrix and uniform distribution throughout the volume of the material.

The influence of the amount of introduced nanofillers with different architectures (BCd nanofibers and CeONPs) on the structure and properties of the composites was investigated. An increase in film stiffness was observed as a result of reinforcing the CS matrix with BCd nanofibers: the Young's modulus increased from 4.55 to 6.3 GPa with the introduction of 5% BCd. A further increase in Young's modulus of 6.7 GPa and a significant increase in film strength (22% increase in yield stress compared to the CS film) were observed when the BCd concentration was increased to 20%. The incorporation of CeONPs into the polysaccharide matrices has no significant effect on the mechanical properties of the films, while the X-ray diffraction data indicate the interaction of CeONPs with the polymeric components of the composite and the nature of this interaction depends on the amount of nanoparticles introduced. The introduction of 4% CeONPs is accompanied by an increase in the hydrophilicity of the nanocomposite, an increase in the CeONP concentration up to 8% contributes to a change in the structural organization of the nanocomposite, accompanied by a decrease in the swelling degree and the formation of a denser structure. The CS(80)-BCd(20)-CeONP(8) composite exhibits good strength in both dry and swollen states.

All nanocomposite matrices were found to be biocompatible. The CS(80)-BCd(20)-CeONP(8) composite showed the best biocompatibility and improved cell adhesion to the scaffold surface. Thus, the positive effect of the introduction of BCd and CeONPs on the properties of the CS-based nanocomposite film was demonstrated, and the resulting films were found to be promising for use as a matrix material for MSC culture. In addition, the results obtained suggest the possibility of successfully using the developed materials for the preparation of wound dressings.

Author Contributions: Conceptualization, V.A.P., I.V.G. and Y.A.S.; methodology, V.A.P., I.V.G. and A.S.G.; formal analysis, I.V.G. and A.E.B.; investigation, V.A.P., I.V.G., A.S.G., A.I.M., N.V.D., A.K.K., E.M.I., E.N.V., D.P.R. and A.V.M.; resources, A.V.Y., V.K.I. and A.E.B.; writing—original draft preparation, V.A.P., I.V.G. and A.S.G.; writing—review and editing, Y.A.S.; supervision, A.V.Y. and V.K.I.; funding acquisition, A.V.Y. All authors have read and agreed to the published version of the manuscript.

Funding: This work was financially supported by the Russian Science Foundation (project 22-13-00068).

Institutional Review Board Statement: The study was conducted according to the guidelines of the Declaration of Helsinki, and approved by the Local Ethics Committee of the Almazov National Medical Research Centre (№ 12.26/2014; 1 December 2014). Written informed consent was obtained from all subjects prior to adipose tissue biopsy.

Informed Consent Statement: Informed consent was obtained from all subjects involved in the study.

Data Availability Statement: Data available upon request.

Conflicts of Interest: The authors declare no conflict of interest.

References

1. Puppi, D.; Chiellini, F. Biodegradable polymers for biomedical additive manufacturing. *Appl. Mater. Today* **2020**, *20*, 100700. [CrossRef]
2. Strnad, S.; Zemljič, L.F. Cellulose–chitosan functional biocomposites. *Polymers* **2023**, *15*, 425. [CrossRef] [PubMed]
3. Roman, D.L.; Ostafe, V.; Isvoran, A. Deeper inside the specificity of lysozyme when degrading chitosan. A structural bioinformatics study. *J. Mol. Graph. Model.* **2020**, *100*, 107676. [CrossRef] [PubMed]
4. Di Martino, A.; Sittinger, M.; Risbud, M.V. Chitosan: A versatile biopolymer for orthopaedic tissue-engineering. *Biomaterials* **2005**, *26*, 5983–5990. [CrossRef]
5. Madni, A.; Kousar, R.; Naeem, N.; Wahid, F. Recent advancements in applications of chitosan-based biomaterials for skin tissue engineering. *J. Bioresour. Bioprod.* **2021**, *6*, 11–25. [CrossRef]
6. Azmana, M.; Mahmood, S.; Hilles, A.R.; Rahman, A.; Arifin, M.A.B.; Ahmed, S. A review on chitosan and chitosan-based bionanocomposites: Promising material for combatting global issues and its applications. *Int. J. Biol. Macromol.* **2021**, *185*, 832–848. [CrossRef]
7. Ling, S.; Chen, W.; Fan, Y.; Zheng, K.; Jin, K.; Yu, H.; Buehler, M.J.; Kaplan, D.L. Biopolymer nanofibrils: Structure, modeling, preparation, and applications. *Prog. Polym. Sci.* **2018**, *85*, 1–56. [CrossRef]
8. Jin, T.; Liu, T.; Lam, E.; Moores, A. Chitin and chitosan on the nanoscale. *Nanoscale Horiz.* **2021**, *6*, 505–542. [CrossRef]

9. Kiroshka, V.V.; Petrova, V.A.; Chernyakov, D.D.; Bozhkova, Y.O.; Kiroshka, K.V.; Baklagina, Y.G.; Romanov, D.P.; Kremnev, R.V.; Skorik, Y.A. Influence of chitosan-chitin nanofiber composites on cytoskeleton structure and the proliferation of rat bone marrow stromal cells. *J. Mater. Sci. Mater. Med.* **2017**, *28*, 21. [CrossRef]
10. Petrova, V.A.; Golovkin, A.S.; Mishanin, A.I.; Romanov, D.P.; Chernyakov, D.D.; Poshina, D.N.; Skorik, Y.A. Cytocompatibility of bilayer scaffolds electrospun from chitosan/alginate-chitin nanowhiskers. *Biomedicines* **2020**, *8*, 305. [CrossRef]
11. Shrivastav, P.; Pramanik, S.; Vaidya, G.; Abdelgawad, M.A.; Ghoneim, M.M.; Singh, A.; Abualsoud, B.M.; Amaral, L.S.; Abourehab, M.A. Bacterial cellulose as a potential biopolymer in biomedical applications: A state-of-the-art review. *J. Mater. Chem. B* **2022**, *10*, 3199–3241. [CrossRef]
12. Petrova, V.A.; Gofman, I.V.; Golovkin, A.S.; Mishanin, A.I.; Dubashynskaya, N.V.; Khripunov, A.K.; Ivan'kova, E.M.; Vlasova, E.N.; Nikolaeva, A.L.; Baranchikov, A.E. Bacterial cellulose composites with polysaccharides filled with nanosized cerium oxide: Characterization and cytocompatibility assessment. *Polymers* **2022**, *14*, 5001. [CrossRef]
13. Carvalho, T.; Guedes, G.; Sousa, F.L.; Freire, C.S.; Santos, H.A. Latest advances on bacterial cellulose-based materials for wound healing, delivery systems, and tissue engineering. *Biotechnol. J.* **2019**, *14*, 1900059. [CrossRef]
14. Dai, L.; Nan, J.; Tu, X.; He, L.; Wei, B.; Xu, C.; Xu, Y.; Li, S.; Wang, H.; Zhang, J. Improved thermostability and cytocompatibility of bacterial cellulose/collagen composite by collagen fibrillogenesis. *Cellulose* **2019**, *26*, 6713–6724. [CrossRef]
15. Petersen, N.; Gatenholm, P. Bacterial cellulose-based materials and medical devices: Current state and perspectives. *Appl. Microbiol. Biotechnol.* **2011**, *91*, 1277–1286. [CrossRef]
16. Kim, J.; Cai, Z.; Chen, Y. Biocompatible bacterial cellulose composites for biomedical application. *J. Nanotechnol. Eng. Med.* **2010**, *1*, 011006. [CrossRef]
17. Müller, F.A.; Müller, L.; Hofmann, I.; Greil, P.; Wenzel, M.M.; Staudenmaier, R. Cellulose-based scaffold materials for cartilage tissue engineering. *Biomaterials* **2006**, *27*, 3955–3963. [CrossRef]
18. Pulkkinen, H.; Tiitu, V.; Lammentausta, E.; Hämäläinen, E.-R.; Kiviranta, I.; Lammi, M.J. Cellulose sponge as a scaffold for cartilage tissue engineering. *Bio-Med. Mater. Eng.* **2006**, *16*, S29–S35.
19. Wang, J.; Wan, Y.; Huang, Y. Immobilisation of heparin on bacterial cellulose-chitosan nano-fibres surfaces via the cross-linking technique. *IET Nanobiotechnol.* **2012**, *6*, 52–57. [CrossRef]
20. Ul-Islam, M.; Subhan, F.; Islam, S.U.; Khan, S.; Shah, N.; Manan, S.; Ullah, M.W.; Yang, G. Development of three-dimensional bacterial cellulose/chitosan scaffolds: Analysis of cell-scaffold interaction for potential application in the diagnosis of ovarian cancer. *Int. J. Biol. Macromol.* **2019**, *137*, 1050–1059. [CrossRef]
21. Cazón, P.; Vázquez, M. Improving bacterial cellulose films by ex-situ and in-situ modifications: A review. *Food Hydrocoll.* **2021**, *113*, 106514. [CrossRef]
22. Gong, T.; Xie, J.; Liao, J.; Zhang, T.; Lin, S.; Lin, Y. Nanomaterials and bone regeneration. *Bone Res.* **2015**, *3*, 15029. [CrossRef]
23. Zienkiewicz-Strzałka, M.; Deryło-Marczewska, A.; Skorik, Y.A.; Petrova, V.A.; Choma, A.; Komaniecka, I. Silver nanoparticles on chitosan/silica nanofibers: Characterization and antibacterial activity. *Int. J. Mol. Sci.* **2019**, *21*, 166. [CrossRef] [PubMed]
24. Fernandes, S.C.; Oliveira, L.; Freire, C.S.; Silvestre, A.J.; Neto, C.P.; Gandini, A.; Desbriéres, J. Novel transparent nanocomposite films based on chitosan and bacterial cellulose. *Green Chem.* **2009**, *11*, 2023–2029. [CrossRef]
25. Phisalaphong, M.; Jatupaiboon, N. Biosynthesis and characterization of bacteria cellulose–chitosan film. *Carbohydr. Polym.* **2008**, *74*, 482–488. [CrossRef]
26. Qian, H.; Liu, J.; Wang, X.; Pei, W.; Fu, C.; Ma, M.; Huang, C. The state-of-the-art application of functional bacterial cellulose-based materials in biomedical fields. *Carbohydr. Polym.* **2022**, *300*, 120252. [CrossRef]
27. Shcherbakov, A.B.; Reukov, V.V.; Yakimansky, A.V.; Krasnopeeva, E.L.; Ivanova, O.S.; Popov, A.L.; Ivanov, V.K. CeO_2 nanoparticle-containing polymers for biomedical applications: A review. *Polymers* **2021**, *13*, 924. [CrossRef]
28. Hirst, S.M.; Karakoti, A.S.; Tyler, R.D.; Sriranganathan, N.; Seal, S.; Reilly, C.M. Anti-inflammatory properties of cerium oxide nanoparticles. *Small* **2009**, *5*, 2848–2856. [CrossRef]
29. Hosseini, M.; Mozafari, M. Cerium oxide nanoparticles: Recent advances in tissue engineering. *Materials* **2020**, *13*, 3072. [CrossRef]
30. Kargozar, S.; Baino, F.; Hoseini, S.J.; Hamzehlou, S.; Darroudi, M.; Verdi, J.; Hasanzadeh, L.; Kim, H.-W.; Mozafari, M. Biomedical applications of nanoceria: New roles for an old player. *Nanomedicine* **2018**, *13*, 3051–3069. [CrossRef]
31. Serebrovska, Z.; Swanson, R.; Portnichenko, V.; Shysh, A.; Pavlovich, S.; Tumanovska, L.; Dorovskych, A.; Lysenko, V.; Tertykh, V.; Bolbukh, Y. Anti-inflammatory and antioxidant effect of cerium dioxide nanoparticles immobilized on the surface of silica nanoparticles in rat experimental pneumonia. *Biomed. Pharmacother.* **2017**, *92*, 69–77. [CrossRef]
32. Petrova, V.A.; Dubashynskaya, N.V.; Gofman, I.V.; Golovkin, A.S.; Mishanin, A.I.; Aquino, A.D.; Mukhametdinova, D.V.; Nikolaeva, A.L.; Ivan'kova, E.M.; Baranchikov, A.E. Biocomposite films based on chitosan and cerium oxide nanoparticles with promising regenerative potential. *Int. J. Biol. Macromol.* **2023**, *229*, 329–343. [CrossRef]
33. Zugenmaier, P. Order in cellulosics: Historical review of crystal structure research on cellulose. *Carbohydr Polym* **2021**, *254*, 117417.
34. Wada, M.; Sugiyama, J.; Okano, T. Native celluloses on the basis of two crystalline phase (iα/iβ) system. *J. Appl. Polym. Sci.* **1993**, *49*, 1491–1496. [CrossRef]
35. Wada, M.; Kondo, T.; Okano, T. Thermally induced crystal transformation from cellulose iα to iβ. *Polym. J.* **2003**, *35*, 155–159. [CrossRef]
36. Baklagina, Y.G.; Klechkovskaya, V.V.; Kononova, S.V.; Petrova, V.A.; Poshina, D.N.; Orekhov, A.S.; Skorik, Y.A. Polymorphic modifications of chitosan. *Crystallogr. Rep.* **2018**, *63*, 303–313. [CrossRef]

37. Gofman, I.V.; Nikolaeva, A.L.; Khripunov, A.K.; Ivan'kova, E.M.; Shabunin, A.S.; Yakimansky, A.V.; Romanov, D.P.; Popov, A.L.; Ermakov, A.M.; Solomevich, S.O. Bacterial cellulose-based nanocomposites containing ceria and their use in the process of stem cell proliferation. *Polymers* **2021**, *13*, 1999. [CrossRef]
38. Ivanova, O.; Shekunova, T.; Ivanov, V.; Shcherbakov, A.; Popov, A.; Davydova, G.; Selezneva, I.; Kopitsa, G.; Tret'yakov, Y.D. One-stage synthesis of ceria colloid solutions for biomedical use. *Dokl. Chem.* **2011**, *437*, 638–641. [CrossRef]
39. Bradbury, A.; Sakai, Y.; Shafmadeh, F. The chemistry of solid wood. *Appl. Polym. Sci.* **1979**, *23*, 3271–3283. [CrossRef]
40. de Britto, D.; Campana-Filho, S.P. Kinetics of the thermal degradation of chitosan. *Thermochim. Acta* **2007**, *465*, 73–82. [CrossRef]
41. López Gómez, F.A.; Mercê, A.; Alguacil, F.J.; López-Delgado, A. A kinetic study on the thermal behaviour of chitosan. *J. Therm. Anal. Calorim.* **2008**, *91*, 633–639. [CrossRef]
42. Bolbasov, E.; Goreninskii, S.; Tverdokhlebov, S.; Mishanin, A.; Viknianshchuk, A.; Bezuidenhout, D.; Golovkin, A. Comparative study of the physical, topographical and biological properties of electrospinning pcl, plla, their blend and copolymer scaffolds. In Proceedings of the IOP Conference Series: Materials Science and Engineering, Amsterdam, The Netherlands, 11–13 December 2017; p. 012012.
43. Chen, L.; Yan, C.; Zheng, Z. Functional polymer surfaces for controlling cell behaviors. *Mater. Today* **2018**, *21*, 38–59. [CrossRef]
44. Horie, M.; Nishio, K.; Kato, H.; Fujita, K.; Endoh, S.; Nakamura, A.; Miyauchi, A.; Kinugasa, S.; Yamamoto, K.; Niki, E.; et al. Cellular responses induced by cerium oxide nanoparticles: Induction of intracellular calcium level and oxidative stress on culture cells. *J. Biochem.* **2011**, *150*, 461–471. [CrossRef] [PubMed]
45. Popov, A.L.; Popova, N.R.; Selezneva, I.I.; Akkizov, A.Y.; Ivanov, V.K. Cerium oxide nanoparticles stimulate proliferation of primary mouse embryonic fibroblasts in vitro. *Mater. Sci. Eng. C Mater. Biol. Appl.* **2016**, *68*, 406–413. [CrossRef] [PubMed]
46. Okano, T.; Yamada, N.; Okuhara, M.; Sakai, H.; Sakurai, Y. Mechanism of cell detachment from temperature-modulated, hydrophilic-hydrophobic polymer surfaces. *Biomaterials* **1995**, *16*, 297–303. [CrossRef]
47. Kononova, S.V.; Kruchinina, E.V.; Petrova, V.A.; Baklagina, Y.G.; Klechkovskaya, V.V.; Orekhov, A.S.; Vlasova, E.N.; Popova, E.N.; Gubanova, G.N.; Skorik, Y.A. Pervaporation membranes of a simplex type with polyelectrolyte layers of chitosan and sodium hyaluronate. *Carbohydr. Polym.* **2019**, *209*, 10–19. [CrossRef]
48. Buyanov, A.L.; Revel'skaya, L.; Kuznetzov, Y.P.; Khripunov, A.K. Cellulose–poly (acrylamide–acrylic acid) interpenetrating polymer network membranes for the pervaporation of water–ethanol mixtures. II. Effect of ionic group contents and cellulose matrix modification. *J. Appl. Polym. Sci.* **2001**, *80*, 1452–1460. [CrossRef]
49. Buyanov, A.L.; Revel'skaya, L.G.; Kuznetzov, Y.P.; Shestakova, A.S. Cellulose–poly (acrylamide or acrylic acid) interpenetrating polymer network membranes for the pervaporation of water–ethanol mixtures. *J. Appl. Polym. Sci.* **1998**, *69*, 761–769. [CrossRef]
50. Gofman, I.V.; Nikolaeva, A.L.; Khripunov, A.; Yakimansky, A.; Ivan'Kova, E.M.; Romanov, D.P.; Ivanova, O.g.S.; Teplonogova, M.A.; Ivanov, V.K. Impact of nano-sized cerium oxide on physico-mechanical characteristics and thermal properties of the bacterial cellulose films. *Nanosyst. Phys. Chem. Math.* **2018**, *9*, 754–762. [CrossRef]
51. Zotkin, M.A.; Vikhoreva, G.A.; Kechek'yan, A.S. Thermal modification of chitosan films in the form of salts with various acids. *Polym. Sci. Ser. B* **2004**, *46*, 39–42.
52. Goreninskii, S.; Guliaev, R.; Stankevich, K.; Danilenko, N.; Bolbasov, E.; Golovkin, A.; Mishanin, A.; Filimonov, V.; Tverdokhlebov, S. "Solvent/non-solvent" treatment as a method for non-covalent immobilization of gelatin on the surface of poly (l-lactic acid) electrospun scaffolds. *Colloids Surf. B Biointerfaces* **2019**, *177*, 137–140. [CrossRef]

International Journal of *Molecular Sciences*

MDPI

Article

Evaluation of Chitosans as Coagulants—Flocculants to Improve Sand Filtration for Drinking Water Treatment

Eleanor B. Holmes [1], Hemali H. Oza [1,2], Emily S. Bailey [3,*] and Mark D. Sobsey [1]

1 Department of Environmental Sciences and Engineering, Gillings School of Global Public Health, The University of North Carolina at Chapel Hill, Chapel Hill, NC 27599, USA
2 Gangarosa Department of Environmental Health, Rollins School of Public Health, Emory University, Atlanta, GA 30322, USA
3 Department of Public Health, College of Pharmacy & Health Sciences, Campbell University, Buies Creek, NC 27506, USA
* Correspondence: ebailey@campbell.edu

Abstract: The World Health Organization (WHO) reports that two billion people worldwide lack access to safely managed water sources, including 1.2 billion who already have access to improved water sources. In many countries, household point-of-use (POU) water-treatment options are used to remove or deactivate microorganisms in water, but not all POU technologies meet WHO performance requirements to achieve safe drinking water. To improve the effectiveness of POU technologies, the use of multiple treatment barriers should be used as a way to increase overall treatment performance. The focus of this research is to evaluate multiple barrier treatment using chitosan, an organic coagulant–flocculant, to improve microbial and turbidity reductions in combination with sand filtration. Bench-scale intermittently operated sand filters with 16 cm layers of sands of two different grain sizes representing slow and rapid sand filters were dosed daily over 57 days with microbially spiked surface water volumes corresponding to household use. *E. coli* bacteria and MS2 coliphage virus reductions were quantified biweekly (N = 17) using culture methods. Bacteria and virus removals were significantly improved over sand filtration without chitosan pretreatment (Wilcoxon Rank-Sum, $p < 0.05$). When water was pretreated at an optimal chitosan dose of 10 mg/L followed by sand filtration, $\log_{10}$ reductions in bacteria and viruses met the two-star WHO performance level of effectiveness. Microbial and turbidity reductions generally improved over the filter operating period but showed no trends with filtration rates.

Keywords: HWTS; POU; chitosan; coagulation; sand filtration; bacterial reduction; virus reduction

1. Introduction

Although access to improved drinking water sources is expanding globally, access alone does not mean the absence of a health risk. An estimated 1.2 million people died as a result of unsafe water sources in 2017 [1]. A critical intervention that improves the safety of both improved and unimproved water sources for at-risk populations is household water treatment and safe storage (HWTS). Various physical, chemical, and biological household treatment options are widely available, including chlorination, solar disinfection, flocculant/disinfectant powders and granules, slow sand filtration, and ceramic filtration [2,3]. Despite extensive availability of multiple-barrier POU treatment technologies, consumers continue to use many single-barrier POU technologies that are less effective and do not meet the protective or highly protective performance criteria of the World Health Organization (WHO) for health risk reductions for all classes if microbes (viruses, bacteria, and protozoan parasites) [4].

In 2014, the WHO published initial results of many POU treatment performance evaluations relative to interim, protective, and highly protective reduction requirements for water-treatment technologies under its newly established International Scheme [5].

Highly protective technologies must achieve log_{10} reductions for bacteria, viruses, and protozoan cysts of greater than 4, 5, and 4, respectively. For protective status, the HWT technology much achieve >2, >3, and >2 log_{10} reductions for bacteria, viruses, and protozoan cysts, respectively. To be considered an interim (minimally protective) technology, protective performance requirements for two of the three pathogens must be met, along with epidemiological evidence of diarrheal disease reduction in credible field studies [4,6,7].

Individual point-of-use (POU) technologies have generally failed to achieve target microbial reductions in field settings, often achieving only half of the reported maximum log_{10} reductions observed in laboratory studies [4]. The WHO now reports baseline and maximum log_{10} reduction values (LRVs) for all POU technologies to illustrate the differences between reductions achieved in field use and those achieved in a controlled laboratory setting. Biosand filtration, for example, can achieve a maximum LRV of three for viruses in laboratory studies; however, in practice only up to a 0.5 log_{10} reduction is typically observed, especially for viruses [4,8,9]. Despite these shortcomings, epidemiological studies suggest that reductions in diarrheal disease attributable to POU technologies is between 30 and 40%, suggesting that these technologies still provide substantial health benefits to users [10–13].

Most single-barrier household water filtration technologies, such as biosand and ceramic filters, are inadequate for virus removal according to WHO performance targets, while household water chlorination is ineffective in reducing the infectivity of the protozoan parasite pathogen *Cryptosporidium*. However, combining technologies or using multiple technologies in series may substantially improve reductions in all classes of microbes, while also maintaining ease of use and accessibility at affordable cost.

In previous laboratory studies the positively charged chitin derivative chitosan, a natural and biodegradable polysaccharide byproduct of the crustacean fishing industry and available as well from other natural sources, such as insect exoskeletons, has been proposed as a coagulant–flocculant POU treatment for use in multiple-barrier methods, such as with cloth and ceramic filters [14–17]. When chitosan is added to water, suspended colloidal material including viruses, bacteria, and spores are coagulated, allowing the coagulated particles to then flocculate together during slow mixing and then settle out of the water or be remove by subsequent filtration. Inorganic coagulants, including ferric sulfate and aluminum sulfate, are commonly used in large water-treatment facilities; however, they are highly pH and dose dependent for optimum performance, which limits their effectiveness for practical use at the household level. Previous studies have found that, unlike inorganic coagulants, pH and dose levels do not significantly affect the efficacy of chitosan as an organic coagulant [17–19]. Furthermore, chitosan is inexpensive, non-toxic, biodegradable, easy-to-use, naturally occurring, and readily available in most places around the world. This makes it an attractive and environmentally friendly supplemental treatment step prior to existing water filtration treatment technologies.

When chitosan was assessed as a coagulant, it effectively reduced bacteria and virus concentrations and turbidity in model and natural waters treated by chitosan coagulation followed by microporous ceramic filtration or cloth filtration, with no appreciable change in pH [16–19]. However, the extent to which chitosan coagulation–flocculation may improve microbial and turbidity reductions achieved via sand filtration has not been reported in model or natural waters, to our knowledge.

The purpose of this laboratory study was to assess the efficacy of chitosan as a coagulant–flocculant in natural waters, followed by treatment with sand filters. Variables such as optimal dosage of chitosan, sand grain size, and flow rate were evaluated for their potential effect on turbidity and microbe reduction. The reductions in turbidity and the indicator microbes of bacteria and viruses were compared to the WHO household water-treatment performance targets. Protozoan parasites were not included in this study because there is documented evidence that physical removals of viruses and bacteria, which are smaller in size than protozoans, would provide sufficient evidence for performance efficacy of these microbial removal processes, including protozoan parasites [20].

2. Results

The mean *E. coli* strain KO11 LRVs for each column over the 57-day duration of filter operation are summarized in Figure 1 for duplicate columns of both sand types. The average MS2 coliphage LRVs are summarized in Figure 2. The average turbidity LRVs for each column are summarized in Figure 3. The error bars represent the standard deviation of the average values.

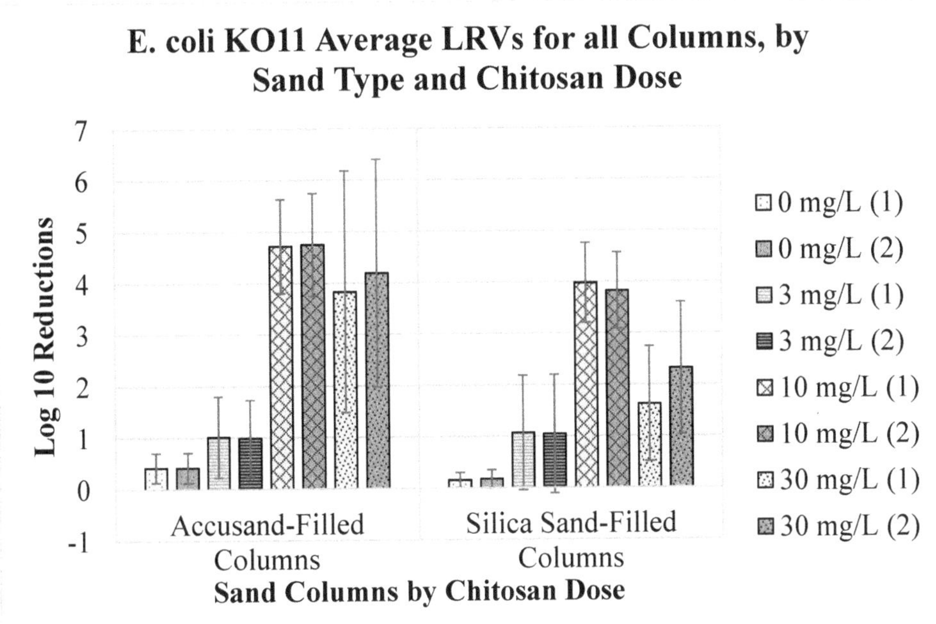

Figure 1. Average $\log_{10}$ reduction values (LRVs) with standard deviation error bars for *E. coli* KO11 in water treated by chitosan coagulation followed by Accusand and silica sand column filtration, based on 17 successive samples collected throughout the 57-day experiment period.

For *E. coli* (strain KO11), the average LRVs by combined chitosan coagulation and sand filtration ranged from less than 0.5-$\log_{10}$ to greater than 4.5 $\log_{10}$ in Accusand columns and from 0.5-$\log_{10}$ to nearly 4-$\log_{10}$ for silica columns (Figure 1). The average LRVs reported for Accusand columns were higher than those reported for silica sand columns for all chitosan doses except 3 mg/L, which were both around 1 LRV (see LRV data below). Control filters not dosed with chitosan (dose = 0 mg/L) did not exceed a 0.5-$\log_{10}$ *E. coli* LRV for both sand types. Filters dosed with water coagulated with 10 mg/L chitosan achieved average LRVs of about 4.5-$\log_{10}$ and nearly 4-$\log_{10}$ for Accusand and silica sand columns, respectively. Although greater variability is observed between duplicate filters of each sand type, those dosed with water coagulated with 30 mg/L chitosan reached average LRVs exceeding 3.5-$\log_{10}$ for Accusand-filled columns and greater than 1.5-$\log_{10}$ for silica sand-filled columns.

The average LRVs of coliphage MS2 by combined chitosan coagulation and sand filtration ranged from less than 0.5-$\log_{10}$ to nearly 4.5-$\log_{10}$ in the duplicate Accusand-filled columns and ranged from less than 0.5-$\log_{10}$ to nearly 4-$\log_{10}$ for the duplicate silica sand-filled columns (Figure 2). As was observed with *E. coli* KO11 LRVs, the MS2 coliphage LRVs were generally higher for Accusand columns compared to silica sand columns across chitosan doses. However, the control filters for both sand types achieved similar MS2 LRVs from waters with no chitosan (dose = 0 mg/L) that did not exceed 0.5 LRV for

MS2 coliphage. Filters dosed with water coagulated with 3 mg/L chitosan achieved on average 1.5-$\log_{10}$ and 1.0-$\log_{10}$ reductions for MS2 coliphage in Accusand and silica sand columns, respectively. Filters dosed with water coagulated with 10 mg/L chitosan achieved LRVs for MS2 coliphage of approximately 4.5-$\log_{10}$ for Accusand columns and greater than 3.5-$\log_{10}$ for silica columns. Filters dosed with water treated with 30 mg/L chitosan achieved average MS2 coliphage LRVs of approximately 4-$\log_{10}$ and 2.5-$\log_{10}$ for Accusand and silica columns, respectively.

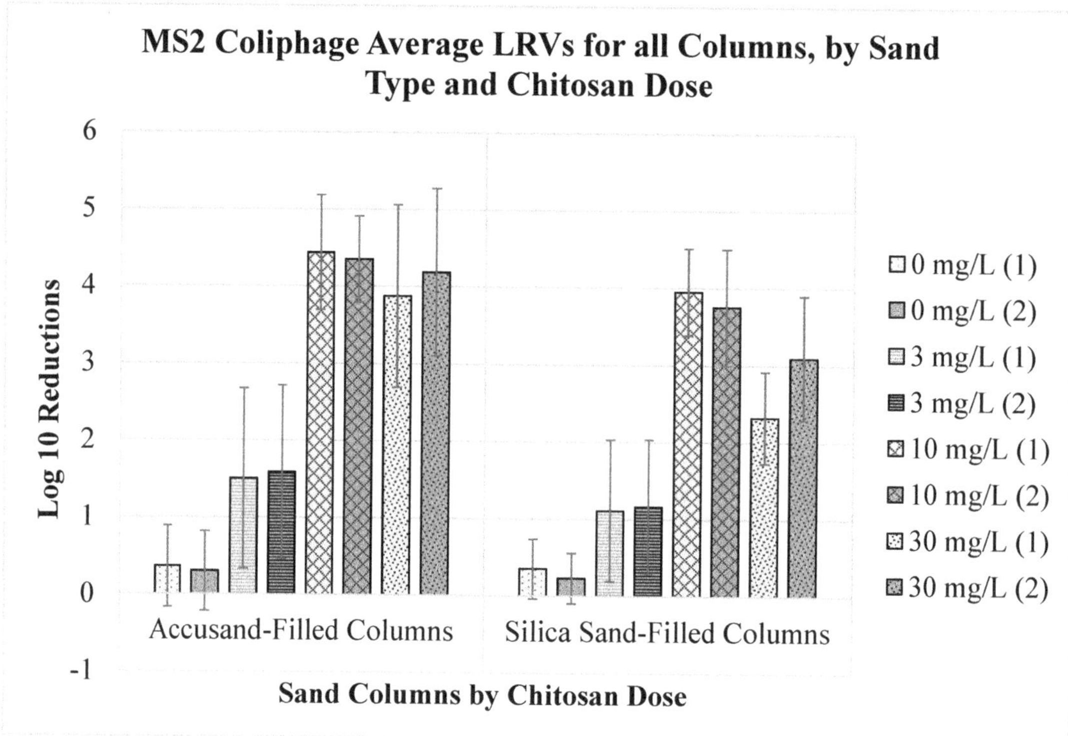

Figure 2. Average $\log_{10}$ reduction values (LRVs) with standard deviation error bars for MS2 coliphage in water treated by chitosan coagulation and Accusand and silica sand column filtration, based on 17 successive samples collected throughout the 57-day experiment period.

The turbidity reductions observed for Accusand columns were typically slightly greater than those observed for silica sand columns across chitosan doses (Figure 3). The turbidity reductions for both sand types ranged from less than 0.2-$\log_{10}$ (<40%) to about 1-$\log_{10}$ (90%). Control filters receiving no chitosan dose (chitosan dose = 0 mg/L) achieved 0.2 LRVs (40%) or less for both sand types. With a 3 mg/L chitosan dose followed by sand filtration LRVs were about 0.6-$\log_{10}$ (75%) for Accusand filters and about 0.5-$\log_{10}$ (68%) for silica sand filters. At a 10 mg/L chitosan dose followed by sand filtration, LRVs were higher, reaching 1.0-$\log_{10}$ (90%) for Accusand filters and about 0.9-$\log_{10}$ (87%) for silica sand filters. However, at the highest dose of chitosan tested, 30 mg/l, followed by sand filtration, turbidity reductions were lower than at 10 mg/L chitosan dose and similar to the 3 mg/L chitosan dose, with LRVs of approximately 0.7-$\log_{10}$ (80%) and 0.6-$\log_{10}$ (75%) for Accusand and silica sand filtration, respectively.

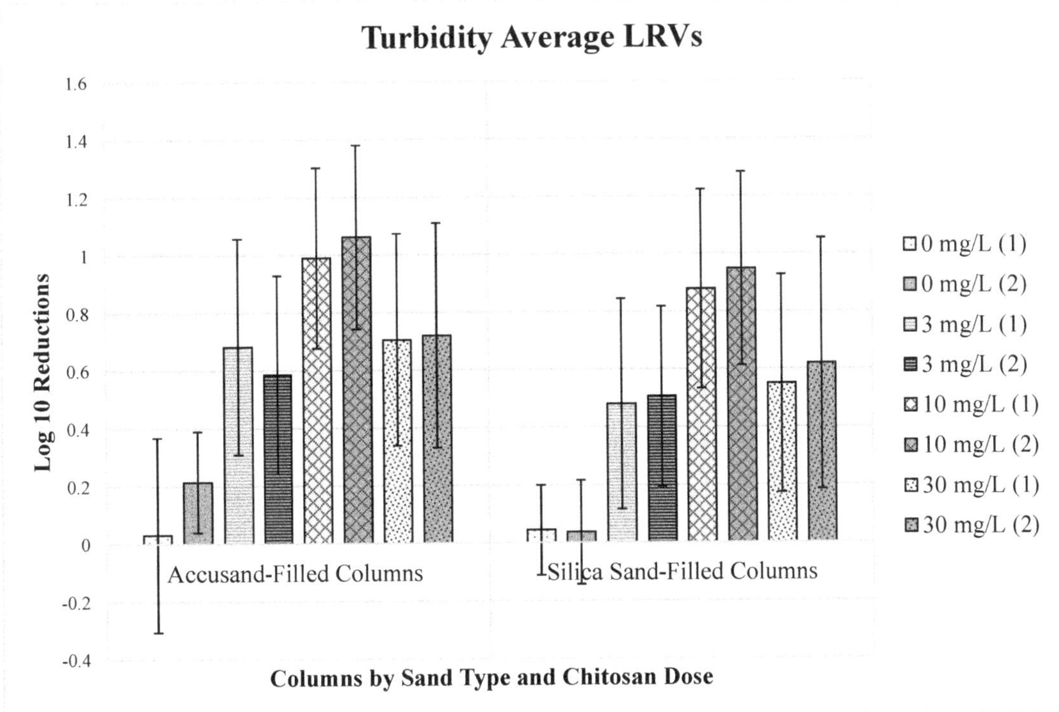

Figure 3. Average $\log_{10}$ reduction values (LRVs) with standard deviation error bars for turbidity in water treated by chitosan coagulation and Accusand and silica sand column filtration, based on 17 successive samples collected throughout the 57-day experiment period.

The average differences in LRV values between the duplicate filters are reported in Table 1 along with the range of differences observed over the 57-day study period. For most conditions, duplicate filters had on average less than 0.5-$\log_{10}$ differences between duplicates for *E. coli* KO11 and MS2 coliphage and less than 0.2-$\log_{10}$ for turbidity. Columns of both sand types dosed with water coagulated with 30 mg/L chitosan had higher average differences between duplicate filters for *E. coli* KO11 and MS2 coliphage compared to all other doses tested. LRV differences were 0.608 and 1.146 for *E. coli* with Accusand and silica sand, respectively, and 0.370 and 0.804 for MS2 with Accusand and silica sand, respectively. *E. coli* KO11 LRVs for water coagulated with 30 mg/L chitosan reported maximum differences between duplicate filters exceeding 3-$\log_{10}$ for both Accusand and silica sand columns. MS2 coliphage LRVs for 30 mg/L chitosan followed by silica sand filtration resulted in an average of about 0.8-$\log_{10}$ difference between the duplicates, with a maximum difference of 1.6-$\log_{10}$. Overall, these results suggest that duplicate columns of the same sand type and dose tend to achieve similar LRVs for bacteria, viruses, and turbidity, with rare exceptions.

Table 1. Average, minimum and maximum differences in $\log_{10}$ reduction values between duplicate filters for each sand type and chitosan dose over the 57-day filter operating time.

		E. coli K011			MS2 Coliphage			Turbidity		
Sand Type	**Dose (mg/L)**	**Average**	**Minimum**	**Maximum**	**Average**	**Minimum**	**Maximum**	**Average**	**Minimum**	**Maximum**
Accusand	0	0.105	0.000	0.356	0.136	0.002	0.363	0.194	0.003	1.024
	3	0.087	0.012	0.247	0.194	0.024	0.632	0.135	0.029	0.472
	10	0.340	0.000	1.041	0.328	0.041	0.829	0.152	0.035	0.550
	30	0.608	0.000	3.395	0.370	0.008	1.322	0.162	0.002	0.449
Silica	0	0.094	0.000	0.321	0.249	0.000	0.698	0.087	0.006	0.304
	3	0.120	0.010	0.293	0.204	0.007	0.509	0.135	0.029	0.472
	10	0.201	0.002	0.662	0.366	0.024	1.031	0.152	0.035	0.550
	30	1.146	0.127	3.046	0.804	0.000	1.600	0.162	0.002	0.449

The Wilcoxon rank-sum test was also used to compare overall median LRVs of *E. coli* KO11 between columns receiving water with the same chitosan dose but treated with the two different sand types. The results are presented in Table 2. Statistically significant differences in achieved LRVs were observed between sand types for columns dosed with 0, 10, and 30 mg/L chitosan-treated water ($p < 0.05$). LRVs achieved by Accusand columns dosed with 3 mg/L chitosan-treated water were not significantly different from those achieved by silica sand columns at the same dose. For MS2 coliphage, there were no significant differences between the LRVs achieved by the two different sand column types dosed with untreated water and 3 mg/L chitosan-treated water. However, LRVs attained with Accusand filter columns were statistically significantly higher for 10 mg/L and 30 mg/L doses of chitosan-treated water than those attained with silica columns ($p < 0.05$). No significant differences in turbidity were observed between the LRVs achieved by Accusand columns compared to silica columns across all doses of chitosan.

Table 2. Results of the Wilcoxon rank-sum test comparing cumulative median $\log_{10}$ reduction values of *E. coli* KO11, MS2, and turbidity by sand type, stratified by dose. Reported *p*-values were adjusted using the Bonferroni correction, m = 4.

	Accusand	**Silica**	**Dose (mg/L)**	***p*-Value** [a]
E. coli K011	Accusand	Silica	0	$\mathbf{5.06 \times 10^{-4}}$ *
	Accusand	Silica	3	1.00
	Accusand	Silica	10	$\mathbf{7.88 \times 10^{-4}}$ *
	Accusand	Silica	30	$\mathbf{2.91 \times 10^{-3}}$ *
MS2	Accusand	Silica	0	1.00 *
	Accusand	Silica	3	0.934
	Accusand	Silica	10	**0.0138**
	Accusand	Silica	30	$\mathbf{3.12 \times 10^{-5}}$ *
Turbidity	Accusand	Silica	0	0.073
	Accusand	Silica	3	0.560
	Accusand	Silica	10	0.690
	Accusand	Silica	30	1.00

[a] Bolded *p*-values are statistically significant at $p < 0.05$; * *p*-values estimated with ties.

3. Discussion

This study reports the first known results evaluating the effectiveness of using chitosans as a coagulation–flocculation pretreatment in natural waters to improve the removal of bacteria, viruses, and turbidity by intermittently operated slow sand filters (IOSSFs). Extensive reductions in bacteria, viruses, and turbidity were achieved by sand columns dosed with 10 mg/L and 30 mg/L chitosan-pretreated water. Sand columns dosed with water treated with 10 mg/L met the protective performance targets for bacteria and viruses specified by the WHO for HWT technologies ($\geq$2 and $\geq$3 $\log_{10}$ reductions, respectively).

The maximum average LRVs of *E. coli* KO11 and MS2 coliphage were achieved with a dose of 10 mg/L of chitosan acetate and Accusand column filtration, reaching 4.75 (+/−0.99) and 4.43 (+/−0.74) log_{10}, respectively. By comparison, Accusand filtration without chitosans achieved maximum average LRVs of only 0.42 (+/−0.29) for bacteria and 0.36 (+/−0.53) for viruses. The average reported reductions at 10 mg/L chitosan doses with Accusand filter columns met the protective LRV targets set by the WHO for HWT technologies, exceeding the 3-log_{10} reduction level for viruses and the 2-log_{10} reduction level for bacteria. These performance targets were also met for 10 mg/L silica sand filter columns in the first half of the study; however, cumulative median LRVs for MS2 coliphage eventually were below the 3-log_{10} target by the end of the 57-day study period (data not shown). Average LRVs for the 30 mg/L chitosan dose did not consistently meet the WHO safe performance targets.

Pretreatment with chitosan salts followed by column sand filtration as a model IOSSF also improved turbidity reductions. All three doses of chitosan tested significantly improved turbidity reductions compared to filtration alone; however, the 10 mg/L chitosan dose was the only dose to achieve <1 NTU for average effluent turbidity with both sand filter column types. Filtration alone without chitosan coagulation pre-treatment produced average filtrate water turbidity levels between 4 and 6 NTU. The 10 mg/L chitosan pretreatment followed by sand filtration achieved on average maximum and minimum filtrate water turbidites of 0.90 NTU (+/−0.67) and 0.56 NTU (+/−0.27), respectively. This chitosan dose coupled with filtration met the 1 NTU level recommended by the WHO GDWQ for drinking water turbidity. All three chitosan doses followed by small-scale sand column filtration exceeded an average of 0.4-log_{10} (60%) turbidity reduction, with 10 mg/L exceeding 0.8-log_{10} (84%) for both sand types. By comparison, filtration alone achieved on average <0.25-log_{10} (<44%) turbidity removal. The addition of chitosan to challenge waters did not significantly change pH, even at the highest chitosan dose.

3.1. Filter Maturation

Previous studies have associated filter maturation or media aging to improved microbial reductions by IOSSF [8,21]. Typically, filtration rate is used as a proxy to indicate media aging. Filtration rates in this study were kept within a target range for each sand type, therefore media aging effects in filtration rate and performance were not directly investigated. Separating LRVs into time intervals may provide some insight into how media aging may impact filter performance. However, because media aging was not directly evaluated as an experimental variable, potential impacts are only speculative. The results suggest that when water is treated with an optimal chitosan dose for the influent water quality, in this case 10 mg/L chitosan, substantially improved reductions in bacteria and viruses are achieved, independent of filter operating time and filter maturation. At non-optimal chitosan doses and for untreated water, media aging is correlated with increased LRVs. However, variability in performance across time intervals suggests improvements in reductions occur at different rates for different chitosan doses and sand filter column types. Additionally, the extent to which filter maturation enhances microbial reductions is probably chitosan dose-dependent and influenced by sand type. Declines in LRVs over time may indicate a plateau effect in terms of the extent to which media aging may improve filter performance, or it may indicate that media aging is a less important indicator of LRVs compared to other experimental parameters such as surface water quality, chitosan dose or type of sand.

Prior research evaluating biosand filters (BSFs) with replicate columns of the same conditions have also experienced some lack of reproducibility for experimental conditions and LRV results [8,21]. In this current study, different rates of maturation or ripening of the filter, including chitosan accumulation, increased biological activity in the sand bed of the filter, and weekly scouring procedures are potentially responsible for variability in LRVs within each column and between duplicate columns. Additionally, bacterial regrowth in stored samples and analytical instrument imprecision may account for the variability observed in turbidity measurements. Without further investigation of these parameters, it is impossible to determine to what extent experimental design limitations and unintended differences in sand filtration system design and operation contribute to observed performance variability.

3.2. Removal Mechanisms

The specific mechanisms by which microorganisms and turbidity were removed via chitosan coagulation–flocculation and slow sand filtration were not directly investigated in this research. General principles and information from the literature may provide some insight into plausible mechanistic considerations, but these explanations and interpretations are speculative and require further testing. The mechanisms by which chitosan acts as a coagulant are documented in the literature, but the interactions between the formed chitosan–colloid floc and the sand media in IOSSFs are not well-characterized. The two primary coagulation processes associated with chitosan are charge neutralization and interparticle bridging [18,22,23]. Negatively charged particles in water, including microorganisms, clay, and other inorganic and organic material, adsorbed to the cationic polyamine sites on the chitosan polymer chain. These attraction forces between the polymer and particles promote coagulation–flocculation. The resulting floc, if neutralized and dense, settles out of solution via sedimentation. The supernatant water, with remaining suspended floc, is dosed into IOSSFs.

The processes by which bacteria and viruses are removed with IOSSFs and BSFs are probably different after water has been pretreated with chitosan. Prior research has suggested the schmutzdecke plays an important role in bacterial reductions either by physical straining or reduced flow rate, resulting in enhanced depth filtration [21,24,25]. Bacteria are more amenable to physical straining than viruses because they are larger. Physical straining through the schmutzdecke has little effect on virus removal, therefore other removal or inactivation mechanisms are probably responsible for virus reductions from slow sand and biosand filters [21,24]. Proposed mechanisms include sorption to the granular media, attachment to biofilms, predation, and biological activity [21]. With the addition of a chitosan coagulation–flocculation pretreatment step, it is unclear to what extent the importance of each mechanism changes. However, in a previous study we observed that chitosan coagulation resulted in large floc particle sizes as determined by particle size analysis (Oza et al., 2022). These mechanisms were not directly studied, but speculative mechanistic considerations are proposed based on how the processes function under typical operating conditions.

Despite weekly cleaning and disruption of the schmutzdecke, high LRVs were still observed for bacteria and viruses in this study. This suggests that the development of the schmutzdecke is not necessarily essential for slow sand filtration if water is pretreated with chitosan. The impact on microbial reduction performance as a result of incorporating a diffuser plate into the bench-scale column design and altering the cleaning procedure to promote schmutzdecke growth requires further investigation. Physical straining and deep-bed filtration may be more important mechanisms for removal when operated without the biological layer and with a coagulant, which makes the particles to be removed from water larger and therefore easier to retain.

Many of the proposed mechanisms for virus removal are also dependent on idle time within the filter media bed. Prior studies have shown that increased idle time within the ISSF improves microbial attenuation within the filter [8,26]. In this study, water was pretreated with chitosan and allowed to mix and flocculate for a 30 min period before it was dosed in the filters. The daily charge volume greatly exceeded the pore volume of the filters, and effluent samples were taken after 300 mL had already passed through the media bed. This means the collected effluent spent little time within the filter column where it would be exposed to the biological processes that enhance removal. This short contact time suggests that biological mechanisms are probably not the primary ones for removal. Short contact and idle times also make this dual-treatment barrier a potentially more convenient (in terms of time before providing water to a consumer), reliable, and sustainable process than traditional BSFs.

3.3. Limitations

This research demonstrates that combining chitosan coagulation–flocculation with ISSFs improves filter performance in microbial and turbidity reductions; however, there were limitations to this study which could be addressed in future research. Filters were designed with graduated cylinders and were not equipped with an upper receptacle to maintain constant head. Manual dosing introduced variability in flow rates, which was dependent on how quickly the columns were refilled to maximum head. They were not designed to meet the specifications of any existing IOSSF or BSF. Maximum head and sand bed depth were determined based on available materials for filter construction. Filter operation was also not optimized based on recommended IOSSF or BSF guidelines. Optimal conditions for schmutzdecke growth were not prioritized. Absence of the diffuser plate and weekly cleaning procedures probably disrupted any biological growth on the top of the sand media bed. Idle time within the filter was also not maximized, as is recommended in BSF operation. Idle time within the filter, which allows for biological processes to occur, accounts for much of the virus attenuation typically observed in BSF use [21]. The sand types used in this study were purposely chosen based on typical grain size ranges for BSFs and RSFs; however, the target filtration rate ranges were a compromise based on column design limitations. Filtration rates were maintained within a specified range over the course of the 57-day evaluation, but that range exceeded recommended rates for BSFs and was far below those recommended for RSFs. Maintaining filtration rate also eliminated a common variable, decline in filtration rate over time, used as a proxy for filter maturation and media aging. Because these variables were effectively eliminated from the experimental design in this study, the mechanisms responsible for enhanced microbial and turbidity reductions with combined chitosan coagulation–flocculation and sand filtration are probably different from those documented for SSFs and BSFs. These mechanisms were not directly evaluated in this study.

This study did not evaluate chitosan coagulation coupled with IOSSF performance in removal of protozoa. For the purposes of this study, it was assumed that because protozoa are larger than bacteria and the removal technology studied was filtration, bacteria can serve as a proxy for protozoa and protozoan LRVs would probably be similar to or greater than the achieved bacteria LRVs [20].

These results were observed in simply designed, intermittently operated, falling-head sand filter setups. Variables such as different filter media characteristics, filtration rate, microbial communities, source water quality, and mechanisms for microbial reduction were not directly investigated in this study. User acceptability of chitosan as a potential water additive was also not systematically evaluated in this study, including potential cultural or religious restrictions that would limit use or application of this product in certain settings. Despite these limitations and further research questions, this study demonstrates that intermittently operated slow sand filtration can be significantly improved in microbial reductions using chitosan as a coagulant–flocculant pretreatment process. this process can potentially be further optimized for a range of granular media filtration technologies, including anthracite, activated carbon, and bone char filters, to further improve these POU filtration systems.

4. Materials and Methods

4.1. Sand Column Design and Operation

A total of 16 sand columns having a sand depth of 16 cm and a column diameter of 3.9 cm were constructed of polypropylene and operated in parallel. This number of experimental columns was not based on power calculations for potential microbial and turbidity reduction performance differences among the experimental variables of sand type and chitosan dose. Instead, the number of columns used was based on what could be managed by the human and materials resources available. Duplicate filters and filtration conditions were used in order to evaluate as many doses across sand types as possible with limited resources. Eight columns contained Accusand silica (Unimin Corp., Le Sueur, MN, USA) and 8 contained silica obtained from a local wastewater-treatment facility (OWASA, Orange Water and Sewer Authority, Chapel Hill, NC, USA) that is used in their full-scale biosand filters and sized for rapid sand filtration (BSFs) (referred to as silica columns hereafter). Accusand was used due to its low organic matter content, chemical purity, and low uniformity coefficient (Schroth, Ahearn, Selker, & Istok, 1996). The Accusand media in the columns was a blend of three sieve fraction sizes (U.S. Standard Mesh 30/40, 40/60, and 50/70) of quartz sand. The combination of these sieve fractions provided a smaller average grain size (d10 = 0.24 mm; d60/d10 = 1.40) compared to the silica sand as described by [8,27]. The silica sand was a larger size than Accusand, with d10 = 0.50 mm, d60/d10 = 1.40 as used in rapid sand filters. Target filtration rates were maintained for each type of column, with a filtration rate for the Accusand-filled columns of 0.4–0.6 m/h and for the silica sand-filled columns 1.0–1.4 m/h. Flow rates for all sand filters were measured twice per week, on water sampling days, over the course of the evaluation period. All columns were pre-washed via 24 h exposure to 10% concentrated HCl and rinsed until effluent water reached a pH of 5 [28]. The underdrain of each column was 1–3 cm layer of poly-fil polyester fiber.

Filters were intermittently operated with a daily charge volume of 500 mL per filter per day. This was determined by comparing the surface area ratio of a household-scale BSF treating 20 L/day to the corresponding dimensions of the sand column filters. A volume of 20 L/day is considered the minimum amount of water required per person per day for basic drinking, hygiene, and food preparation needs [29].

Weekly scouring of the top 3 cm layer of each column was conducted to disrupt any growth of a schmutzdecke. The weekly scouring procedure involved: (1) disrupting the top 3 cm of sand in each filter column with a sterile 5 mL pipette for 30 s, (2) then adding 25 mL of deionized (DI) water to the top of the columns in order to suspend the material released from this scoured top layer of the sand bed, and (3) then aspirating this resulting mixture from the top of the sand filter column into the pipette and discharging it to waste. This cleaning procedure was repeated twice in succession for each sand column filter, and then the filters were returned to daily use.

4.2. Challenge Water and Microbial Detection

The challenge water for chitosan coagulation–flocculation–sedimentation and then sand filter column dosing was obtained by periodic surface grab sampling from University Lake in Carrboro, NC, USA. University Lake is a protected reservoir supplying drinking water to the residents of Chapel Hill and Carrboro, NC, USA. University Lake does not receive any identifiable wastewater discharges. A diagram of the experimental design is presented in Figure 4 and average water quality parameters are described in Supplementary Table S1.

Test waters were spiked with *E. coli* KO11 (ATCC# 55124) and male-specific (F+) coliphage (bacteriophage) MS2 (ATCC# 15597-B1). Microbial stocks were prepared to achieve at least a 6 $\log_{10}$ per 100 mL spiking concentrations in test water so that reductions of 99.9999% or 6 $\log_{10}$ could be quantified. A one mL volume of frozen suspension of overnight logarithmic phase growth culture of *E. coli* KO11 was added to 200 mL tryptic soy broth (TSB) with 1% V/V chloramphenicol stock solution (100× stock concentration, 3.4 g/L chloramphenicol dissolved in ethanol, filtered through 0.22 µm pore size membrane filter) in a shaker flask. The culture was incubated at 37 °C on a shaker table at 100 rpm for 18–24 h. The resulting culture was distributed into four 50 mL polystyrene tubes and centrifuged at 3000 rpm for 15 min at 4 °C in a Sorvall refrigerated centrifuge with H6000a swing bucket rotor. Approximately 45 mL of the supernatant was decanted and discarded, then the equivalent volume of phosphate buffer was added (Standard Methods buffer, with 0.4 M $MgCl_2$) and vortex mixed. The suspension was centrifuged and washed three times with this buffer composition. The final suspension of *E. coli* was vortex mixed in phosphate buffer (Standard Methods buffer, with 0.4 M $MgCl_2$) until the pellet was completely dispersed in solution. The resulting *E. coli* concentration of this suspension was approximately 10^6 CFU/mL. Each 2.5 L batch of test water received 15 mL of this concentrated *E. coli* KO11 suspension per day. Washed *E. coli* KO11 cells were prepared each sample day, and unused washed cells were used on non-sampling days. A 1 mL volume of propagated and chloroform extracted MS2 bacteriophage stock at a titer of 1×10^{11} PFU/mL, stored in −80 °C, was added to each 2.5 L batch of challenge water daily.

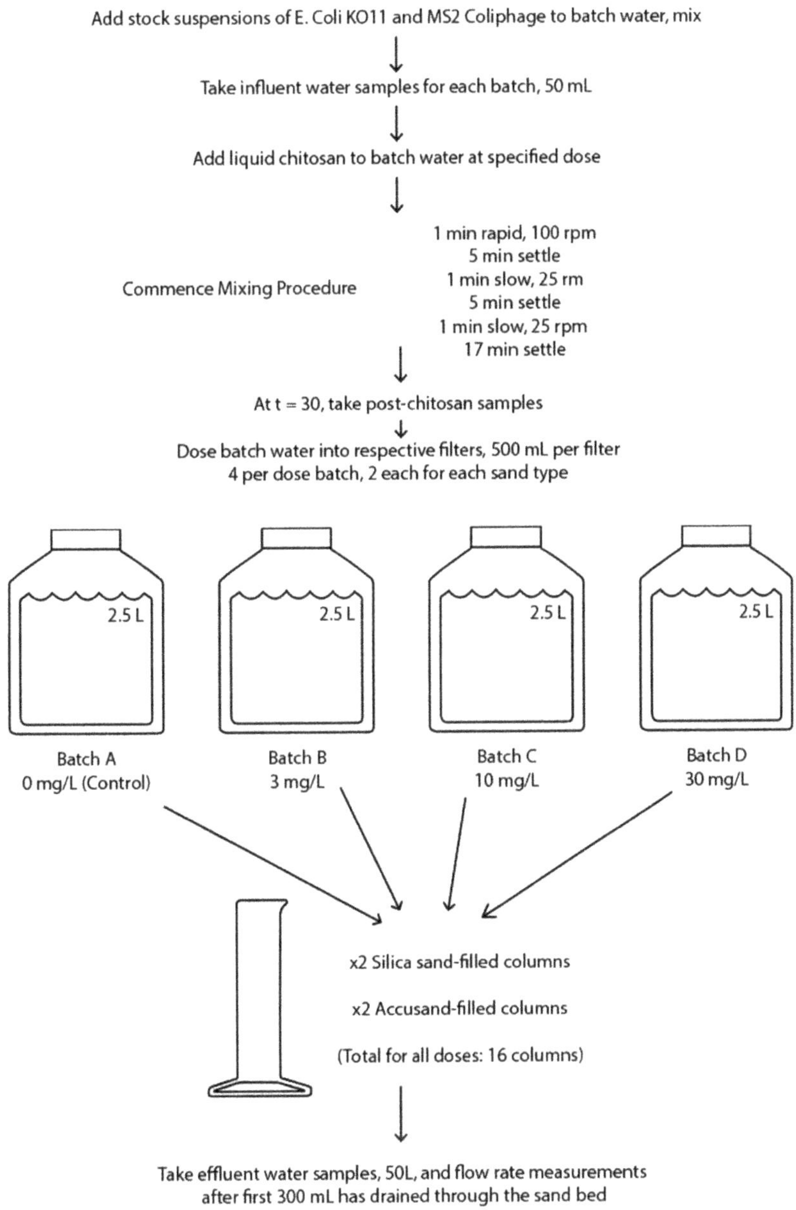

Figure 4. Experimental design for batch water preparation and dosing into sand filters.

The spread plate method was used to quantify *E. coli* KO11 concentrations as colony forming units (CFU) per mL. Selective/differential agar medium, consisting of 40 g/L tryptic soy agar plus 30 mg/L neutral red and 10 g/L lactose, amended with 1% (V/V) chloramphenicol stock (100 × stock concentration, 3.4 g/L chloramphenicol dissolved in ethanol, filtered through 0.22 μm pore size membrane filter), was used to limit background organism interference and discern *E. coli* colonies. The double agar layer (DAL) plaque assay method (EPA 1601) on tryptic soy agar (TSA) plates amended with 1% (V/V) streptomycin/ampicillin stock (100× stock concentration, 1.5 g/L ampicillin sodium salt and 1.5 g/L streptomycin sulfate dissolved in deionized water, filtered through 0.22 μm pore size membrane filter) and *E. coli* F_{amp} host bacteria to quantify MS2 as plaque forming units (PFU)/mL [30].

4.3. Chitosan Dosing

Chitosan acetate (CH3COO−) was selected based on previous studies evaluating microbial removal from water [18] and purchased from Sarchem Labs in powder form. A 2 g/L solution of liquid chitosan was prepared using 2 g chitosan acetate and 1 L of autoclaved lab grade deionized water. Three doses of chitosan were tested in the 57-day bench-scale treatment study along with a control condition with no chitosan treatment: 0 mg/L, 3 mg/L, 10 mg/L, and 30 mg/L. The choice of these chitosan doses was based on previous studies by our laboratory on the range of effective chitosan doses for microbial and turbidity reductions [14–16,18].

Test water with added chitosan was rapidly mixed at approximately 100 rpm for 1 min, then the water was left to settle for 5 min. The water was then slowly mixed for 1 min at approximately 25–30 rpm, left to settle for 5 min, and then slow mixed for a final minute before settling for a final 17 min. Total coagulation–flocculation–settling time for all test waters was 30 min. At 30 min, post-chitosan-treated samples were taken for microbial analysis and 500 mL of the challenge water was dosed to each sand filter column.

4.4. Statistical Analysis

Data analysis and graphical representations were created in R Studio version 4.2.0. Log_{10} reductions were calculated by subtracting the log_{10} concentration of microorganisms in the effluent water from the concentration in the influent water. Non-parametric statistics were used to compare median log_{10} reductions achieved across chitosan doses and the two sand types. The Wilcoxon rank-sum test was used to compare reductions in bacteria and viruses between chitosan doses for the same sand type, and between sand types with the same chitosan dose. An alpha level of 0.05 was used as the significance level.

5. Conclusions

This study reports the first known results evaluating the effectiveness of using chitosans as a coagulation–flocculation pretreatment in natural waters to improve the removal capacity of bacteria, viruses, and turbidity by intermittently operated slow sand filtration for 57 successive days. Sand filter columns with two different sand types achieved extensive reductions in bacteria, viruses, and turbidity when applied water was pre-treated by dosing with 10 mg/L chitosan and then flocculated. Sand columns dosed with water treated with 10 mg/L chitosan met the protective performance targets of the WHO for the HWT technologies program, specifically, 2 log_{10} for bacteria and 3 log_{10} for viruses. Filter performance varied over time, possibly due to scouring procedures, variable source water quality, and inconsistent flow rates. These results were observed in simply designed, intermittently operated, falling-head sand filter setups indicating that simple adjustments to pretreatment prior to the operation of BSF/IOSSF can result in adequately treated drinking water meeting the WHO two-star performance level of health risk reduction for bacteria and viruses.

Supplementary Materials: The following supporting information can be downloaded at: https://www.mdpi.com/article/10.3390/ijms24021295/s1.

Author Contributions: Data collection, methodology, validation, writing—original draft preparation, E.B.H.; data collection, methodology, writing—review and editing, H.H.O.; writing—review and editing, E.S.B. and M.D.S.; study design, research proposal submission and funding acquisition, M.D.S. All authors have read and agreed to the published version of the manuscript.

Funding: This research was funded by the Gillings Innovation Labs Awards of the Gillings School of Global Public Health, University of North Carolina at Chapel Hill.

Institutional Review Board Statement: Not applicable.

Informed Consent Statement: Not applicable.

Data Availability Statement: Data included in this study are available from the corresponding author upon reasonable request.

Conflicts of Interest: The authors declare no conflict of interest.

References

1. World Health Organization (WHO); The United Nations Children's Fund (UNICEF). *Progress on Household Drinking Water, Sanitation and Hygiene 2000–2020: Five Years into the SDGs*; World Health Organization (WHO): Geneva, Switzerland, 2021.
2. Sobsey, M.D.; Stauber, C.E.; Casanova, L.M.; Brown, J.M.; Elliott, M.A. Point of Use Household Drinking Water Filtration: A Practical, Effective Solution for Providing Sustained Access to Safe Drinking Water in the Developing World. *Environ. Sci. Technol.* **2008**, *42*, 4261–4267. [CrossRef]
3. Clasen, T.F.; Alexander, K.T.; Sinclair, D.; Boisson, S.; Peletz, R.; Chang, H.H.; Majorin, F.; Cairncross, S. Interventions to improve water quality for preventing diarrhoea. *Cochrane Database Syst. Rev.* **2015**, *10*, CD004794. [CrossRef]
4. World Health Organization. *Evaluating Household Water Treatment Options: Health-Based Targets and Microbiological Performance Specifications*; World Health Organization: Geneva, Switzerland, 2011.
5. World Health Organization. *WHO International Scheme to Evaluate Household Water Treatment Technologies*; World Health Organization: Geneva, Switzerland, 2014.
6. World Health Organization. *Results of Round I of the WHO International Scheme to Evaluate Household Water Treatment Technologies*; World Health Organization: Geneva, Switzerland, 10 December 2022.
7. World Health Organization. *Results of Round II of the WHO International Scheme to Evaluate Household Water Treatment Technologies*; World Health Organization: Geneva, Switzerland, 2019.
8. Elliott, M.A.; Stauber, C.E.; Koksal FA TM, A.; DiGiano, F.A.; Sobsey, M.D. Reductions of *E. coli*, echovirus type 12 and bacteriophages in an intermittently operated household-scale slow sand filter. *Water Res.* **2008**, *42*, 2662–2670. [CrossRef]
9. Elliott, M.; Stauber, C.E.; DiGiano, F.A.; Fabiszewski de Aceituno, A.; Sobsey, M.D. Investigation of *E. coli* and Virus Reductions Using Replicate, Bench-Scale Biosand Filter Columns and Two Filter Media. *Int. J. Environ. Res. Public Health* **2015**, *12*, 10276–10299. [CrossRef]
10. Fewtrell, L.; Kaufmann, R.B.; Kay, D.; Enanoria, W.; Haller, L.; Colford, J.M., Jr. Water, sanitation, and hygiene interventions to reduce diarrhoea in less developed countries: A systematic review and meta-analysis. *Lancet Infect. Dis.* **2005**, *5*, 42–52. [CrossRef]
11. Clasen, T.; Schmidt, W.P.; Rabie, T.; Roberts, I.; Cairncross, S. Interventions to improve water quality for preventing diarrhoea: Systematic review and meta-analysis. *BMJ* **2007**, *334*, 782. [CrossRef]
12. Tintle, N.; Heynen, A.; Van De Griend, K.; Ulrich, R.; Ojo, M.; Boven, E.; Brokus Wade, R.; Best, A.A. Evaluating the efficacy of point-of-use water filtration units in Fiji. *Trop. Med. Health* **2019**, *47*, 48. [CrossRef]
13. Tamene, A.; Habte, A.; Woldeyohannes, D.; Tamrat, H.; Endale, F.; Eajo, T.; Afework, A. Water treatment at the point-of-use and treatment preferences among households in Ethiopia: A contemporaneous systematic review and meta-analysis. *PLoS ONE* **2022**, *17*, e0276186. [CrossRef]
14. Abebe, L.S.; Chen, X.; Sobsey, M. Chitosan Coagulation to Improve Microbial and Turbidity Removal by Ceramic Water Filtration for Household Drinking Water Treatment. *Int. J. Environ. Res. Public Health* **2016**, *13*, 269. [CrossRef]
15. Coleman, C.K.; Mai, E.; Miller, M.; Sharma, S.; Williamson, C.; Oza, H.; Holmes, E.; Lamer, M.; Ly, C.; Stewart, J.; et al. Chitosan Coagulation Pretreatment to Enhance Ceramic Water Filtration for Household Water Treatment. *Int. J. Mol. Sci.* **2021**, *22*, 9736. [CrossRef]
16. Oza, H.H.; Holmes, E.B.; Bailey, E.S.; Coleman, C.K.; Sobsey, M.D. Microbial reductions and physical characterization of chitosan flocs when using chitosan acetate as a cloth filter aid in water treatment. *PLoS ONE* **2022**, *17*, e0262341. [CrossRef]
17. Fabris, R.; Chow, C.; Drikas, M. Evaluation of chitosan as a natural coagulant for drinking water treatment. *Water Sci. Technol.* **2010**, *61*, 2119–2128. [CrossRef]
18. Soros, A.; Amburgey, J.E.; Stauber, C.E.; Sobsey, M.D.; Casanova, L.M. Turbidity reduction in drinking water by coagulation-flocculation with chitosan polymers. *J. Water Health* **2019**, *17*, 204–218. [CrossRef]

19. Christensen, E.; Håkonsen, T.; Robertson, L.J.; Myrmel, M. Zirconium and chitosan coagulants for drinking water treatment—A pilot study. *J. Water Supply: Res. Technol.-AQUA* **2016**, *65*, 635–644. [CrossRef]
20. LeChevallier, M.W.; Au, K.K. *Water Treatment and Pathogen Control—Process Efficiency in Achieving Safe Drinking-Water*; World Health Organization: Geneva, Switzerland, 2004.
21. Elliott, M.A.; Di Giano, F.; Sobsey, M. Virus attenuation by microbial mechanisms during the idle time of a household slow sand filter. *Water Res.* **2011**, *45*, 4092–4102. [CrossRef]
22. Rinaudo, M. Chitin and chitosan: Properties and applications. *Prog. Polym. Sci.* **2006**, *31*, 603–632. [CrossRef]
23. Kumar, M.R.; Muzzarelli, R.; Muzzarelli, C.; Sashiwa, H.; Domb, A.J. Chitosan Chemistry and Pharmaceutical Perspectives. *Chem. Rev.* **2004**, *104*, 6017–6084. [CrossRef]
24. Hijnen, W.A.; Schijven, J.F.; Bonne, P.; Visser, A.; Medema, G.J. Elimination of viruses, bacteria and protozoan oocysts by slow sand filtration. *Water Sci. Technol.* **2004**, *50*, 147–154. [CrossRef]
25. Unger, M.; Collins, M. Assessing Escherichia coli removal in the schmutzdecke of slow-rate biofilters. *J. AWWA* **2008**, *100*, 60–73. [CrossRef]
26. Jenkins, M.W.; Tiwari, S.; Darby, J. Bacterial, viral and turbidity removal by intermittent slow sand filtration for household use in developing countries: Experimental investigation and modeling. *Water Res.* **2011**, *45*, 6227–6239. [CrossRef]
27. Stauber, C.E.; Elliott, M.A.; Koksal, F.; Ortiz, G.M.; Di Giano, F.A.; Sobsey, M.D. Characterisation of the biosand filter for *E. coli* reductions from household drinking water under controlled laboratory and field use conditions. *Water Sci. Technol.* **2006**, *54*, 1–7. [CrossRef]
28. Litton, G.M.; Olson, T.M. Colloid deposition rates on silica bed media and artifacts related to collector surface preparation methods. *Environ. Sci. Technol.* **1993**, *27*, 185–193. [CrossRef]
29. World Health Organization. *Domestic Water Quantity, Service, Level and Health*; World Health Organization: Geneva, Switzerland, 2003.
30. United States Environmental Protection Agency. *Method 1601: Male-specific (F+) and Somatic Coliphage in Water by Two-Step Enrichment Procedure*; United States Environmental Protection Agency: Washington, DC, USA, 2001.

International Journal of
Molecular Sciences

MDPI

Article

Development of Inhalable Chitosan-Coated Oxymatrine Liposomes to Alleviate RSV-Infected Mice

Jianqing Peng [1,2], Qin Wang [1,2], Mingyang Guo [3], Chunyuan Liu [3], Xuesheng Chen [3], Ling Tao [1,2], Ke Zhang [1,3,*] and Xiangchun Shen [1,2,*]

1 The High Efficacy Application of Natural Medicinal Resources Engineering Center of Guizhou Province, School of Pharmaceutical Sciences, Guizhou Medical University, Guiyang 550025, China
2 The Key Laboratory of Optimal Utilization of Natural Medicine Resources, School of Pharmaceutical Sciences, Guizhou Medical University, Guiyang 550025, China
3 The Key and Characteristic Laboratory of Modern Pathogenicity Biology, School of Basic Medical Sciences, Guizhou Medical University, Guiyang 550025, China
* Correspondence: kezhang@gmc.edu.cn (K.Z.); shenxiangchun@126.com (X.S.); Tel.: +86-0851-884-16022 (K.Z.); +86-0851-881-74180 (X.S.)

Abstract: Human respiratory syncytial virus (RSV) infection is the most important cause of acute lower respiratory tract infection in infants, neonates, and young children, even leading to hyperinflation and atelectasis. Oxymatrine (OMT), originating from natural herbs, possessed potential antivirus activity against influenza A virus, Coxsackie B3 virus, and RSV, whereas the absence of an in vivo study indicated the difficulties in overcoming the physiological obstacles. Since RSV basically replicated in lung tissue, in this study, we fabricated and characterized a chitosan (CS)-coated liposome with OMT loaded for the treatment of lethal RSV infection via inhalation. The results uncovered that OMT, as a hydrophilic drug, was liable to diffuse in the mucus layer and penetrate through the gas–blood barrier to enter systemic circulation quickly, which might restrict its inhibitory effect on RSV replication. The CS-coated liposome enhanced the distribution and retention of OMT in lung tissue without restriction from mucus, which contributed to the improved alleviative effect of OMT on lethal RSV-infected mice. Overall, this study provides a novel inhalation therapy for RSV infection, and the CS-coated liposome might be a potential inhalable nanocarrier for hydrophilic drugs to prevent pulmonary infections.

Keywords: oxymatrine; RSV; chitosan; inhalation; liposomes

1. Introduction

Human respiratory syncytial virus (RSV) is a human-originated orthopneumovirus that circulates in winter and spring worldwide. RSV presents a high tropism to the pulmonary system of preterm infants, neonates, and young children, it being the number-one cause of acute lower respiratory tract infection (ALRTI), i.e., pneumonia or bronchiolitis, and severe conditions such as asthma [1]. RSV infection is also a significant cause of morbidity and mortality in the elderly, immunocompromised individuals, and patients with chronic underlying illnesses [2]. Moreover, the specific mechanism for the pulmonary disease is poorly dissected; direct viral cytotoxic injury companied with the pathogenesis of RSV was thought to be a main cause for necrosis of the epithelial cells of the small airways and systemic extrinsic apoptosis of $CD4^+$ and $CD8^+$ T cells [3,4]. Consequently, such sloughed cells and mucus plug the bronchioles, and thereafter, hyperinflation and atelectasis occur [5].

Matrine-type alkaloids are a group of monomeric compounds that were isolated from *Sophora flavescens* Ait (*S. flavescens* Ait, "Kushen" in Chinese) by a Japanese group in 1895 [6]. They are well known, with a wide range of pharmacological effects such as antiviral, antioxidant, antibacterial, antitumor, antifibrotic, and insecticidal properties [7]. Oxymatrine

(OMT) is the representative alkaloid of *S. flavescens* Ait, as investigated in several reports, including against the influenza A virus (IAV), by inactivating the promoters of *TLR3*, *TLR4*, *TLR7*, *MyD88*, and *TRAF6* of host cells and deregulating IAV infection-induced activations of NF-κB, Akt, ERK1/2, and p38 MAPK pathways [8], and against the Coxsackie B3 virus by ameliorating myocarditis in vivo described by our group [9]. Impressively, Ma et al. presented the anti-RSV infection effect of OMT in vitro, which preliminarily demonstrated the anti-RSV activity with promising selectivity index in comparison with ribavirin [10]; however, there is an absence of an in vivo studies. Effective in vivo delivery of OMT to pulmonary infection lesions might be a promising strategy to alleviate RSV-infected mice.

Pulmonary administration has the priority in the treatments of pulmonary infectious diseases, due to it being free from first-pass effect and direct delivery of a loaded drug to lung lesions [11]. However, challenges remain in the pulmonary delivery of therapeutics to infected regions. OMT must overcome numerous physiologic barriers, including enzymatic degradation, mucociliary clearance, pulmonary surfactant absorption and macrophage phagocytosis to achieve pulmonary accumulation [12–14]. In the exploration of inhalable drug delivery system against infectious lung diseases, liposomes with great biocompatibility and wide applicability hold a strong promise in improving the stability and cellular uptake of loaded drugs in the respiratory tree, especially for hydrophilic drugs [15,16]. OMT liposomes have been reported for several decades, and they were mainly applied on liver diseases, including acute liver injury [17], liver fibrosis [18], and hepatocellular carcinoma [19]. In addition, OMT liposomes were applied in the treatment of ulcerative colitis and intervertebral disc degeneration as well [20,21]. In our previous work, OMT liposomal preparations have been developed for the treatment of emphysema via pulmonary administration [22]. Although carboxymethyl chitosan (CMCS) coating slightly enhanced the pulmonary retention of positively charged liposomes, the fast systemic distribution of OMT liposomes, regardless of the CMCS coating, indicated the necessity for further optimization of the prescription.

Enlightened by the constitution of endogenous pulmonary surfactants, we prepared OMT liposomes (Lipo/OMT) by 1,2-dipalmitoyl-sn-glycero-3-phosphocholine (DPPC), hydrogenated soybean phosphatidylcholine (HSPC), and 1,2-dipalmitoyl-sn-glycero-3-phospho-(1′-rac-glycerol) (DPPG) in this study instead of soyabean lecithin, in order to improve its stability in vivo. To further enhance the pulmonary delivery, a surface modification strategy was used to prevent the degradation and immune recognition of liposomes and facilitate the mucus penetration [23]. 1,2-distearoyl-sn-glycero-3-phosphoethanolamine-N-methoxy (polyethylene glycol) (DSPE-PEG) has been widely used to escape from immune surveillance and facilitate mucus penetration via hydration shell, whereas as a double-edged sword, PEGylation hindered the interactions with cells at the same time [24]. Given that chitosan (CS) has been widely used in the construction of nanocarrier for pulmonary delivery as both matrix and shell due to its high biocompatibility and biodegradability [25,26], it is reasonable to circumvent the "PEG dilemma" via coating of CS on the periphery of PEGylated OMT liposomes. Both high-molecular-weight (30 kDa) and low-molecular-weight (800~1000 Da) CS were adopted to prepare the surface-coated liposomes (Lipo/OMT@CS). The optimal type of CS and the weight ratio of CS/liposome was optimized in terms of the characteristics of liposomal preparations. Both Lipo/OMT and Lipo/OMT@CS were used for the treatment of RSV-infected mice via inhalation.

In this study, we developed inhalable CS-coated OMT liposomes for the treatment of RSV-infected mice. The optimal CS-coated OMT liposomes were evaluated in vitro and applied on the RSV-infected mice model in which the anti-RSV mechanisms of OMT might be attributed to its preventive effects on RSV (Figure 1). This study provides an effective liposomal platform for pulmonary delivery of hydrophilic therapeutics against viral infection.

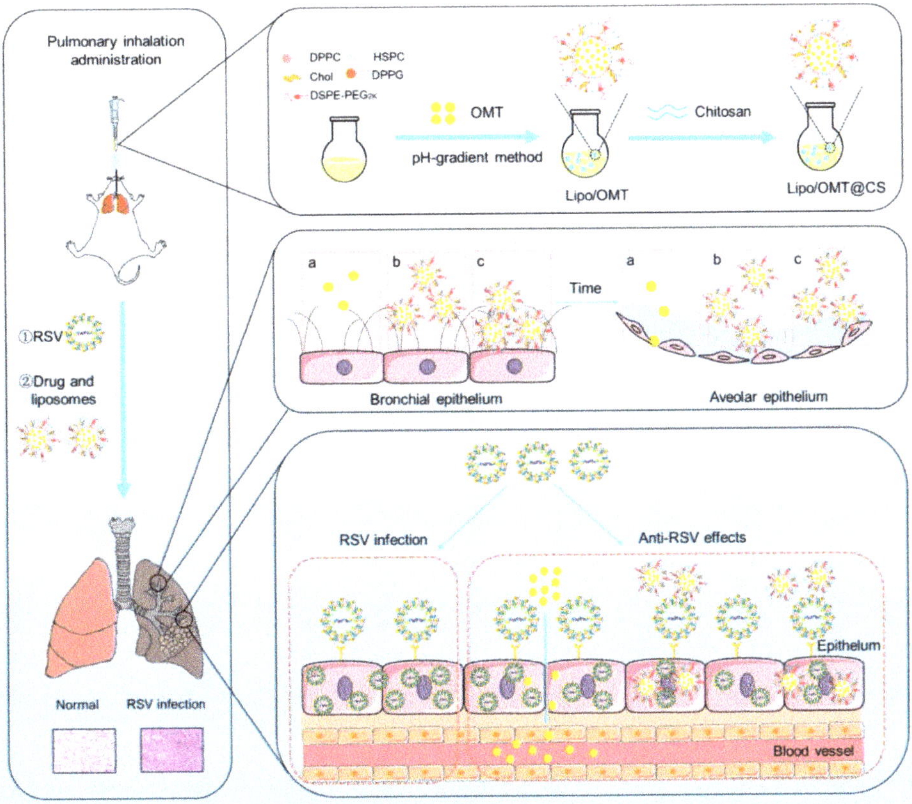

Figure 1. Schematic illustration of the preparation process of CS-coated OMT liposomes, the distribution after inhalation and the inhibitory effect on RSV infection (a: free OMT, b: Lipo/OMT, c: Lipo/OMT@CS).

2. Results and Discussion

2.1. Inhibitory Effects of OMT on RSV-Infected HEp-2 Cell

Based on the previous report that related to the inhibitory effect of OMT on RSV in vitro. Above all, the cytotoxicity of OMT and ribavirin on HEp-2 cells were evaluated (Figure S1), and the non-cytotoxic concentrations of OMT and ribavirin were used in the following study. We evaluated the inhibitory effect of OMT and ribavirin on RSV from 4 aspects. They were virus biosynthesis (Figure 2A), virus penetration (Figure 2B), virus attachment (Figure 2C), and drug pre-treatment (Figure 2D). Since RSV was capable to enter cells quickly, it was incubated with HEp-2 cells, at 37 °C, for 1 h when no drug was presented in the medium (Figure 2A,D). In the study of virus penetration, RSV was pre-incubated with HEp-2 cells, at 4 °C, for 2 h to achieve sufficient binding before penetration (Figure 2B). In terms of inhibition on the virus attachment, RSV was mixed with drugs ahead of co-incubation with HEp-2 cells, and thus, a longer incubation time of 2 h at 37 °C was used (Figure 2C). Based on the $TCID_{50}$ assay result, measurements of 150 μg/mL, 75 μg/mL, 37.5 μg/mL, and 18.75 μg/mL OMT were found to significantly decrease viral titers and viral gene copies indicating the intervention effects of OMT on the virus biosynthesis (Figure 2A). Ribavirin showed more excellent anti-RSV effect than OMT as virus biosynthesis inhibitor. According to the $TCID_{50}$ result, 150 μg/mL, 75 μg/mL, and 37.5 μg/mL OMT significantly blocked the penetration of virus into cells (Figure 2B), which was confirmed by the result of viral gene copies. Ribavirin showed more excellent anti-RSV effect than OMT as a virus penetration inhibitor. In the study of the virus attachment, OMT and ribavirin showed weak anti-RSV effect. No statistical differences in both OMT and ribavirin was observed in terms of viral gene copies (Figure 2C). After pre-treatment of OMT, the viral titer and gene copies were slightly reduced, which was inferior to the inhibitory effects of ribavirin (Figure 2D). Taken together, OMT displayed different ways to alleviate RSV infection in HEp-2 cells. RSV replication was restrained by OMT especially via inhibiting virus biosynthesis and virus penetration. Taken together, OMT displayed different ways to alleviate RSV infection in HEp-2 cells. RSV replication was restrained by OMT, especially via inhibiting virus biosynthesis and virus penetration.

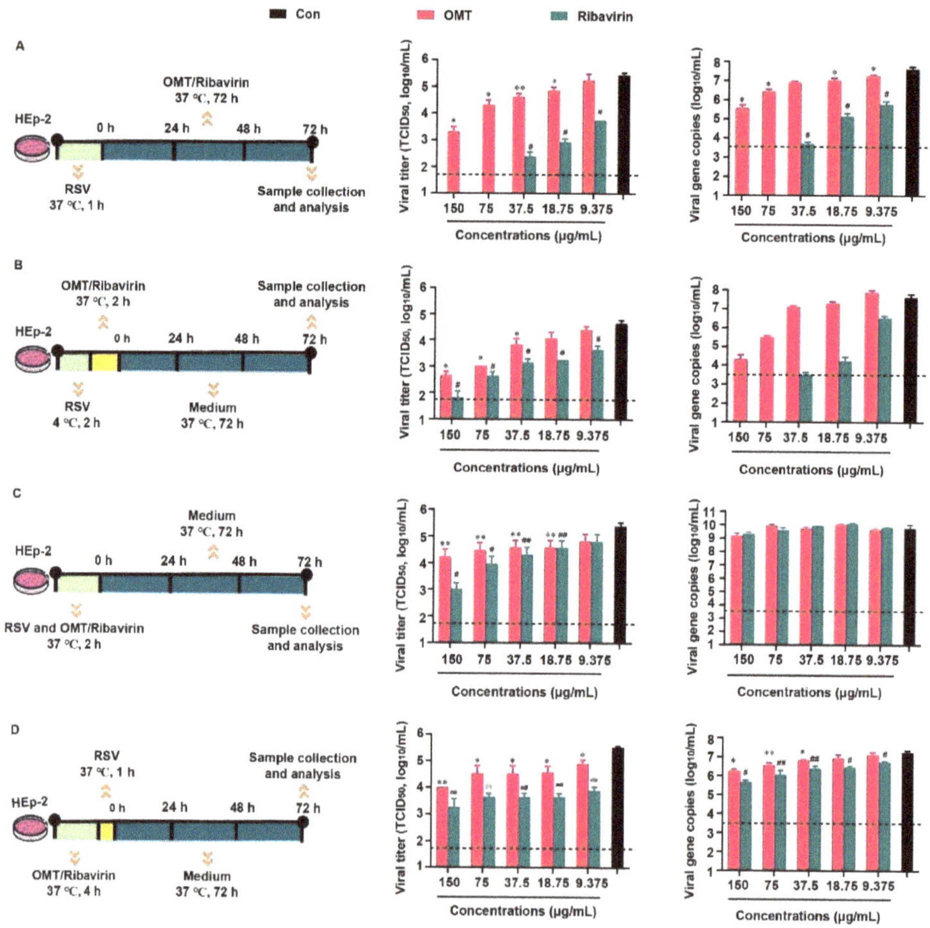

Figure 2. Inhibitory effects of OMT on RSV infection and replication in HEp-2 cell. The experimental scheme, $TCID_{50}$ and qRT-PCR result of (**A**) virus biosynthesis assay, (**B**) virus penetration assay, (**C**) virus attachment assay, and (**D**) drug pre-treatment assay. All the data were expressed as mean + SD ($n = 3$). * and ** represent OMT compared with virus control (black bar) as $p < 0.05$ and $p < 0.01$; # and ## represent ribavirin compared with virus control (black bar) as $p < 0.05$ and $p < 0.01$. The dash line indicates the limit of detection in $TCID_{50}$ assay and qRT-PCR assay, respectively. Con: virus control.

2.2. Preparation and Characterization of CS-Coated OMT Liposomes

In our previous reports, we prepared a CMCS-coated OMT liposomes via a pH-gradient method that constituted soybean phospholipid (SPC), cholesterol (Chol), polyoxyethylene stearate (SPEG) and octadecyl amine (ODA) for pulmonary emphysema treatment [22]. However, a fast release of OMT was detected regardless of CMCS coating. The instability of SPC was proposed to contribute to the leakage of OMT, and the usage of saturated lipids might be a solution [17,27]. Therefore, the OMT liposomes prepared by SPC, DPPC/SPC and DPPC/HSPC were formulated and characterized (Figure S2). The particle size of liposome increased along with the percentage of saturated lipids-DPPC and HSPC. An obviously retarded drug release was observed in the OMT liposomes prepared by DPPC/HSPC. Hence, DPPC/HSPC/DPPG/Chol/DSPE-PEG at the weight ratio of 20:1:5:10:1 was used to prepare Lipo/OMT and used in the following studies.

CS coating was suggested to confer muco-adhesivity to nanoparticles and mediate cellular uptake [28–31]. Both high-molecular-weight CS (HCS) and low-molecular-weight CS (LCS) were coated on the liposomes at various weight ratios (Figures S3 and S4). As a matter of fact, HCS showed a strong combination with liposomes, leading to increased particle size. A large amount of precipitation and a liquid–solid interface were observed after standing still for 4 h until the HCS:liposome weight ratio reached 2.5:1. Intriguingly, the particle size of LCS-coated liposomes reached the highest particle size (337 nm) at a LCS:liposome ratio of 1:10 (*w*/*w*) and sharply decreased to 234 nm at 1:5 (*w*/*w*), and then a slightly variation was observed. Given that the high ratio of HCS resulted in a high viscosity of the solution that failed in effective inhalation, the LCS was used in the following study. The zeta potential was decreased until the LSC:liposome ratio reached 1:2.5 (*w/w*), and a larger SD value was measured at a higher LSC:liposome ratio, which suggested that LCS reached saturated adsorption on liposomes. Therefore, LCS was coated on the surface of Lipo/OMT at the weight ratio of 1:2.5 to receive Lipo/OMT@CS.

The physiochemical properties of OMT liposomal preparations were characterized and the stability was evaluated for one week (Figure 3). Lipo/OMT and Lipo/OMT@CS are 287.28 $\pm$ 2.02 nm and 246.16 $\pm$ 3.03 nm, respectively (Figure 3A). Small PDI value indicated good polydispersity of OMT liposomes. After CS coating, the zeta potential of liposomes increases from -19.63 ± 2.92 mV to -14.54 ± 3.04 mV, which remains negative (Figure 3B). The morphology of OMT liposomes were observed by transmission electron microscope (TEM). Lipo/OMT possesses spheric morphology with a clear layer outside that divides the particle into two parts, whereas irregular particles without an interface were observed in the Lipo/OMT@CS sample (Figure 3D), confirming the successful coating of CS on the liposomes. The loading efficiency (LE) and drug loading capacity (DL) of OMT liposomes were approximate 40% and 6%, respectively (Figure 3C). Drug release behavior was studied in the pH 7.4 medium (Figure 3E). More than 80% of OMT was released from liposomes at 12 h, and CS slightly retarded drug release from 2 h. The stability of OMT liposomes was evaluated in terms of particle size and zeta potential, at 4 °C, for 7 days (Figure 3F,G). No obvious fluctuation on the size and zeta potential was detected, which indicated the good stability of OMT liposomal preparations in vitro. Collectively, the CS coating on the periphery of liposomes decreased the surface charge without affecting the stability, while it exhibited a limited effect on retarding the release of OMT, which was likely due to the huge intermolecular channels of CS.

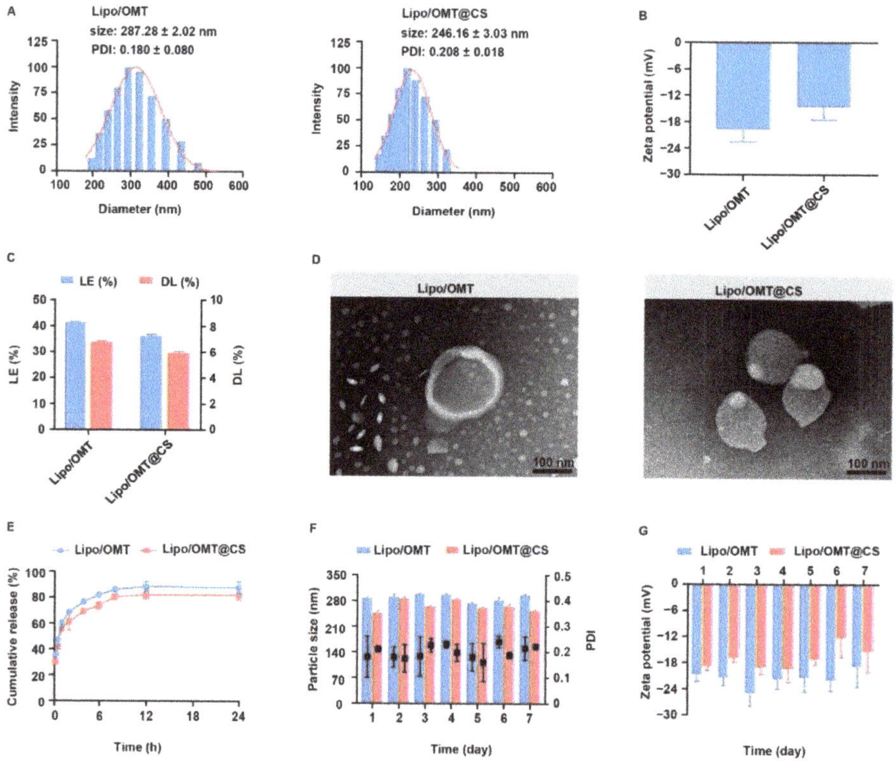

Figure 3. Characterization of OMT liposomal preparations. (**A**) Evolution of particle size, PDI and (**B**) zeta potential of OMT liposomal preparations. (**C**) Drug loading efficiency (LE) and drug loading capability (DL) of OMT liposomal preparations. (**D**) Representative TEM images of Lipo/OMT and Lipo/OMT@CS (scale bar 100 nm). (**E**) In vitro release behavior of OMT from liposomal preparations in pH 7.4 PBS. (**F**) The variation in particle size, PDI and (**G**) zeta potential of OMT liposomal preparations, at 4 °C, for 7 days. All the data were expressed as mean + SD (n = 3).

2.3. Stability and Penetration Behavior of CS-Coated Liposomes in Mucin Media

The high stability of OMT-loaded liposomes in pulmonary-relevant fluids is the prerequisite of subsequent cellular uptake and drug release by the epithelial cells, since the aggregation or dissociation of liposomes would seriously restrict the pharmacological effects of OMT. As mucin play an important role in the mucus, the size variation of OMT-loaded liposomes in mucin media was monitored. As shown in Figure 4A, a slight fluctuation on diameter was observed in Lipo/OMT within 48 h, while a sharp decrease in the diameter of Lipo/OMT@CS was detected after 12 h incubation with mucin, thereafter re-increasing at 24 h. CS has been reported to show muco-adhesivity via the interaction between its amino groups and sialic acid and sulfated sugar in the mucin chains [28,32,33]. It was reasonable to suggest that the negatively charged Lipo/OMT prevented the interference from mucin, and the CS coating mediated the electrical compression and adhesion from mucin. To investigate the interaction between mucin and liposomes, the absorbance of liposomes in mucin at 650 nm was detected (Figure 4B). A similar absorbance of mucin and mucin + OMT was observed. Upon incubation, the absorbance of liposomes + mucin approximately equals to the sum of the absorbance of liposomes and mucin, indicating the poor tendency in the interaction between liposomes and mucin regardless of the CS coating. Correspondingly, no variation in the zeta potential of the liposomes was exhibited after incubation with mucin (Figure 4C).

For further assessment on the potential diffusion of liposomes through the airway mucus, two in vitro models were used to evaluate the penetration of rhodamine B (RB) liposomal preparations and OMT liposomal preparations. A two-layer model that consisted of a mucin upper layer and a gelatin sublayer was used to directly evaluate the penetration capability of the hydrophilic drug and its liposomes (Figure 4D,E). Both the visual observation and the quantitation of the percentage of RB in gelatin demonstrated a fast penetration at the initial 0.5 h, and no obvious differences were detected among all the groups. Meanwhile, a transwell model consisted of upper chamber filled with mucin and lower chamber filled with PBS was used to evaluate the penetration capability of OMT liposomal preparations (Figure 4F). Approximately 50% OMT was detected in the well at 0.5 h, and 100% penetration was achieved at 12 h. Intriguingly, free OMT showed even faster penetration than its liposomal preparations. It is reasonable to suggest that the hydrophilic drug might not be restricted by the mucus layer after inhalation, whereas the three-dimensional network formed by mucin will be an obstacle to nanosized particles. Although several reports have been pointing to the advantage of PEGylation on the deep and fast penetration of nanoparticles across mucus via reduction in hydrophobic and electrostatic interactions with mucin and other proteins, the high negative potential and strong steric hindrance of PEG shell might impede the cellular uptake after mucus penetration [34,35]. Hence, LCS was used in this study to increase the interactions between OMT liposomes and targeted cells without alteration on the surface electrical property, since positively charged liposomes might strongly induce interactions with mucin to form aggregation. In addition, the different penetration percentage between the two models might result from the resistance of high concentration of gelatin.

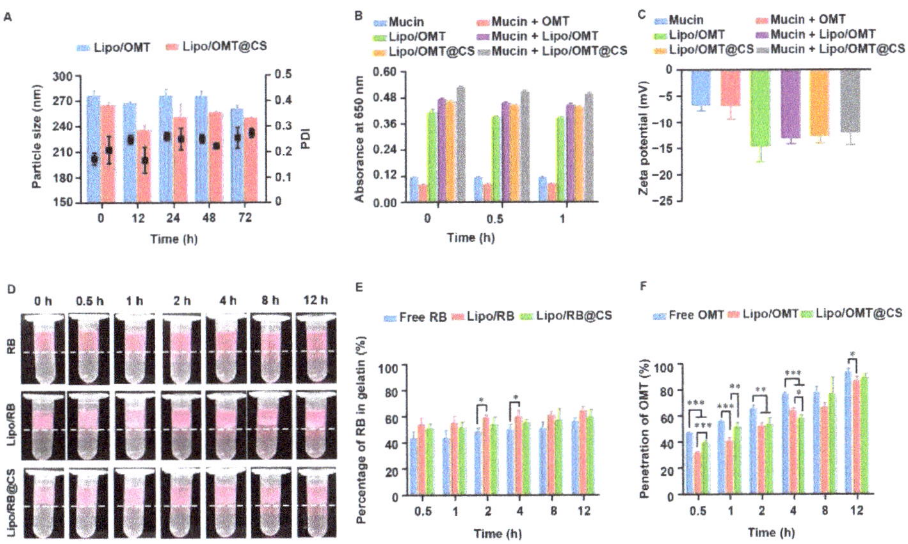

Figure 4. Stability and penetration capability of OMT/RB liposomal preparations in mucin media. (**A**) Particle size of OMT liposomal preparations in mucin. (**B**) Absorbance at 650 nm and (**C**) zeta potential of OMT liposomal preparations alone or in a mucin solution. (**D**) Visual observation of RB liposomes penetrations through the mucin layer at different time point (dotted line indicated the interface between mucin and gelatin layer). (**E**) The percentage of RB in gelatin layer. (**F**) Penetration percentage of OMT through mucin layer in transwells. All the data were expressed as mean + SD ($n = 3$). * $p < 0.05$, ** $p < 0.01$, *** $p < 0.001$.

2.4. Biodistribution of RB Liposomal Preparations after Inhalation

For lung infection treatment, pulmonary distribution and retention closely related to the therapeutic effects and systemic escape of drugs might lead to undesirable side effects. To discover the effects of liposomal carrier and CS coating on the biodistribution of hydrophilic drug, RB liposome (Lipo/RB) and CS-coated Lipo/RB (Lipo/RB@CS) were prepared, and the concentration of RB in lung, heart, liver, spleen, kidney, bronchoalveolar lavage fluid (BALF) and plasma were quantified at 0.5 h, 1 h, 2 h, 4 h and 8 h after inhalation (Figure 5). Half an hour after inhalation, the free RB group showed significantly higher concentration of RB in liver and kidney, and lower concentration in lung and BALF than that of the liposomal groups. Lipo/RB@CS showed the highest concentration of RB in both BALF and lung tissue within 1 h. No significant difference was detected between the Lipo/RB and Lipo/RB@CS after 2 h in lung. We can conclude that liposomes enhanced the accumulation of RB in lung, especially with CS coating.

Given that faster mucus penetration behavior of hydrophilic drug was observed in vitro (Figure 4), liposomal carrier and CS coating might contribute to the retention of RB in BALF. Furthermore, the higher accumulation of RB in lavaged lung of the Lipo/RB@CS group might result from the increased cellular uptake mediated by CS. The low BALF retention and high plasma concentration of RB in free RB group is closely related to fast diffusion of the hydrophilic drug though mucus layer, which might facilitate the penetration of RB through blood–gas barrier at the trachea and the alveolar sacs resulting in higher plasma concentration. As previously reported, the large surface area, highly permeable biomembrane, and good blood supply of the alveolar region are favorable for rapid absorption of drugs [36]. However, it is not favorable for the local treatments on pulmonary infection. CS-coated liposomes have the potential to increase the lung accumulation and reduce the systemic and extrapulmonary exposure of the hydrophilic drug, which provides the foundation for the alleviative effects of OMT on RSV-induced lung infection.

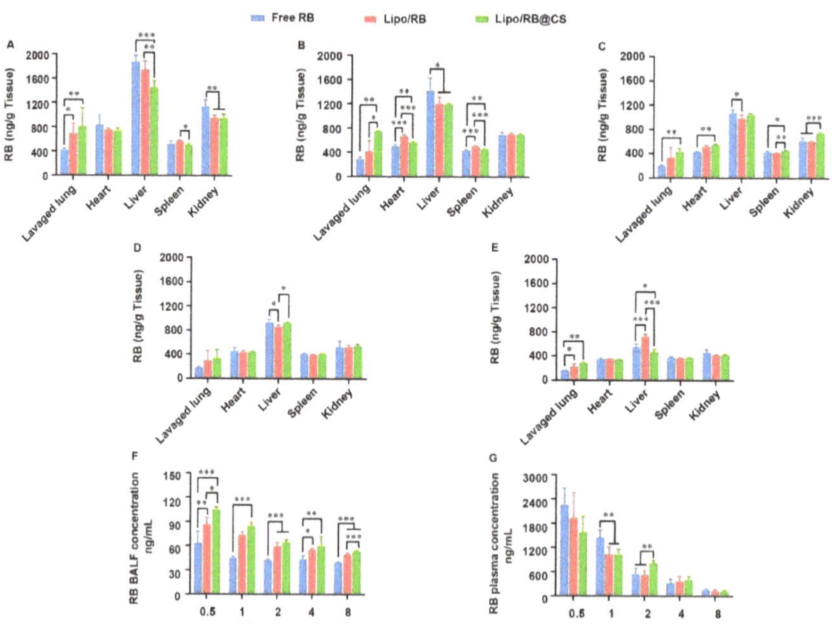

Figure 5. Biodistribution of RB in mice at (**A**) 0.5 h, (**B**) 1 h, (**C**) 2 h, (**D**) 4 h and (**E**) 8 h after inhalation of RB formulations and the concentration of RB in (**F**) BALF and (**G**) plasma. All the data were expressed as mean + SD ($n = 6$). * $p < 0.05$, ** $p < 0.01$, *** $p < 0.001$.

2.5. Alleviative Effects and Mechanisms of OMT Liposomes on RSV-Infected Mice

In order to evaluate the potential alleviative effects of OMT liposomal preparations in vivo, we introduced the lethal RSV infection mice model developed by our group [37]. As shown in Figure 6A, except for the Mock mice, the rest of the mice received a 300 μL GZ08-18 inoculation via inhalation within 2 days to establish the lethal RSV infection model. In the meantime, Blank Carrier mice were given CS-coated liposome. Free OMT mice were given 5 mg/kg or 20 mg/kg OMT solution as low dosage group and high dosage group, respectively. Lipo/OMT and Lipo/OMT@CS mice were inoculated with OMT liposomal preparations at an equivalent OMT dosage of 5 mg/kg. Mock and Model mice were given saline instead of drugs. The treatments were all conducted intra tracheal once per day for 5 consecutive days. The sampling of lungs for virus titration, hematoxylin and eosin (H&E) staining, and immunofluorescence assay was implemented at the 2nd day post-inoculation (DPI), and the rest of mice were kept till 21 DPI for daily weight and survival recording. The safety of OMT liposomes and blank carriers on the normal mice was implemented before the treatment. No mortality was observed after treatment by the OMT liposomes and blank carriers followed the same schedule in Figure 6A without RSV infection.

Combining the results of Figure 6B,C, Blank Carrier mice showed the worst inhibitory effect against the virus challenge, indicated by the loss of weight from −4 to 5 DPI, and the entire group died off at 6 DPI. However, two of nine Model mice survived the RSV lethal infection. The rest of the groups all had survivor (survivors) and weight curves that demonstrated the recovery after sharp declines which almost happened from 2 to 6 DPI. The survival rates of the Model, Blank Carrier, Free OMT (High), Free OMT (Low), Lipo/OMT, and Lipo/OMT@CS groups were 22.22% (2/9), 0 (0/9), 22.22% (2/9), 11.11 (1/9), 27.27% (3/11), and 37.5% (3/8), respectively. Compared with the Blank Carrier group, the Free OMT, Lipo/OMT, and Lipo/OMT@CS showed a protective effect against lethal RSV infection to mice, especially for Lipo/OMT@CS. However, no statistical difference was shown between Lipo/OMT@CS and other groups. Although a higher mortality and weight loss in the Blank Carrier group was observed than that in the Model group, fewer than three survivors in the Model, Blank carrier and Free OMT groups might be seen as a survivorship bias. Hence, a larger sample size is necessary in future studies.

Furthermore, the H&E staining confirmed that Model and Blank Carrier mice suffered from severe acute viral pneumonia indicated by alveolar structure destruction, leukocytes and cytokines infiltrations, congestion and edema of small vessels, etc., (Figure 6D). Although partial destruction of alveolar structure and enhanced thickness of alveolar wall were observed in free OMT, Lipo/OMT, and Lipo/OMT@CS groups, the basic alveolar structure was maintained, which was essential for O_2/CO_2 exchange, and fewer leukocytes and cytokines infiltrations were shown, resulting in a lower possibility of fatal cytokine storm. The H&E staining slides were further evaluated by pathologists (Figure 6E) according to the inflammation and tissue damage scoring system as previously reported [38]. In addition, the qRT-PCR results presented the significantly lower viral gene copies in BALF samples of Lipo/OMT@CS mice in comparison with Model and Blank Carrier. The Free OMT (Low) group showed significantly decreased viral gene copies compared with Model group as well. However, no significant difference was observed between Lipo/OMT@CS and other OMT preparations (Figure 6F).

Since F protein is one of viral membrane proteins which fuse the virus and cell membrane or infected cell membranes together, the expression of F protein demonstrates the attachment and entry capability of the virus. Hence, the immunofluorescence assay was conducted to show the expression of RSV F protein (Figure 7A). A huge amount of viral protein was found in the lumen of lung tissue in Model mice. The pulmonary lumen and interstitial were filled with RSV F protein in Blank Carrier mice as well. Similarly to Blank Carrier, scatter distribution of RSV F protein was also found in free OMT groups. However, less viral protein was shown in Lipo/OMT, especially in Lipo/OMT@CS mice. Compared with the results of gene copies (Figure 6F), a remarkably low level of viral protein was found in Lipo/OMT@CS, it being the lowest level among all the groups ($p < 0.001$, Figure 7B). In accord with the in vitro anti-RSV effects on HEp-2 cells (Figure 2), the main therapeutic mechanisms of OMT on RSV-infection might be attributed to its potential inhibitory effects on virus penetration and biosynthesis. Collectively, the alleviative effects of OMT on the lethal RSV infection of mice were endorsed, and Lipo/OMT@CS improved the preventive and therapeutic effects of OMT to some extent. The development and optimization of more effective inhalable liposomal carriers for pulmonary delivery of OMT is highly in need.

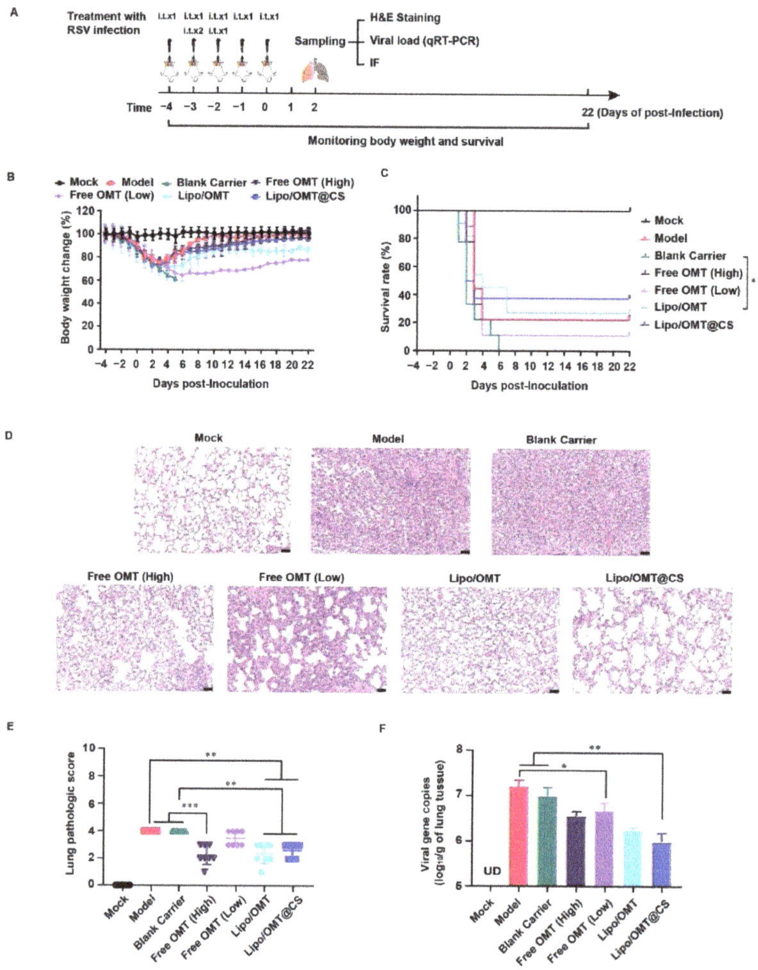

Figure 6. Alleviative effect of OMT liposomal preparations on RSV-infected mice. (**A**) Schematic illustration for the evaluation of alleviative effect of OMT liposomal preparations on RSV-infected mice. (**B**,**C**) Body weight change and survival rate of Mock (n = 8 at beginning), Model (n = 9 at beginning), Blank Carrier (n = 9 at beginning), Free OMT (High) (n = 9 at beginning), Free OMT (Low) (n = 9 at beginning), Lipo/OMT (n = 11 at beginning), and Lipo/OMT@CS (n = 8 at beginning). The timepoint of 22 DPI was set as end point of recording. Data for body weight change are expressed as mean ± SD. (**D**) H&E staining of pathologic sections of lung tissues for each group at 2 DPI (scale bar = 50 μm). (**E**) The lung pathologic scores basic on the H&E staining were double-blinded and accessed by anonymous pathologists and statistically processed. Data are expressed as mean ± SD (n = 8). (**F**) The qRT-PCR results for viral gene copies of lung BALF samples. Data are expressed as mean + SD (n = 3). UD: under detectable. * $p < 0.05$, ** $p < 0.01$, *** $p < 0.001$.

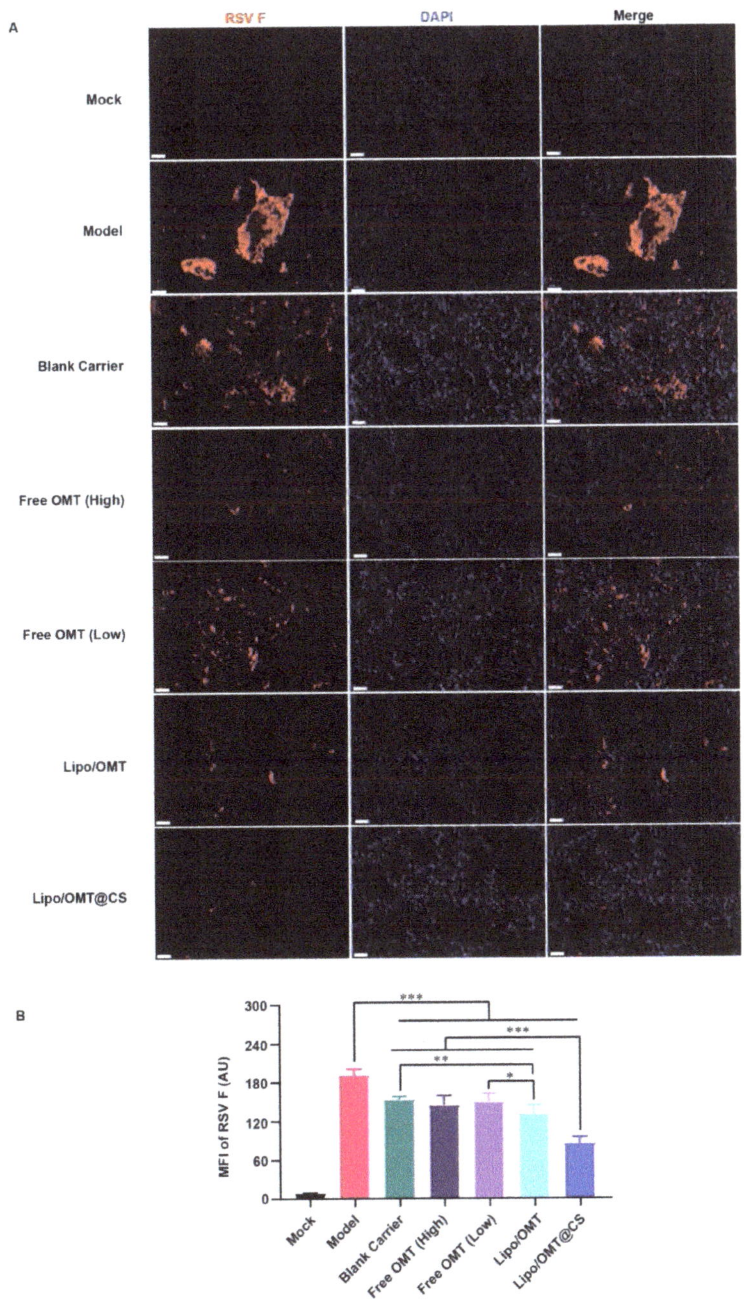

Figure 7. Immunofluorescence assay for OMT liposomal preparations on RSV-infected mice. (**A**) The representative images of each group were captured by confocal microscope (Scale bar = 50 μm). (**B**) Statistical result for the mean fluorescence intensity (MFI) of RSV F that calculated by Image J software from 8 equal sized images. All the data were expressed as mean + SD ($n = 8$). * $p < 0.05$, ** $p < 0.01$, *** $p < 0.001$.

3. Materials and Methods

3.1. Materials and Reagents

Oxymatrine (OMT, purity ≥ 98%, lot no. C10572744), rhodamine B (RB, purity ≥ 99%, lot no. C10006604), and chitosan (CS, 800~1000 Da, lot no. C13207154) were offered by Macklin Biochemical Co., Ltd. (Shanghai, China). Cholesterol (Chol, purity ≥ 92.5%, lot no. WXBB1238) was purchased from VetecTM reagent grade (Shanghai, China). 1,2-dipalmitoyl-sn-glycero-3-phosphocholine (DPPC, lot no. CC0657) was provided by Corden pharma Switzerland LLC (Liestal, Switzerland). Hydrogenated soybean phosphatidylcholine (HSPC, lot no. B60455) and 1,2-dipalmitoyl-sn-glycero-3-phospho-(1′-rac-glycerol) (DPPG, lot no. B90497) were supported by A.V.T Pharmaceutical Co., Ltd. (Shanghai, China). 1,2-distearoyl-sn-glycero-3-phosphoethanolamine-N-methoxy (polyethylene glycol) (DSPE-PEG_{2K}, lot no. RS0200412) was purchased from Xi'an ruixi Biological Technology Co., Ltd. (Xi'an, China).

3.2. Cells, Virus, and Animals

HEp-2 cells (CCL-23, ATCC, United States) were cultured under 5% fetal bovine serum (FBS) and 1% penicillin/streptomycin (P/S) supplemented with DMEM/F12 GlutaMax-I (10565, Gibco Ltd., Grand Island, NY, USA). GZ08-18, a highly virulent mouse-adapted RSV strain, was developed and stocked in our lab [37], and it was cultured under 2% FBS serum supplemented with DMEM/F12 GlutaMax-I. The cultured GZ08-18 was stock under −80 °C and viral titer was titrated by $TCID_{50}$ assay and calculated by Spearman–Karber method.

Eight-month-old retired breeder BALB/c female mice (25–35 g) were supplied by the Experimental Animal Center of Guizhou Medical University. All animal care and experiments have been approved by the Animal Welfare and Ethics Committee of Guizhou Medical University (No: 1800020). Mice were raised at room temperature (23 ± 2 °C) and constant humidity (45 ± 10%) in an SPF condition. One week before starting experiments, animals were placed for adaptation.

3.3. Effects of OMT on Coculture of RSV and HEp-2 Cell

To evaluate the intervention of OMT on virus biosynthesis, GZ08-18 dilution with a multiplicity of infection (MOI) of 0.05 was added to HEp-2 cells for 1 h at 37 °C, then we discarded the virus dilution, washed the cells once with PBS, removed unadsorbed viruses, and added cell maintenance solutions containing different concentrations of OMT to culture for 72 h. To evaluate the inhibitory effects of OMT on virus penetration, GZ08-18 dilution with MOI of 0.05 was added to Hep-2 cells and cultured, at 4 °C, for 2 h, then we discarded the virus solution, washed the cells once with pre-cooled PBS, added different concentrations of OMT solutions, and cultured at 37 °C for 2 h. The OMT solution was discarded, and the cell maintenance solution was added to incubate cells for 72 h. To evaluate the inhibitory effects of OMT on virus attachment, pre-cooled OMT solutions and GZ08-18 diluent with MOI of 0.05 were added to Hep-2 cells, for 2 h, at 4 °C, then we discarded the OMT solution and virus diluent, washed the cells once with PBS, and removed unabsorbed OMT/virus. The cell maintenance solution was added to the cell and cultured for 72 h. To evaluate the preventive protective effects of OMT, different concentrations of OMT were added to the cells, incubated at 37 °C for 4 h, then we washed the cells once with PBS, added GZ08-18 dilution with MOI 0.05 for 1 h, at 37 °C, and then we washed the cells with PBS once and added the cell maintenance solution to incubate cells for 72 h. Meanwhile, ribavirin was used as the positive drug to coculture with virus and HEp-2 cells according to above experiments as the substitution for OMT.

3.4. Preparation of CS-Coated OMT Liposomal Preparations

OMT liposomal preparations were prepared using the pH-gradient method as reported previously [19]. In brief, DPPC, HSPC, DPPG, Chol, and DPEG-PEG2K were dissolved in a mixture of methanol and trichloromethane at a mass of 20:15:5:10:1 and transferred to a round bottom flask. Then, the solvent was removed by rotary evaporation (RE52CS, Shanghai Yarong biochemical instrument, Shanghai, China) to form thin films. The lipid film was resuspended with 0.15 mol/L citric acid solution (pH 1.9), and then incubated in water bath for 30 min, at 50 °C. Next, film was hydrated using a water bath ultrasonic apparatus for 2 min. The obtained liposomes were filtered through 0.45 μm filter. Additionally, the pH was then adjusted to 8.0 by sodium carbonate solution (0.3 mol/L), followed by coculture of OMT and liposomes for 10 min, at 50 °C, which were then cooled with cold water immediately. For the preparation of Lipo/OMT@CS, CS was coated on liposomal surface through electrostatic adsorption. Appropriate amount of CS (100 mg/mL) was dissolved under stirring in distilled water and added dropwise to the freshly prepared Lipo/OMT solution under stirring at 500 rpm. Ultimately, the free OMT was removed via dialysis against PBS to obtain Lipo/OMT and CS-coated OMT liposomal preparations.

The RB-encapsulated liposomes were prepared via a film dispersion method. Briefly, a thin lipid film was cocultured with 4 mL RB solution at 0.5 mg/mL and ultrasonicated. The Lipo/RB was obtained after passing through 0.45 μm filter. Lipo/RB@CS were obtained after modification of CS via the same methods described above.

3.5. Characterization of OMT Liposomal Preparations

The morphology of Lipo/OMT and Lipo/OMT@CS were observed by transmission electron microscope (TEM, Tecnai 12, Philips, Holland, Amsterdam, The Netherlands). The particle diameter, polydispersity (PDI), and zeta potential of Lipo/OMT and Lipo/OMT@CS were measured by NanoBrook 90Plus PALS (Brookhaven, GA, USA). The concentration of OMT in liposomal preparations was measured by high performance liquid chromatography (HPLC), which used Shimadzu LC-16 (Suzhou, China) with a UV detector (SPD-16, Shimadzu, Suzhou, China) using a reversed phase column (Hanbon, 4.6 × 250 mm, 5 μm, Huaian, China), at 30 °C. A mobile phase consisted of acetonitrile and 0.2% triethylamine (8: 92, *v/v*) at a flow rate of 1 mL/min and detecting at the wavelength of 210 nm. The loading efficiency (LE) and drug loading capacity (DL) were calculated according to Equations (1) and (2).

$$\text{LE}\ (\%) = \frac{W_{\text{drug loaded}}}{W_{\text{total Liposome}}} \times 100 \tag{1}$$

$$\text{DL}\ (\%) = \frac{W_{\text{drug loaded}}}{W_{\text{drug added}}} \times 100 \tag{2}$$

where $W_{\text{drug loaded}}$ is the drug encapsulated into liposomes, $W_{\text{total liposome}}$ is the mass of liposome with drug encapsulated, and $W_{\text{drug added}}$ is the drug added in the preparation of liposomes.

In vitro release profile was measured using the dialysis method. Dialysis bag (MWCO, 14 KDa) holding 1 mL OMT liposomal preparations were immersed in 15 mL of PBS in a capped bottle and incubated, at 37 °C, with stirring at a speed of 50 rpm. At predetermined time point, 500 μL medium was withdrawn and replenished with fresh PBS at the same volume. The amount of OMT released into the medium was determined using the HPLC method described above. To determine the stability of formulations, OMT liposomal preparations were kept in an incubator, at 4 °C, for one week. The particle size, PDI and zeta potential were measured over the storage periods.

3.6. In Vitro Assessment of the Interactions between Liposomal Preparations and Mucus

3.6.1. Stability of OMT Liposomal Preparations in Mucus

Stability of OMT liposomal preparations in mucin solution was evaluated. Mucin powder was dispersed in PBS and stirred overnight, then the dispersion was centrifuged at 3500× *g* for 10 min to receive a 0.08% (*w/v*) mucin supernatant solution. For the analyses, 200 μL of OMT liposomal preparations were added to 2 mL of medium and incubated, at 37 °C, for 72 h. At selected time intervals, the particle size of OMT liposomal preparations were monitored. The variation on particle size were considered as indicative of disassembly/aggregation state.

3.6.2. Mucoadhesive Tendency

The mucoadhesive tendency of OMT liposomal preparations were assessed by turbidimetric measurement as previously reported [39,40]. Briefly, mucin solutions (0.08%, *w/v*) and OMT liposomal preparations were mixed at volume ratio of 1:1 and vortexed for 1 min. The absorbance at 650 nm of the mixtures was measured at 0, 30 and 60 min after incubation, at 37 °C, by the ultraviolet spectrophotometer (UV-2700, Shimadzu, Japan). The mucin solution and the liposomal preparations dispersions in PBS were analyzed as controls. For the measurement of zeta potential, the mucin solutions (0.08%, *w/v*) and mixture samples were diluted in PBS (1:4, *v/v*) and measured.

3.6.3. Penetration of RB Liposomal Preparations through Mucus

Mucus penetration of RB-encapsulated liposomal preparations were evaluated by the in vitro mucus model [40,41]. One milliliter of 10% (*w/v*) hot gelatin solution was prepared, placed into a centrifuge tube, hardened at room temperature, and placed in a refrigerator, at 4 °C, until use. For the mucus penetration study, 0.5 mL of mucin solution (0.08%, *w/v*) was added above the gelatin solution and 125 μL of free RB solution and RB liposomal preparations were added on the mucus layer and maintained, at 37 °C. After 0.5 h, 1 h, 2 h, 4 h, 8 h, and 12 h incubation, the mucus layer was transferred to 1 mL volumetric flask. RB in the mucus layer was extracted with methanol and measured by a fluorescence spectrophotometer (Cary Eclipse, Varian, USA) at $\lambda_{Ex}/\lambda_{Em}$ = 550/580 nm. The percentage of RB penetrated through mucus was calculated according to Equation (3).

$$Penetration\ percentage\ (\%) = \frac{Drug_{\mathrm{added}} - Drug_{\mathrm{mucus}}}{Drug_{\mathrm{added}}} \times 100\% \quad (3)$$

3.6.4. Transwell Penetration of OMT Liposomal Preparations through Mucus

The transwell multiwall plates were applied to evaluate the penetration capacity of OMT liposomal preparations across mucus layer [42]. Briefly, 150 μL of mucin solution (0.08%, *w/v*) was transferred to the insert of transwell (polycarbonate membrane, 3 μm, 0.33 cm^2). Then, 100 μL of free OMT and OMT liposomal preparations were gently added above the mucus layer. The insert of transwell was immersed in the acceptor chambers that were filled with 1 mL of PBS and incubated, at 37 °C. At the scheduled time intervals, 100 μL of samples were collected from acceptor chambers, and an equal volume of fresh PBS were replenished. The amount of OMT in the acceptor chambers was determined using the HPLC method, and the percentage of OMT penetrated through mucus was calculated according to Equation (3).

3.7. In Vivo Biodistribution of RB Liposomal Preparations

The distribution of RB liposomal preparations in organs were investigated according to the method reported previously [22]. Tracheal intubation (22G surflo®. I. V. Catheter, Terumo Corp, Biñan, Philippines) was used for the inhalation administration of free RB and RB liposomal preparations. Briefly, at specific time points, mice were anesthetized by isoflurane via inhalation anesthesia. Free RB solution and RB liposomal preparations were inhalation administered at an equivalent RB dose of 0.75 mg/kg, respectively. At 0.5 h, 1 h, 2 h, 4 h and 8 h after treatment and under anesthesia, blood samples were taken from the aorta abdominalis, and the plasma was obtained by centrifugation (3000 rpm, 10 min) and stored, at 4 °C, until analysis. After the euthanasia treatment, BALF and organs (heart, liver, spleen, lung, kidney) were obtained. The tissues were homogenized in PBS. Afterwards, ethanol was added to the plasma, BALF and organ homogenates at certain ratio to extract

RB. Thereafter, the samples were centrifugated at 3000× g for 10 min, and the supernatants were used to measure the concentration of RB via a fluorescence spectrophotometer.

3.8. Treatments on RSV-Infected Mice

Based on lethal RSV-infected mouse models established by our laboratory as one regimen, 100 μL GZ08-18 culture medium (containing 1×10^{10}TCID$_{50}$ virus) was inoculated into mice via trachea, and three regimen (with 10–12 h interval) should be inoculated into pulmonary system of mice to establish the lethal RSV-infected mouse model. A total of 63 BALB/c mice were involved in the body weight change and survival experiments. They were randomly divided into 7 groups (n = 8–11 for each group): Mock, Model (virus control), Blank Carrier (blank preparation plus virus), Free OMT (High) (free OMT 20 mg/kg plus virus), Free OMT (Low) (free OMT 5 mg/kg plus virus), Lipo/OMT (5 mg/kg Lipo/OMT plus virus), and Lipo/OMT@CS (5 mg/kg Lipo/OMT@CS plus virus). The Mock mice were administered with saline as negative control. The Model mice were administered with saline plus virus. The Blank Carrier mice were administered with Lipo@CS solution.

The schematic experiment arrangement for exploration of the alleviated effect of OMT liposomal preparations was illustrated in Figure 6A. Briefly, the Mock, Model, Blank Carrier, Free OMT (High), Free OMT (Low), Lipo/OMT, and Lipo/OMT@CS were received saline, free OMT, or OMT liposomal preparations treatments for 6 consecutive days. All the groups except the Mock were infected with 300 μL GZ08-18 for 2 consecutive days. We define the day after last inoculation as the 1st day post-inoculation (1 DPI). The body weight and survival of each mouse were documented daily till 22 DPI.

3.9. Analysis of Bronchoalveolar Lavage Fluid (BALF)

At 2 DPI, the mice were properly anaesthetized and euthanized. Subsequently, the right lobe of lung was fully perfused with 800 μL DMEM/F12 GlutaMax-I (containing Vancomycin 20 μg/mL, Ciprofloxacin 20 μg/mL, Amikacin 0.05 μg/mL, Nystatin 50 μg/mL, and P/S 1%), and the BALF was obtained. After centrifugating the BALF sample under 4 °C, at 12,000 rpm for 10 min, the cell-free supernatant is transferred to sterilized enzyme-free 1.5 mL centrifuge tubes for subsequent qRT-PCR assay for viral load analysis. Briefly, the viral RNA was firstly extracted by using the TaKaRa MiniBEST Viral RNA/DNA Extraction Kit (9766, TaKaRa Ltd., Beijing, China) from BALF-derived cell-free supernatant. cDNA was generated by PrimeScript™ 1st Strand cDNA Synthesis Kit (6110A, TaKaRa Ltd., Bejing, China). The qRT-PCR was then performed with the TB Green® Fast qPCR Mix kit (RR430A, TaKaRa Ltd., Beijing, China). We performed the RSV qRT-PCR absolute quantitative detection method established by our laboratory to quantitatively measure the viral copies via a standard curve of RSV *N* gene. Since *N* is the most conserved RSV gene across all known genetic clades, and is also expressed most abundantly during viral replication [43].

3.10. Histopathology Evaluation of Lung Tissue

At 2 DPI, the mice were properly anaesthetized and euthanized. The left lung lobe was collected and fixed immediately in 4% formaldehyde under 0.01 M phosphate buffer (pH 7.4). After dehydration and embedding in paraffin, the lung tissue was cut into 5.0 μm-thick sections and stained with hematoxylin and eosin (H&E) for evaluation of histopathology and of the severity of pneumonia. The slides were scored with a semi-quantitative system according to the relative degree of inflammation and tissue damage (Table S1).

3.11. Immunofluorescence Experiment

The left lung lobe harvested at 2 DPI was processed into paraffin sections, and then the sections were then deparaffinized and rehydrated. After treatment, immersed the slides in EDTA antigen retrieval buffer for antigen retrieval. Because F protein is one of viral membrane proteins and the function of F protein is fusing RSV and cell membrane or infected cell membranes together. So its expression would demonstrate the attachment and entry capability of virus. Therefore, slides were blocked by 1% BSA and incubated with primary anti-RSV F antibody (sc-101362, Santa Cruz Biotechnology Ltd., Dallas, TX, USA). Then, slides were incubated with fluorescent secondary antibody. 4′,6-diamidino-2-phenylindole (DAPI) counterstained nucleus for 15 min, and then images were detected and collected by fluorescent microscope (ECLIPSE C1, Nikon, Tokyo, Japan) and confocal microscope (DS-U3, Nikon, Tokyo, Japan). Nucleus is blue by labeling with DAPI, and positive signals are red according to the fluorescent labels used.

3.12. Statistical Analysis

Data are presented as mean ± or + SD. The one-way ANOVA followed by post hoc Tukey test was used in the statistical analysis among multiple groups using GraphPad Prism 6. Comparisons between Mock and Model groups were assessed by unpaired, two-tailed Student's t test. Variance test (F test) was implemented before the one-way ANOVA analysis and t-test to check whether the data met the normal distribution. The Kaplan–Meier method was used for survival analysis, and the Log-rank test was performed to analyze the differences among the survival curves. The Mann–Whitney test was applied for the pathologic scores analysis. Significance was defined as p values of less than 0.05.

4. Conclusions

In this study, we developed a CS-coated liposome to facilitate the alleviative effect of OMT on lethal RSV infection via improving the lung accumulation and retention of OMT to exert its anti-virus effects by inhibiting the penetration and biosynthesis of virus. We revealed that effective penetration through the blood–gas barrier leads to fast systemic distribution and short exposure at lung, instead of pulmonary mucus being the main barrier for OMT in the treatment of RSV infection. OMT liposomal preparations, especially Lipo/OMT@CS, reduced the virus replication and increased the survival rate of RSV-infected mice. Further investigation is warranted to elaborate the exact mechanisms of OMT on the alleviation of lethal RSV infection and to optimize the liposomal carriers.

Supplementary Materials: The following supporting information can be downloaded at: https://www.mdpi.com/article/10.3390/ijms232415909/s1.

Author Contributions: Conceptualization, J.P. and K.Z.; methodology, Q.W. and M.G.; software, C.L.; validation, J.P. and K.Z.; investigation, C.L. and X.C.; resources, L.T.; writing—original draft preparation, J.P., K.Z., Q.W. and M.G.; writing—review and editing, J.P. and K.Z.; project administration, Q.W. and M.G.; supervision and funding acquisition, J.P., K.Z. and X.S.; J.P., Q.W. and M.G. contributed equally to this work. All authors have read and agreed to the published version of the manuscript.

Funding: This research was funded by the National Natural Science Foundation of China (No. 81860752), the High-level Overseas Talent Fund for Home Work from the Ministry of Human Resources and Social Security (RSBLXHGZ202001), the Guizhou Provincial Natural Science Foundation (No. [2020]1Z069), the Excellent Young Talents Plan of Guizhou Medical University (No. 2020-102, 2021-103), the Scientific and Technologic Project of Guiyang (No. 2020-16-9), the Science and Technology Foundation of Health Commission of Guizhou Province (gzwkj2021-463, gzwkj2022-230), the Cultivation Project of National Natural Science Foundation of China of the Affiliated Hospital of Guizhou Medical University and Guizhou Medical University (I-2020-06, 20NSP010), and the Doctor Training Program of the Affiliated Hospital of Guizhou Medical University (gyfybsky-2022-09).

Institutional Review Board Statement: The animal study protocol was approved by the Animal Welfare and Ethics Committee of Guizhou Medical University (No: 1800020).

Informed Consent Statement: Not applicable.

Data Availability Statement: Not applicable.

Acknowledgments: The authors extend their appreciation to Yusi Luo from the Department of Emergency at the Affiliated Hospital of Guizhou Medical University for technical support.

Conflicts of Interest: The authors declare no conflict of interest.

References

1. Smith, D.; Seales, S.; Budzik, C.J.A.f.p. Respiratory Syncytial Virus Bronchiolitis in Children. *Am. Fam. Physician* **2017**, *95*, 94–99. [PubMed]
2. Shaw, C.A.; Galarneau, J.-R.; Bowenkamp, K.E.; Swanson, K.A.; Palmer, G.A.; Palladino, G.; Markovits, J.E.; Valiante, N.M.; Dormitzer, P.R.; Otten, G.R. The role of non-viral antigens in the cotton rat model of respiratory syncytial virus vaccine-enhanced disease. *Vaccine* **2013**, *31*, 306–312. [CrossRef] [PubMed]
3. Meissner, H.C. Viral Bronchiolitis in Children. *N. Engl. J. Med.* **2016**, *374*, 62–72. [CrossRef] [PubMed]
4. Roe, M.F.; Bloxham, D.M.; White, D.K.; Ross-Russell, R.I.; Tasker, R.T.; O'Donnell, D.R. Lymphocyte apoptosis in acute respiratory syncytial virus bronchiolitis. *Clin. Exp. Immunol.* **2004**, *137*, 139–145. [CrossRef]
5. Welliver, T.; Garofalo, R.; Hosakote, Y.; Hintz, K.; Avendano, L.; Sanchez, K.; Velozo, L.; Jafri, H.; Chavez-Bueno, S.; Ogra, P.; et al. Severe human lower respiratory tract illness caused by respiratory syncytial virus and influenza virus is characterized by the absence of pulmonary cytotoxic lymphocyte responses. *J. Infect. Dis.* **2007**, *195*, 1126–1136. [CrossRef]
6. Plugge, P.C. Matrin, das Alkalid von Sophora angustifolia. *Arch. Pharm.* **1895**, *233*, 441–443. [CrossRef]
7. Lan, X.; Zhao, J.; Zhang, Y.; Chen, Y.; Liu, Y.; Xu, F. Oxymatrine exerts organ- and tissue-protective effects by regulating inflammation, oxidative stress, apoptosis, and fibrosis: From bench to bedside. *Pharmacol. Res.* **2020**, *151*, 104541. [CrossRef]
8. Dai, J.P.; Wang, Q.W.; Su, Y.; Gu, L.M.; Deng, H.X.; Chen, X.X.; Li, W.Z.; Li, K.S. Oxymatrine Inhibits Influenza A Virus Replication and Inflammation via TLR4, p38 MAPK and NF-kappaB Pathways. *Int. J. Mol. Sci.* **2018**, *19*, 965. [CrossRef]
9. Jiang, Y.; Zhu, Y.; Mu, Q.; Luo, H.; Zhi, Y.; Shen, X. Oxymatrine provides protection against Coxsackievirus B3-induced myocarditis in BALB/c mice. *Antiviral Res.* **2017**, *141*, 133–139. [CrossRef]
10. Ma, S.C.; Du, J.; But, P.P.H. Deng, X.L.; Zhang, Y.W.; Ooi, V.E.C.; Xu, H.X.; Lee, S.H.S.; Lee, S.F. Antiviral Chinese medicinal herbs against respiratory syncytial virus. *J. Ethnopharmacol.* **2002**, *79*, 205–211. [CrossRef]
11. Jain, H.; Bairagi, A.; Srivastava, S.; Singh, S.B.; Mehra, N.K. Recent advances in the development of microparticles for pulmonary administration. *Drug Discov. Today* **2020**, *25*, 1865–1872. [CrossRef]
12. Douafer, H.; Andrieu, V.; Brunel, J.M. Scope and limitations on aerosol drug delivery for the treatment of infectious respiratory diseases. *J. Control. Release* **2020**, *325*, 276–292. [CrossRef] [PubMed]
13. He, S.; Gui, J.; Xiong, K.; Chen, M.; Gao, H.; Fu, Y. A roadmap to pulmonary delivery strategies for the treatment of infectious lung diseases. *J. Nanobiotechnol.* **2022**, *20*, 101. [CrossRef] [PubMed]
14. Wang, W.; Huang, Z.; Huang, Y.; Zhang, X.; Huang, J.; Cui, Y.; Yue, X.; Ma, C.; Fu, F.; Wang, W.; et al. Pulmonary delivery nanomedicines towards circumventing physiological barriers: Strategies and characterization approaches. *Adv. Drug Deliv. Rev.* **2022**, *185*, 114309. [CrossRef] [PubMed]
15. Loira-Pastoriza, C.; Todoroff, J.; Vanbever, R. Delivery strategies for sustained drug release in the lungs. *Adv. Drug Deliv. Rev.* **2014**, *75*, 81–91. [CrossRef]
16. Tahara, K.; Tomida, H.; Ito, Y.; Tachikawa, S.; Onodera, R.; Tanaka, H.; Tozuka, Y.; Takeuchi, H. Pulmonary liposomal formulations encapsulated procaterol hydrochloride by a remote loading method achieve sustained release and extended pharmacological effects. *Int. J. Pharm.* **2016**, *505*, 139–146. [CrossRef]
17. Liu, M.; Jin, S.; Yan, H.; Du, S. Effect of oxymatrine HSPC liposomes on improving bioavailability, liver target distribution and hepatoprotective activity of oxymatrine. *Eur. J. Pharm. Sci.* **2017**, *104*, 212–220. [CrossRef]
18. Zhang, S.; Wu, J.; Wang, H.; Wang, T.; Jin, L.; Shu, D.; Shan, W.; Xiong, S. Liposomal oxymatrine in hepatic fibrosis treatment: Formulation, in vitro and in vivo assessment. *AAPS PharmSciTech* **2014**, *15*, 620–629. [CrossRef]
19. Guo, J.; Zeng, H.; Shi, X.; Han, T.; Liu, Y.; Liu, Y.; Liu, C.; Qu, D.; Chen, Y. A CFH peptide-decorated liposomal oxymatrine inactivates cancer-associated fibroblasts of hepatocellular carcinoma through epithelial-mesenchymal transition reversion. *J. Nanobiotechnol.* **2022**, *20*, 114. [CrossRef]
20. Tang, Q.; Zhang, W.; Zhang, C.; Guan, Y.; Ding, J.; Yuan, C.; Tan, C.; Gao, X.; Tan, S. Oxymatrine loaded nitric oxide-releasing liposomes for the treatment of ulcerative colitis. *Int. J. Pharm.* **2020**, *586*, 119617. [CrossRef]
21. Wang, H.; Ding, Y.; Zhang, W.; Wei, K.; Pei, Y.; Zou, C.; Zhang, C.; Ding, J.; Fang, H.; Tan, S. Oxymatrine Liposomes for Intervertebral Disc Treatment: Formulation, in vitro and vivo Assessments. *Drug Des. Devel. Ther.* **2020**, *14*, 921–931. [CrossRef]
22. Peng, J.; Cai, Z.; Wang, Q.; Zhou, J.; Xu, J.; Pan, D.; Chen, T.; Zhang, G.; Tao, L.; Chen, Y.; et al. Carboxymethyl Chitosan Modified Oxymatrine Liposomes for the Alleviation of Emphysema in Mice via Pulmonary Administration. *Molecules* **2022**, *27*, 103402. [CrossRef] [PubMed]
23. De Leo, V.; Milano, F.; Agostiano, A.; Catucci, L. Recent Advancements in Polymer/Liposome Assembly for Drug Delivery: From Surface Modifications to Hybrid Vesicles. *Polymers* **2021**, *13*, 1027. [CrossRef] [PubMed]
24. Huang, X.; Chisholm, J.; Zhuang, J.; Xiao, Y.; Duncan, G.; Chen, X.; Suk, J.S.; Hanes, J. Protein nanocages that penetrate airway mucus and tumor tissue. *Proc. Natl. Acad. Sci. USA* **2017**, *114*, 6595–6602. [CrossRef]

25. Manconi, M.; Manca, M.L.; Valenti, D.; Escribano, E.; Hillaireau, H.; Fadda, A.M.; Fattal, E. Chitosan and hyaluronan coated liposomes for pulmonary administration of curcumin. *Int. J. Pharm.* **2017**, *525*, 203–210. [CrossRef] [PubMed]
26. Rasul, R.M.; Tamilarasi Muniandy, M.; Zakaria, Z.; Shah, K.; Chee, C.F.; Dabbagh, A.; Rahman, N.A.; Wong, T.W. A review on chitosan and its development as pulmonary particulate anti-infective and anti-cancer drug carriers. *Carbohydr. Polym.* **2020**, *250*, 116800. [CrossRef]
27. Hao, Y.L.; Deng, Y.J.; Chen, Y.; Wang, X.M.; Zhong, H.J.; Suo, X.B. In Vitro and In Vivo Studies of Different Liposomes Containing Topotecan. *Arch. Pharm. Res.* **2005**, *28*, 626–635. [CrossRef]
28. Mohammed, M.A.; Syeda, J.T.M.; Wasan, K.M.; Wasan, E.K. An Overview of Chitosan Nanoparticles and Its Application in Non-Parenteral Drug Delivery. *Pharmaceutics* **2017**, *9*, 53. [CrossRef]
29. Murata, M.; Nakano, K.; Tahara, K.; Tozuka, Y.; Takeuchi, H. Pulmonary delivery of elcatonin using surface-modified liposomes to improve systemic absorption: Polyvinyl alcohol with a hydrophobic anchor and chitosan oligosaccharide as effective surface modifiers. *Eur. J. Pharm. Biopharm.* **2012**, *80*, 340–346. [CrossRef]
30. Naveed, M.; Phil, L.; Sohail, M.; Hasnat, M.; Baig, M.; Ihsan, A.U.; Shumzaid, M.; Kakar, M.U.; Mehmood Khan, T.; Akabar, M.D.; et al. Chitosan oligosaccharide (COS): An overview. *Int. J. Biol. Macromol.* **2019**, *129*, 827–843. [CrossRef]
31. Netsomboon, K.; Bernkop-Schnurch, A. Mucoadhesive vs. mucopenetrating particulate drug delivery. *Eur. J. Pharm. Biopharm.* **2016**, *98*, 76–89. [CrossRef] [PubMed]
32. Boroumand, H.; Badie, F.; Mazaheri, S.; Seyedi, Z.S.; Nahand, J.S.; Nejati, M.; Baghi, H.B.; Abbasi-Kolli, M.; Badehnoosh, B.; Ghandali, M.; et al. Chitosan-Based Nanoparticles Against Viral Infections. *Front. Cell. Infect. Microbiol.* **2021**, *11*, 643953. [CrossRef]
33. Mura, S.; Hillaireau, H.; Nicolas, J.; Kerdine-Romer, S.; Le Droumaguet, B.; Delomenie, C.; Nicolas, V.; Pallardy, M.; Tsapis, N.; Fattal, E. Biodegradable nanoparticles meet the bronchial airway barrier: How surface properties affect their interaction with mucus and epithelial cells. *Biomacromolecules* **2011**, *12*, 4136–4143. [CrossRef] [PubMed]
34. Ensign, L.M.; Schneider, C.; Suk, J.S.; Cone, R.; Hanes, J. Mucus penetrating nanoparticles: Biophysical tool and method of drug and gene delivery. *Adv. Mater.* **2012**, *24*, 3887–3894. [CrossRef] [PubMed]
35. Gubernator, J. Active methods of drug loading into liposomes Recent strategies for stable drug entrapment and increased in vivo. *Expert Opin. Drug Deliv.* **2011**, *8*, 565–580. [CrossRef]
36. Patton, J.S.; Fishburn, C.S.; Weers, J.G. The lungs as a portal of entry for systemic drug delivery. *Proc. Am. Thorac. Soc.* **2004**, *1*, 338–344. [CrossRef]
37. Zhang, K.; Li, C.; Luo, Y.S.; Wen, L.; Yuan, S.; Wang, D.; Ho-Yin Wong, B.; Zhao, X.; Chiu, M.C.; Ye, Z.W.; et al. Establishment of a lethal aged mouse model of human respiratory syncytial virus infection. *Antiviral Res.* **2019**, *161*, 125–133. [CrossRef]
38. Ye, Z.-W.; Yuan, S.; Poon, K.-M.; Wen, L.; Yang, D.; Sun, Z.; Li, C.; Hu, M.; Shuai, H.; Zhou, J. Antibody-dependent cell-mediated cytotoxicity epitopes on the hemagglutinin head region of pandemic H1N1 influenza virus play detrimental roles in H1N1-infected mice. *Front. Immunol* **2017**, *8*, 317. [CrossRef]
39. d'Angelo, I.; Casciaro, B.; Miro, A.; Quaglia, F.; Mangoni, M.L.; Ungaro, F. Overcoming barriers in Pseudomonas aeruginosa lung infections: Engineered nanoparticles for local delivery of a cationic antimicrobial peptide. *Colloids Surf. B Biointerfaces* **2015**, *135*, 717–725. [CrossRef]
40. Pellosi, D.S.; d'Angelo, I.; Maiolino, S.; Mitidieri, E.; d'Emmanuele di Villa Bianca, R.; Sorrentino, R.; Quaglia, F.; Ungaro, F. In vitro/in vivo investigation on the potential of Pluronic(R) mixed micelles for pulmonary drug delivery. *Eur. J. Pharm. Biopharm.* **2018**, *130*, 30–38. [CrossRef]
41. Costabile, G.; d'Angelo, I.; d'Emmanuele di Villa Bianca, R.; Mitidieri, E.; Pompili, B.; Del Porto, P.; Leoni, L.; Visca, P.; Miro, A.; Quaglia, F.; et al. Development of inhalable hyaluronan/mannitol composite dry powders for flucytosine repositioning in local therapy of lung infections. *J. Control. Release* **2016**, *238*, 80–91. [CrossRef] [PubMed]
42. Yu, Y.; Li, S.; Yao, Y.; Shen, X.; Li, L.; Huang, Y. Increasing stiffness promotes pulmonary retention of ligand-directed dexamethasone-loaded nanoparticle for enhanced acute lung inflammation therapy. *Bioact. Mater.* **2023**, *20*, 539–547. [CrossRef] [PubMed]
43. Do, L.A.H.; van Doorn, H.R.; Bryant, J.E.; Nghiem, M.N.; Van, V.C.N.; Vo, C.K.; Nguyen, M.D.; Tran, T.H.; Farrar, J.; De Jong, M.D. A sensitive real-time PCR for detection and subgrouping of human respiratory syncytial virus. *J. Virol. Methods* **2012**, *179*, 250–255. [CrossRef] [PubMed]

MDPI AG
Grosspeteranlage 5
4052 Basel
Switzerland
Tel.: +41 61 683 77 34

International Journal of Molecular Sciences Editorial Office
E-mail: ijms@mdpi.com
www.mdpi.com/journal/ijms

www.ingramcontent.com/pod-product-compliance
Lightning Source LLC
Chambersburg PA
CBHW061605210926
51850CB00009B/2219